Intermolecular Interactions and Biomolecular Organization

Intermolecular Interactions and Biomolecular Organization

A. J. HOPFINGER
Case Western Reserve University

A WILEY-INTERSCIENCE PUBLICATION

JOHN WILEY & SONS, New York · London · Sydney · Toronto

Library of Congress Cataloging in Publication Data:

Hopfinger, A J

Intermolecular interactions and biomolecular organization.

"A Wiley-Interscience publication."
Includes bibliographies. and index.
1. Molecular biology. 2. Molecular dynamics.
I. Title.
QH506.H67 574.8′8 76-26540
ISBN 0-471-40910-3

Printed in the United States of America.

10 9 8 7 6 5 4 3 2 1

To my sons

Tim and Tony

. . . to give us insight, not numbers.
C. A. Coulson, 1972

Preface

In writing this text I have attempted to centralize and summarize the results and their interpretations of investigations of *interacting* biomolecular systems. The text focuses on, and is limited to, those studies which yield information concerning structural chemistry and associated molecular thermodynamics. Two classes of interacting biomolecular systems are neglected in this work: enzyme–substrate structure and the organization of viruses. Both systems have been the subjects of many recent reviews. In the case of viruses, it is also questionable if sufficient structural information is available to meet the goals of this text. Very recently, a few of the topics reported herein have become popular review topics. I have dutifully attempted to note such reviews and to slant my discussions along different lines. The interaction of water with proteins is an example of an area recently reviewed.

In the first part of Chapter 1, I attempt to establish the importance of biomolecular interactions and consequential molecular organization in the life cycle. The latter part of Chapter 1 provides a historical summary of multimolecular systems. This is embodied in the various theories of molecular fluids. A comparison of solution and crystal structures of small, biologically active molecules is presented in Chapter 2. Relationships between preferred solution and solid-state structures for flexible molecules are discussed. Chapter 3, dealing with "drug design," may seem out of place in this book to some readers. However, it is my opinion that many of the structural and free energy variables used to establish empirical structure–activity relationships in the pharmacologic sciences can be helpful in explaining the structural properties of other biomolecular systems. Therefore, I have included it for the reader's consideration.

Synthetic polypeptides can be considered as simplistic protein models. Consequently, a large body of structural data on these biopolymers in solution and in the solid state has accumulated in the many attempts to understand protein structure. In Chapter 4, I try to isolate information that deals with the interaction of solvent molecules with polypeptides. Chapters 5 and 6 can be considered as extensions of Chapter 4. Chapter 5 deals with the structural organization of water molecules with and about proteins. Attention is given to comparing solution and crystal protein structures. Chapter 6 reviews the very limited molecular structural data available on the interaction of drugs with proteins. Still, we can expect the elucidation of drug-receptor geometry to provide major advances in molecular pharmacology in the near future.

Chapters 7, 8, and 9 discuss molecular interactions involving DNA. The interaction of water molecules with DNA is first presented in Chapter 7. Although the data for DNA hydration structure is less extensive than for proteins with water, it appears, overall, that the interactions of water molecules with two types of biologic macromolecules are quite similar. The associative binding of a certain class of drugs with DNA, those leading to DNA intercalation, on the other hand, are among the better understood intermolecular interactions. DNA intercalation processes are considerably better understood in terms of structural chemistry than the corresponding protein–drug interactions. Chapter 8 discusses DNA intercalation. Chapter 9 focuses on the complex formation between polypeptides and histones with DNA. There seems to be a rather disjoint set of data regarding the nature and properties of these complexes. Hence, in the survey of the field, I try to compensate by providing a large number of references to the original studies.

Chapter 10 and 11 deal with systems involving polysaccharides. The first part of Chapter 10 summarizes our limited knowledge of glycosaminoglycan (mucopolysaccharide) structure in solution and the solvated crystalline state. Amylose has been studied extensively both in solution and in the solid state (as compared to the glycosaminoglycans) and is discussed in the latter half of Chapter 10 to aid in understanding the solvation and crystallization processes of polysaccharides. Given the background in glycosaminoglycan structure laid in Chapter 10, we proceed to discuss the interactions of glycosaminoglycans with polypeptides, collagen, and some proteoglycan fragments in dilute solutions in Chapter 11. These multimacromolecular systems are considered as connective tissue models.

Chapter 12 attempts to characterize the nature of ion interactions with biomacromolecules. The very exhaustive, yet fragmented literature in this area has made the review task quite difficult. Chapter 13 is related to Chapter 12 in that most biomolecular aggregates involve at least pseudo-molecular electrolytes whose structure and organizational makeup is governed largely by ionic interactions. The entire area of molecular aggregation seems understudied. This is probably a consequence of the difficulty in probing the structure and thermodynamic properties of these diffuse molecular entities. Still, the organizational state of living systems on the molecular level seems to be closely akin to flexible, yet organized, aggregrates. Consequently, I feel this state of molecular organization deserves much more study.

Chapter 14 deals with the theories and models of interacting molecular species. I think it is safe to say that experimental work is considerably far ahead of theoretical advances in the field of interacting molecular systems. One can ask how much further experimental investigations can proceed before the corresponding experimental "pictures" of molecular organization will no longer retain any reasonable theoretical viability say, for example, in terms of molecular energetics.

Chapter 15 looks at a few specific systems whose higher-order molecular organization, ultrastructure, has an experimental basis from the structures of the constituent molecular units. Some readers may feel that the molecular assemblies reported in this chapter could be included in Chapter 13. However, I feel the structural organizations discussed in Chapter 15 contain a higher information content than those of Chapter 13. By putting Chapter 14 between these two sections—biomolecular aggregration and ultrastructural organization—I hope to emphasize their respective unique aspects.

Chapter 16 represents my attempt to overview intermolecular interactions and organization. The generalizations and deductions are speculative and are not always grounded in hard-core evidence. They are also my interpretation of available data, and I am solely responsible for their merits and faults. My only hope is that they will be accepted as at least reasonable viewpoints.

The topics in this book obviously cover a wide spectrum of distinct research areas. Many types of molecular species are considered. Consequently, it is not only possible, but probable, that I have overlooked some significant studies in a few of the areas reported. I must apologize for these oversights and hope that the theme of describing the molecular structure and thermodynamics of the corresponding interacting molecular systems compensates for my lack of familiarity with certain research literatures.

Many individuals have contributed to the development of this book. Dr. Herschel Weintraub provided me with an impetus to study structure–activity relationships in drugs. Chapter 2 and 3 are a result of these interests. Mr. Deepak Malhotra gave valuable assistance in the preparation of Chapter 8. Professor John Blackwell, Dr. John Cael, Ms. Kathleeen Schodt, and Mr. Rudolph Potenzone, Jr., provided considerable help and discussion in the development of Chapters 10, 11, and 13. I must also thank the Alfred P. Sloan Foundation for a Sloan Fellowship that provided me the opportunity to explore the range of intermolecular interactions reported.

In the area of technical preparation of the text I sincerely thank Ms. Terri Moughan for typing and proofreading, Ms. Paula Jacobs for typing, and Mrs. Barbara Leach for typing and editing. My wife Kathy also spent several evenings proofreading this work, and it is most appreciated.

Last, I would be very much remiss if I did not extend my apologies and thanks to my family, Kathy, Tim, and Tony, for their patience with me during the long development of this text. For the second time in three years they have had to give up many evenings and weekends of our all being together in order for me to do something I wanted. I do appreciate it guys!

A. J. HOPFINGER

Cleveland, Ohio
June 1976

Contents

Intermolecular Interactions and Biomolecular Organization

CHAPTER 1

General Concepts and Background

Intermolecular interactions in biologic systems are governed largely by molecular structure. We can roughly envision biomolecular reactions as taking place between species with complementary structures that "fit together." Such complementary fittings lead to organization in the overall biologic process, which, in turn, implies the transmittal or distribution of information. The transfer of information provides the basis for a self-sustaining system: life. In this chapter, we discuss these concepts and outline the history and development of molecular theories of interacting systems.

I INTRODUCTION

Intermolecular interactions have often been used as phenomenologic sources to rationalize the shortcomings of investigations of molecular structure. Although the existence and importance of intermolecular interactions on molecular structure and action has long been realized, even the immense qualitative complexity of describing these events has led to treating such intermolecular processes as relatively mystical happenings. To some degree, an unwritten axiom adopted by many scientists is, perhaps, overly stated: when in doubt about explaining minor data discrepancies, blame the errors on intermolecular interactions. This attitude is reflected in the following quotes from a journal of our departmental seminars kept during the preparation of this text—names have been withheld to protect the guilty:

Differences in the CD spectra, . . . , can be assigned to interactions between water molecules and the polypeptide chains.

September, 1974

Packing forces are probably the cause for the unusual valence angle . . . ,

December, 1974

Polar solvents promote non-theta behavior in this class of macromolecules.

December, 1974

Glycosaminoglycans interact with ionized polypeptides to form a complex.

February, 1975

The protein is partly denatured by the solvent, . . . , resulting in a different activity.

March, 1975

The substrate force field apparently induces a high-energy packing mode.

May, 1975

Solvent composition, that is, water to weak acid concentration, controls the conformational transition.

July, 1975

Ion-counterion interactions apparently change hydration structure, . . . , leading to structural changes in the biopolymer.

October, 1975

Most readers can probably recall similar general statements they have heard made about intermolecular interactions, as well as appreciate why such comments have been necessary.

Over the last two decades or so, however, there has been a growing awareness of both the need and the technical possibility of investigating *in detail* intermolecular processes. This awareness has come from workers with a variety of interests, but perhaps nowhere has it been more pronounced than in the biochemical sciences. The emphasis to explore a fundamental biochemical principle—biologic processes are carried out through chemical actions and reactions *between* molecular species—has probably been a main driving force in studying intermolecular processes. For example, the pharmacologist has long accepted drug-receptor interactions as the basis for drug activity. The enzyme chemist's early postulates about protein-substrate interactions has been substantiated by the detailed X-ray studies of enzyme and enzyme-substrate crystals, as beautifully illustrated by the work of Phillips and colleagues on lysozyme (Blake et al., 1967; Phillips, 1966).

Even further, the X-ray studies of enzyme-substrate interactions reveal what can be taken as a molecular foundation for the biologic process—chemical event principle. Biochemical actions are governed by the molecular structure of the reactants. Thus we must ultimately focus on the organization of molecular structure in biochemically interacting systems. In turn, molecular organization implies the transfer and/or acquisition of information. It is the controlled distribution of information, *through molecular structure*, that recreates, propagates, and regulates living systems. The molecular biologist is making significant headway in understanding organization-information relationships at the ultrastructural and cellular levels. However, a void exists at the molecular plane in regard to precisely how the structural chemistry of interacting systems and resulting molecular thermodynamics control the organization and information networks in biochemical processes. This book attempts to centralize the findings of studies relevant to this question of molecular structure, interaction, and control.

II CLASSIC MOLECULAR THEORIES OF SOLUTIONS

Before proceeding to discuss in detail some molecular interactions, it is worthwhile to mention a few of the early theories of interacting molecular species. As one might guess, the original models for interacting molecular units focused on multicomponent fluids, most notably, liquids. That is, early wokers wished to describe the molecular origins of bulk solution properties. To avoid confusion, it is best to divide the discussion of solutions into two parts: first, to deal with "small" molecules exclusively, and second, macromolecular solutions.

A "Small" Molecule Solutions

When two or more liquids form a solution, there can be an enthalpy and volume change attending the mixing process. The enthalpy and volume changes are consequences of the interaction and resultant reorganization of constituent molecular species. Even if the observed enthalpy and volume changes are both zero, that is, we have an ideal solution that obeys Rauolt's law, we cannot postulate that the various species have not interacted and reorganized. We can only conclude that the net observable consequences of the total set of interactions as reflected by the enthalpy and volume of mixing are unchanged. The classic means of describing the liquid state on the molecular level in the hope of predicting heats and volumes of mixing has been to adopt cell and hole models. In these theories, the liquid state is considered to be composed of a three-dimensional spanning network of cells that can contain constituent molecular species. If a cell does not contain a molecular species, it is termed a hole. Each species in each occupied cell is subjected to some potential field whose complexity and form differentiate one theory from another. Clearly, when only one type of molecular species can occupy the cells, we are dealing with a simple liquid. When two or more unique species can habitate the cells, we are dealing with solution. Examples of the former include the Bragg-Williams theory (Hill, 1962), the Hirschfelder, Stevenson, and Eyring model (1937), the cell theory of Lennard-Jones and Devonshire (LJD) (1937, 1938), and the Cernuschi and Eyring theory (Hill, 1962). Among the solution theories are the McMillan-Mayer theory (1945), Kirkwood and Buff theory (1951), modified Bragg-Williams and LJD theories (Hill, 1962), and Prigogine models (Prigogine, 1957). Separate, but similar theories for dilute electrolyte solutions have also been proposed. Most well known of these is the Debye-Hückel theory (Fowler and Güggenheim, 1939).

Individual solution theories are valid over specific ranges of component concentrations. This is analogous to the limited applicability of gaseous equations of state over P-V-T space. Solutions are often classified as X-type in concentration range C, where X is the name of a particular solution theory whose calculated thermodynamic behavior is most similar to those actually observed. The implicit inference in such a classification is that the molecular interactions and organization assigned by the theoretical model are realistically representative of the actual solution under study. On this basis, it has been possible to estimate characteristic enthalpies and entropies of interaction as well as molecular dimensions of interacting solvent species. Still, the limited applicability of any theory over a range of concentrations suggests that the nature and/or extent of intermolecular interactions changes as the number of molecular species of a given type increases or decreases.

At one limit we might ask how a solution would behave if one set of molecular species were bound to one another in long linear arrays for all concentrations. This question was asked in the early forties in regard to polymer solutions, and the findings are the basis for the next section.

B Macromolecular Solutions

The study of macromolecular solutions, biologic or not, is, as already noted, of relatively recent vintage. A major obstacle to the initiation of a molecular description of solutions of macromolecules was the nonapplicability of existing theories of the liquid state. The most notable breakthrough in the development of a theory of polymer solutions was made nearly simultaneously, but independently, by Flory (1942) and Huggins (1942a,b,c). They reasoned that the entropy of a polymer solution must be small in comparison to a solution of nonpolymer liquids based on the same volume, or weight concentrations, because both the size and connectivity of the polymer chain keep the chain entropy small in comparison to that of a nonpolymeric species. Flory and Huggins, again independently, were able to derive approximate expressions for the number of configurational states a polymer chain could adopt in solution using a lattice model. This laid the basis for what is now called the Flory-Huggins theory of polymer solutions. This theory introduced a most basic parameter, χ_1*, which, in the Flory-Huggins theory, characterizes the total interaction energy of the solution per solvent molecule divided by kT times the volume fraction of the polymer.

The first polymer solution for which experimental results were compared with the Flory-Huggins theory was rubber in benzene (Gee and Orr, 1946). Outside of fair agreement between theory and experiment for the free entropy of mixing, the theory was not particularly successful. Discrepancies between theory and experiment grew larger with increasing dilution. Additional experiments on different systems supported the supposition that the Flory-Huggins theory could only be successful for reasonably dense polymer solutions, and was not at all adequate for dilute systems.

The reason for this is that the lattice model used in the Flory-Huggins model does not consider that a very dilute polymer solution must be discontinuous in structure, consisting of domains or clusters of polymer chains separated, on the average, by regions of bulk solvent. Flory and Krigbaum (1950) modeled such a discontinuous structure. They assumed a model in which each cloud of polymer chain segments is spherical, with a density that is maximum at the center of the sphere and decreases in an approximately Gaussian fashion with distance from the center. The Flory-Krigbaum

* Huggins adopted μ_h^o instead of χ_1.

theory, as this model has come to be known, is reasonably successful in describing some dilute polymer solutions. Just as with the Flory-Huggins theory, one is never quite sure when the Flory-Krigbaum theory can be validly applied.

Furthermore, both the Flory-Huggins and Flory-Krigbaum theories assume that polymer solutions differ from "regular" solutions only through the entropic term. Detailed studies of the deviations between theory and experiment for the interaction parameter χ_1 have led to its reinterpretation as a combined entropy and enthalpy parameter. The major contribution to the χ_1 parameter for polymer-solvent systems arises through the differences in free volumes.

Recognition of a volume-dependent enthalpy contribution in the characterization of a polymer solution has led to the development of several elegant "free-volume" theories of polymer solutions. The pioneering work in this area was carried out by Prigogine and his co-workers (Prigogine, 1957) and extended and put into practice by Flory (Flory, et al., 1968; Eichinger et al., 1968a–d) as well as by Patterson (1967; 1969). Such theories recognize the dissimilarity in the free volumes of the polymer and the solvent as a result of their great difference in size; the usual solvent is more expanded than the polymer. These free-volume theories are usually more accurate than prior polymer-solution theories in that volume changes on mixing can be predicted to within 10–15% of the experimental values, and the correct variation of χ_1 with concentration in the Flory-Huggins theory is also predicted.

In 1959, Maron (1959) introduced a semiempirical theory of polymer solutions based on the *effective* volume of the solute molecule in the solvent medium. In this theory, a characteristic interaction parameter, μ, which takes into account both enthalpy and entropy contributions from solvent-solvent, solvent-solute, and solute-solute interactions, takes the place of χ_1. Because μ is determined from all three types of interactions in a two-component solution, the Maron Theory is applicable over a wide range of concentrations and for a large number of solution types. It is *not* restricted to polymer solutions. Maron and Daniels (1968) and Maron and Lee (1973) contain composite listings of solutions to which the Maron theory has been applied. Although not used, or known, extensively, this may be one of the better polymer-solution theories at present.

III CONTENT OF THE TEXT

Succintly, this book focuses on the results of studies that have attempted to dissect, on the molecular level, χ_1 (or μ) for a variety of biomolecular systems. We seek to understand how nature has evolved sets of interactions between molecular species that are ultimately governed by the constituent structural chemistry. Implicit in this is unraveling how biomolecular species "fit together" to produce characteristic interactive energetics that translate into bulk thermodynamics and kinetics. Very little is said in the text about the temporal make-up of biomolecular interactions. This is because only a limited amount of information, or even conceptual understanding, of time-dependent processes is available. Thus we use the word kinetics only in the broadest fashion.

The published proceedings of the Eighth Jerusalem Symposium (Pullman, 1975) has a theme similar to that generated herein. This book, however, covers a wider range of topics and contains a larger set of references. Still, the two works do, to some degree, complement one another and emphasize the high interest in intermolecular structural organization that exists at this time.

REFERENCES

Blake, C. C. F., L. N. Johnson, G. A. Mair, A. C. T. North, D. C. Phillips, and V. R. Sarna (1967). *Proc. Roy. Soc. (Lond.)*, **B167**, 378.

Eichinger, B. E. and P. J. Flory (1968*a*). *Trans. Faraday Soc.*, **64**, 2035.

Eichinger, B. E. and P. J. Flory (1968*b*). *Trans. Faraday Soc.*, **64**, 2053.

Eichinger, B. E. and P. J. Flory (1968*c*). *Trans. Faraday Soc.*, **64**, 2061.

Eichinger, B. E. and P. J. Flory (1968*d*). *Trans. Faraday Soc.*, **64**, 2066.

Flory, P. J. (1942). *J. Chem. Phys.*, **10**, 51.

Flory, P. J., J. L. Ellenson, and B. E. Eichinger (1968). *Macromolecules*, **1**, 279.

Flory, P. J. and W. R. Krigbaum (1950). *J. Chem. Phys.*, **18**, 1086.

Fowler, R. H. and E. A. Güggenheim (1939). *Statistical Thermodynamics*, Cambridge Press.

Gee, G. and W. J. C. Orr (1946). *Trans. Faraday Soc.*, **42**, 507.

Hill, T. L. (1962). *Introduction to Statistical Thermodynamics*, Addison-Wesley: Reading, Mass.

Hirschfelder, J., D. Stevenson, and H. Eyring (1937). *J. Chem. Phys.*, **5**, 896.

Huggins, M. L. (1942*a*). *Ann. N.Y. Acad. Sci.*, **43**, 1.

Huggins, M. L. (1942*b*). *J. Phys. Chem.*, **46**, 151.

Huggins, M. L. (1942*c*). *J. Am. Chem. Soc.*, **64**, 1712.

Kirkwood, J. G. and F. P. Buff (1951). *J. Chem. Phys.*, **19**, 774.

Lennard-Jones, J. E. and A. F. Devonshire (1937). *Proc. Roy. Soc. (Lond.)*, **A163**, 53.

Lennard-Jones, J. E. and A. F. Devonshire (1937). *Proc. Roy. Soc. (Lond.)*, **A165**, 1.

Maron, S. H. (1959). *J. Polymer Sci.*, **38**, 329.

Maron, S. H. and C. A. Daniels (1968). *J. Macromol. Sci., Phys.*, **B2(4)**, 743.

Maron, S. H. and Min-Shiu. Lee (1973). *J. Macromol. Sci., Phys.*, **B7(1)**, 61.

McMillan, W. G. and J. E. Mayer (1945). *J. Chem. Phys.*, **13**, 276.

Patterson, D. (1967). *Rubber Chem. Tech.*, **40**, 1.

Patterson, D. (1969). *Macromolecules*, **2**, 672.

Phillips, D. C. (1966). *Scientific American*, **215(5)**, 78.

Prigogine, I. (1957). *Molecular Theory of Solutions*, North-Holland: Amsterdam.

Pullman, B. (1975). *Environmental Effects on Molecular Structure and Properties*. Reidel-Dordrecht, Holland.

CHAPTER 2

Conformation of Small Biologically Active Molecules in Solution and Crystals

A classic question in structural chemistry with important biologic implications is: How similar are the conformations of a molecule in solution and in its crystal? Many proposed mechanisms of biologic action are premised on identical conformations in solution and in crystals for the consitituent reactants. Certain comformational features of a molecule in solution can be quantitatively estimated from its characteristic PMR vicinal coupling constants. This chapter reports and compares such solution data to solid-state information on several classes of biologically active molecules. In general, the preferred solution conformation, irrespective of solvent, is similar to the crystal comformation.

I VICINAL COUPLING CONSTANTS AND MOLECULAR CONFORMATION IN SOLUTION

At least part of the conformational behavior of a solute molecule can often be determined in solution from measurement of the proton vicinal coupling constants seen in its PMR spectrum.

This is a recently developed method of solution conformational analysis whose origins date back to the very late 1950s. Today it is probably the most often-employed procedure to determine the distribution of conformational states of small molecules in solution. Coupling-constant analyses are usually restricted to small molecules, since, in macromolecules, the resonances are generally sufficiently broadened so that the splitting cannot be resolved. One important exception to this general rule in many instances is a polymer in the random-coil conformation. Bovey (1974) discusses coupling-constant behavior in random-coil polypeptides and can be consulted for further information. The majority of conformational studies in solution discussed in this chapter, as well as several others throughout the text, are based on NMR-vicinal coupling-constant analyses. Consequently, it is worthwhile to begin this chapter by looking at this technique in some detail and paying particular attention to its current limitations of applicability.

It should be noted immediately that, although the estimation of a vicinal coupling constant is a direct experimental measurement, the relationship between the magnitude of a coupling constant and the corresponding dihedral angle, which defines conformation, is empirical and, in part at least, is based on theoretical calculations as well as experimental data. Since the vicinal coupling constant measures the time-averaged distribution of conformational states about a bond, and the number of spatially different types of preferred conformational states about bonds is small, only a narrow range of dihedral angle values, that is, conformational states, are available in experimental studies. Moreover, the precise value of the dihedral angle is difficult to measure even in the simple model compounds used in experimental calibrations. Representation of the coupling constant/dihedral angle relationship for large excursions from preferred conformational states must be "guessitmated" from empirical curve fitting or from theoretical calculations. In short, determination of vicinal couplings to determine solution conformation is not a true experimental technique in the same sense that X-ray crystallography is completely experimental. Coupling-constant analysis requires an intermediate, empirical relationship between conformation and scalar coupling. A discussion of some of these relationships, generally termed Karplus-type functions, is given below.

Around 1960, both experimental and theoretical investigations of the vicinal coupling constant for ethanic, ethylenic, and related systems was found to vary in an ordered fashion with the dihedral angle, θ, about the carbon-carbon bond. Molecular orbital (MO) calculations were carried out in order to determine the functional relationship between θ and $J_{HH'}$ (Karplus, 1959, 1963; McConnell, 1957; Dewar and Fahey, 1963). Karplus suggested that the relationship could be expressed as;

$$J_{HH'} = A + B \cos \theta + C \cos 2\theta \tag{2-1}$$

For ethane, in which the C—C bond length is 1.543*A* and the bond energy is 9 ev, the constants are $A = 4.22$, $B = -0.5$, and $C = 4.5$. Kopple et al. (1973) have suggested the relationship

$$J_{\alpha\beta} = 11.0 \cos^2 \theta - 1.4 \cos \theta + 1.6 \sin^2 \theta \tag{2-2}$$

to describe the variations in the coupling constant of the

N　　　　　　　H
H—C^{α} —— θ —— C^{β}
C′　　　　　　　H

bond in a polypeptide sidechain. Figure 2-1 is a plot of this two-fold symmetric function. Deber et al. (1971) proposed a nearly equivalent relationship

$$J_{\alpha\beta} = \begin{cases} 8.5 \cos^2 \theta + 1.4 & 0^\circ \leq \theta \leq 90^\circ \\ 10.5 \cos^2 \theta + 1.4 & 90^\circ \leq \theta \leq 180^\circ \end{cases} \tag{2-3}$$

in the analysis of cyclo-triproline.

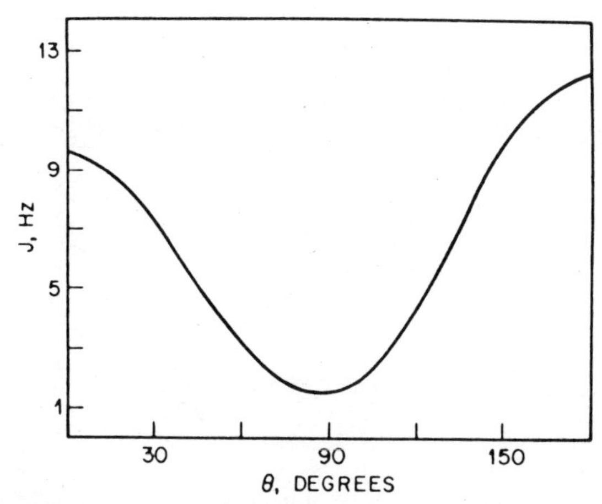

Fig. 2-1 Dependence of α—β J coupling (Hz) for amino acid side chains upon dihedral angle θ, according to Kopple et al. (1973).

Theoretical considerations suggest that serious errors can result if Karplus-like relationships are assumed in systems in which one or more of the chemical species bonded to one or both of the carbons is different from hydrogen. Such assumptions have been made in constructing equations 2-2 and 2-3. An even bolder step has been taken in developing a Karplus-like relationship for HC$\overset{\phi}{\circlearrowright}$NH. This relationship has proved very useful in estimating solution conformations about the N—C backbone bond in oligopeptides. The establishment of this relationship has been the object of a number of studies, both experimental (Balashova and Ovchinnikov, 1973; Bystrov et al., 1969; Ramachandran et al., 1971; Schwyzer, 1971; Thong et al., 1969) and theoretical (Barfield and Gearhart, 1973). A widely used equation is that of Bystrov et al. (1969):

$$J_{N\alpha} = 8.9 \cos^2 \phi' - 0.9 \cos \phi' + 0.9 \sin^2 \phi' \tag{2-4}$$

which is plotted in Figure 2-2. This group has also redetermined the empirical coefficients of equation 2-3 (Balashova and Ovichinnikov, 1973), correcting for substituent electronegativity.

Abraham and Gatti (1969) employed the isomeric state concept for the preferred conformational states about X—C(H)(H)$\overset{\theta}{\circlearrowright}$C(H)(H)—Y bonds in order to determine rotamer populations for the trans state, n_t and gauche states, n_g.

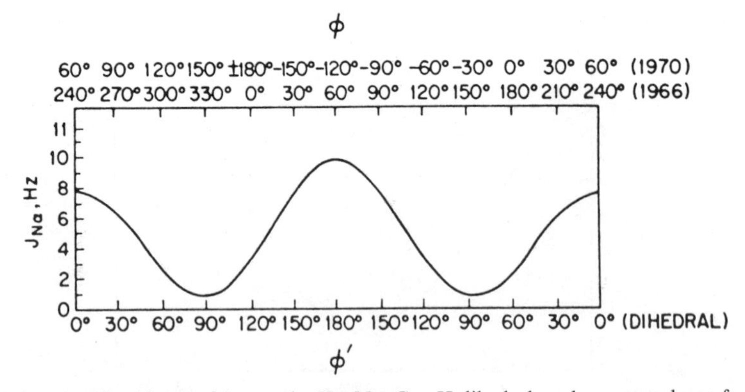

Fig. 2-2 The dependence of $J_{N\alpha}$ on the H—N—C$_\alpha$—H dihedral angle expressed as a function of ϕ in the 1966 and 1970 conventions (upper scales) and as the dihedral angle ϕ' (bottom scale) according to the relationship proposed by Bystrov et al. (1969).

The working equations in this analysis are

$$n_g = \frac{J_{AB'}J_t^g - J_{AB'}J_t^t}{J_g^g(J_t^g - J_t^t)} \tag{2-5a}$$

$$n_t = \frac{J_{AB} + J_{AB'} - n_g(0.5[J_g^t + J_g^{g'}] + J_g^g)}{(J_t^g + J_t^t)} \tag{2-5b}$$

where J_{AB} and $J_{AB'}$ are the observed vicinal proton-proton coupling constants for the protons defined as,

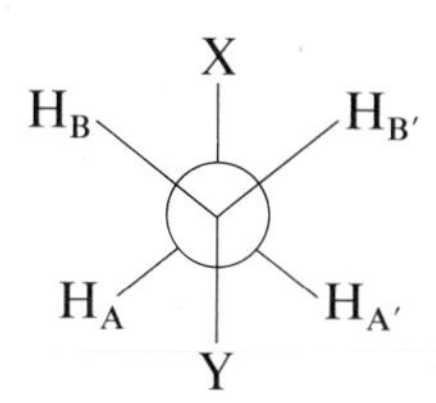

and J_t^t, J_t^g, J_g^t, J_g^g, and $J_g^{g'}$ are individual vicinal couplings estimated by Abraham and Gatti (1969) in terms of substituent-group electronegativities E_x and E_y:

$$J_t^t = 18.07 - 0.88(E_x + E_y) \tag{2-6}$$

$$J_t^g = 1.35 + 0.63(E_x + E_y) \tag{2-7}$$

$$J_g^g = 8.94 - 0.94(E_x + E_y) \tag{2-8}$$

$$J_g^t + J_g^{g'} = 26.92 - 2.03(E_x + E_y) \tag{2-9}$$

Clearly, the quantitative reliability of predicting substituted ethanic rotamer populations depends on the reasonableness of the additive electronegativity correction.

Other types of magnetic resonance vicinal couplings involving nonprotonic nuclei are not considered in this book. For example, some workers (Karplus and Karplus, 1972; Sogn et al., 1973; Lichter and Roberts, 1970) have attempted to use couplings involving the ^{15}N nucleus to determine peptide conformers. Nuclear relaxation studies involving ^{13}C nuclei can also provide conformational information (Levy, 1973; Stothers, 1972).

II ACETYCHOLINE AND CONGENERS

Acetylcholine, Ach, as the endogenous synoptic chemical transmitting agent in cholingergic neural systems, is active at parasympathetic postganglionic sites, voluntary neuromuscular junctions, and autonomic ganglia (Katz,

1966; Brimblecome, 1974). The structure of Ach with defined torsional rotational angles is shown in Figure 2-3. The conformation of Ach and a variety of its molecular congeners have been determined in the solid state (for reviews see, Martin-Smith et al., 1967; Dangoumaw, et al., 1969; Baker, et al., 1971; Pauling, 1973). The crystal structures of a variety of these compounds are defined as (τ_1, τ_2) points on the conformational energy map in Figure 2-4 (Pullman and Courrière, 1972). Several NMR conformational investigations of Ach and congeners in aqueous solution have been performed (Mautner, 1974; Culvenor and Ham, 1966; Shefter, 1971; Cushley and Mautner, 1970; Makriyannis et al., 1973; Behr and Lehn, 1972; Casy et al., 1971; Partingon et al., 1972; Chynoweth et al., 1973; Lichtenberg et al., 1974). The work of Partington et al. (1972) considered 22 compounds from which it was possible to assign the congeners into four conformational classifications with respect to rotation. Table 2-1 lists the rotamer populations about τ_2 for the 22 compounds as well as some characteristic coupling constants. Class I compounds strongly prefer the gauche conformation for τ_2. The J_{NH} coupling constants fall in the range of 2.2–2.8 Hz, and one of the vicinal proton coupling constants, J_{13}, falls into the narrow range of 2.3–3.9 Hz. Class II compounds prefer the trans conformation. The J_{NH} is small, approximately 0.7 Hz, and one of the vicinal proton coupling constants, J_{13}, is large, 11.0–12.6 Hz. Class III compounds are a mixture of gauche and trans conformations, as shown by the intermediate value for J_{13} of 6.4–6.5 Hz and the value of J_{NH} for chlorocholine. The distribution of gauche and trans is close to the 2:1 ratio expected when there is no conformational preference. In Class IV compounds, the gauche degeneracy is destroyed, and one of the two distinct

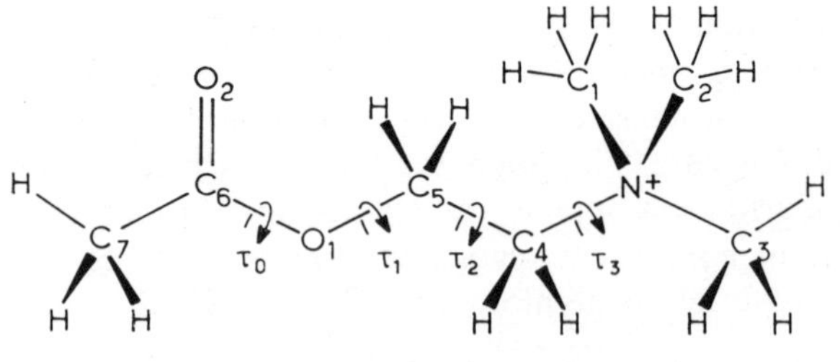

I. ACETYLCHOLINE

$$\tau_1 = \tau(C_6-O_1-C_5-C_4)$$

$$\tau_2 = \tau(O_1-C_5-C_4-N^+)$$

Fig. 2-3 Torsion angles τ_1 and τ_2 in acetycholine. Pullman and Courriere (1972).

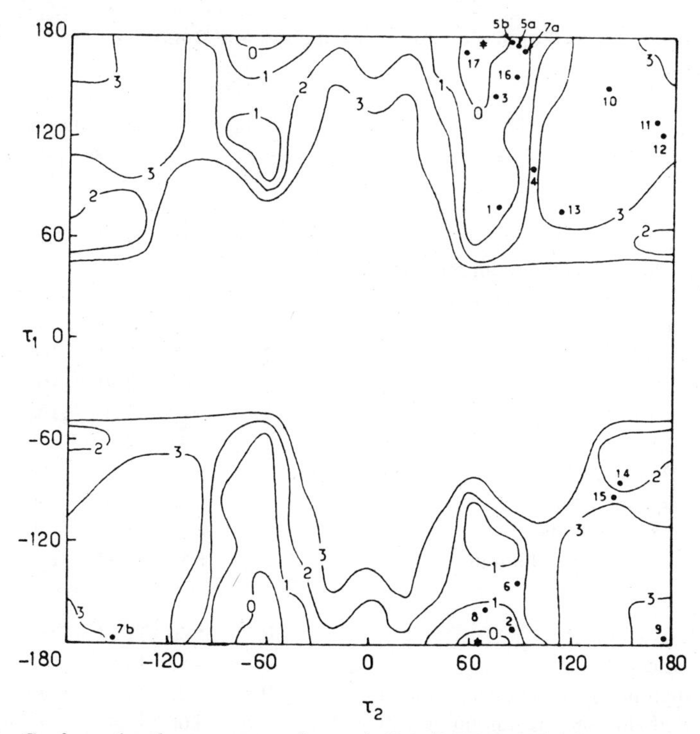

Fig. 2-4 Conformational energy map of acetycholine (PCILO method) Isoenergy curves in kilocalories per mole with respect to the global minimum, $\neq$, taken as zero energy. Shown are experimental conformations (●) in crystals of: 1, acetycholine bromide; 2, acetycholine chloride; 3, L(+)-muscarine iodide; 4, L(+)-*cis*-2(S)-methyl-4(R)-trimethylammonium methyl-1,3-dioxolane iodide; 5,5-methylfurmethide iodide (a and b); 6, L(+)-S-acetyl-β-methylcholine iodide; 7, D(+)-R-acetyl-α-methylcholine iodide (a and b); 8, erythro-acetyl-α(R), β(S)-dimethylcholine iodide; 9, carbamoylcholine; 10, (+)-*trans*-2(S) acetoxycyclopropyl-1(S)-trimethylammonium iodide; 11, acetylthiocholine bromide; 12, acetyselenocholine iodide; 13, (−)-R-3-acetoxyquinuclidine methiodide; 14, 2(S)-trimethylammonium-3(S)-acetoxy-*trans*-decahydronaphthalene iodide; 15, theo-acetyl-α(S),β(S)-dimethylcholine iodide; 16, lactoylcholine iodide; 17, dimethylphenyl-piperazine. Pullman and Courriere (1972).

gauche conformations is preferred over the other. There is a good agreement for the 22 compounds between the preferred solution conformation and the crystal structure, except for carbamylcholine, which in solution has a gauche conformation and in the crystal structure has a trans conformation. Partington et al. (1972) were not able to identify any simple correlation between the predominant solution conformation and the potency of the drugs at either the nicotinic or muscarinic receptor.

Currently there is a popular proposition that biologic activation of "small" drugs like Ach and its analogs may require a conformational change from

Table 2-1 Rotamer Populations of Acetylcholine and Congeners in Solution for τ_2.

Compound	J_{14NH}	J_{14}	J_{13}	*Gauche* %	*Trans* %
Class I					
Choline	2.7	6.56	3.55	89	11
N-Benzyldimethylethanolammonium	2.8	6.6	3.6	88	12
N-Dimethylethanolammonium		6.68	3.93	85	15
Acetylcholine	2.5	6.93	2.35	100	
Butyrylcholine	2.6	7.19	2.27	100	
Suxamethonium	2.5	7.17	2.29	100	
Carbamylcholine	2.7	6.98	2.20	100	
Benzilylcholine		6.98	2.83	96	4
Choline methyl ether	2.7	6.55	3.57	88	12
Choline phenyl ether	2.8	6.44	2.83	96	4
Choline 2,6-xylyl ether	2.7	6.6	3.5	89	11
N,N-Dimethylmorpholinium	2.2	6.57	3.54		
Class II					
Acetylthiocholine[a]	0.75	5.00	11.59	16	84
Acetylselenocholine[a]	<0.7	4.83	12.63	5	95
β-Methylthioethyltrimethylammonium	0.75	5.02	11.43	17	83
Phenylethyltrimethylammonium	0.7	4.68	11.49	17	83
Cyclohexylethyltrimethylammonium		5.00	11.00	22	78
β-Aminoethyltrimethylammonium	<0.7	4.70	11.50	13	87
3,3-Dimethylbutan-1-ol(carbocholine)		5.70	10.06	32	68
4-Hydroxylphenylethyldimethylammonium	0.7	5.00	11.00	22	78
Class III					
β-Chloroethyltrimethylammonium (chlorocholine)	1.8	6.42	6.53	64	36
N-Dimethylethanolamine		6.5	6.5	64	36
Class IV					
Methacholine	1.3	8.8	1.4	77/23[b]	0
Carbamyl-β-methylcholine	1.3	9.4	1.5	79/21[b]	0
Acetyl-α-methylcholine	2.2			80	20

[a] Data of Cushley and Mautner (1970).
[b] Rotamer with O, N^+, and Me *gauche* to each other is least populated.
Error in rotamer populations is ±10%.
Partington et al. (1972).

the preferred solution conformation to fit the shape of the receptor site. Thus the barrier height of such a transition might be the most critical factor to specifying activity in a compound. Lichtenberg et al. (1974) have estimated the barrier height of the trans/gauche conformational interversion for τ_2 to be about 1 kcal/mole for both Ach and Choline. This is in agreement with some theoretical calculations (Kier, 1967; Liquori et al., 1968), but in disagreement with some molecular orbital calculations (Genson and Christofferson, 1973; Port and Pullman, 1973). Recent ^{13}C and ^{14}N NMR studies by Behr and Lehn (1972) suggest trans/gauche barrier heights of about 1.5 kcal/mole for Ach and aldo-choline. The correlation times and the activation parameters for the two methylene carbons of the carbons attached to the nitrogen were found to be very similar both in water and in methanol, suggesting that internal motion within the $(CH_3)_3N^+—CH_2—CH_2$ grouping is relatively slow compared to the tumbling of molecules as a whole. On the other hand, the $—CH_3$ portion of the acetyl group displays a very short correlation time. This suggests a very fast rotation about the $CH_3—\overset{\overset{\displaystyle O}{\|}}{C}—$ bond. Makriyannis (1974) performed spin-lattice relaxation measurements on Ach deuterated either in the α-position or the β-position. Activation energies of relaxation for the $—NCH_2—$ protons were found to be 3.8 kcal in Ach, 3.5 kcal in carbamylcholine, and 2.85 kcal in choline. Rotation of the methyl group attached to the carbonyl carbon was seen to be considerably faster than that of the molecule as a whole.

It is not at all difficult to rationalize interactions between Ach (and homologs) and the receptors that can account for the energy needed to stream over the Ach trans/gauche barrier. Formation of a receptor-agonist hydrogen bond, for example, could provide 3–5 kcals/mole. Hence, for Ach, and probably many of it analogs, it is energetically reasonable to picture the agonist molecule as essentially flexible with respect to the receptor site.

However, Weinstein et al. (1973) have shown that the preferred aqueous solution conformation of Ach is likely very close to its active conformation for muscarinic interactions and probably the precise conformation for enzymatic hydrolysis. This was accomplished by a structure-activity analysis of a rigid Ach analog, 3-acetoxyquinuclidine (3-AcQ), which has a rigidly fixed conformation and electron-charge distribution very nearly identical to the solution conformation and charge distribution, respectively, of Ach. In Figure 2-5 is (*a*) a ball-stick illustration of Ach in its solution conformation, (*b*) a ball-stick representation of 3-AcQ in its fixed conformation, and (*c*) superposition of Ach and 3-AcQ as shown in *a* and *b*. The formal representation of 3-AcQ's ring system consists of three boat-like structures of piperidine, in which each boat shares the nitrogen and three other carbon

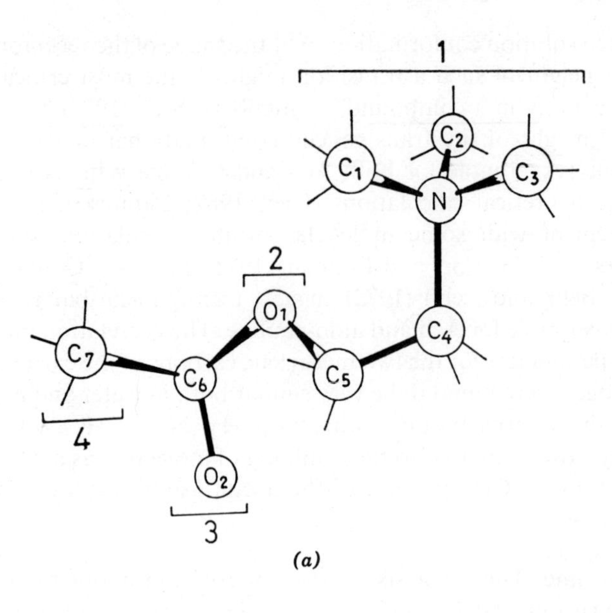

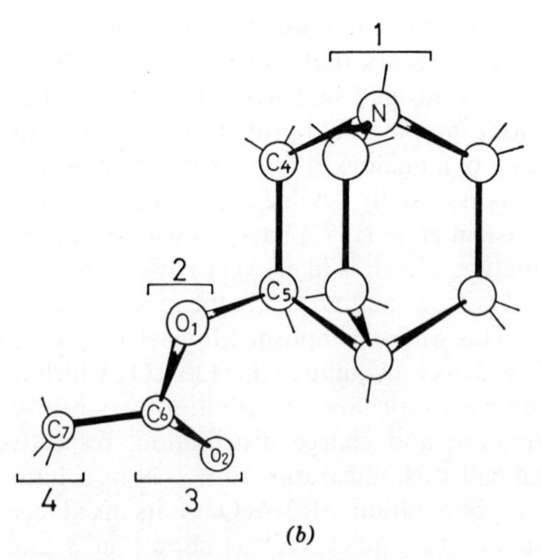

Fig. 2-5 *a*. Notation and active groups in acetylcholine. *b*. Notation and active groups in 3-acetyoxyquinuclidine. *c*. Proposed superposition of the configuration of 3-AcQ and ACh corresponding to similar active group patterns. Weinstein et al. (1973).

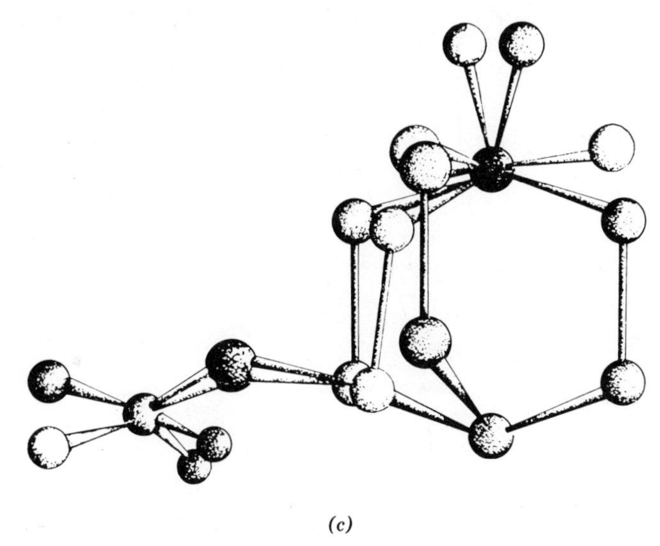

(c)

Fig. 2-5 (*Continued*)

atoms with the others' boats. The (−) enantiomer has the (R) configuration and the stronger muscarinic activity. The high activity of the tertiary compound indicates that the parts of the quinuclidine ring, which are not present in Ach, only marginally disturb the approach to, or the reaction with, the muscarinic receptor, and have virtually no effect on enzymatic hydrolysis interactions. This is particularly interesting because the quaternary salt (Mashkovsky, 1963; Weinstein et al., 1973) has a reduction in affinity by a factor of 200, showing that this class of structural changes, which are smaller than those involving ring geometry, has large effects on the receptor interaction. A direct measure of the muscarinic activity of Ach, (±) 3-AcQ and (±) 3-AcQ-CH_3·I, may be derived from the magnitude of the response evoked in the isolated perfused guinea pig ileum by a given concentration of the compound. Weinstein et al. (1973) have reported the dose-response curves shown in Figure 2-6. The curves in this figure are also similar in shape, suggesting that the receptor uptake of the drug molecule is probably the same for all three compounds. The stringent requirements imposed by the muscarinic receptor site are not necessarily shared by other systems involving interaction with Ach, or the 3-AcQ and 3-AcQ-CH_3·I congeners. Both (±) 3-AcQ and its methiodide are good substrates for the enzymes acetylcholinesterase and butyrylcholinesterase, as shown in Table 2-2. The activity differences of (±) 3-AcQ interacting with muscarinic and enzymatic receptors lends support to the theory that these two types of receptors are dissimilar (Belleau, 1970). However, the moderate activity of (±) 3-AcQ with

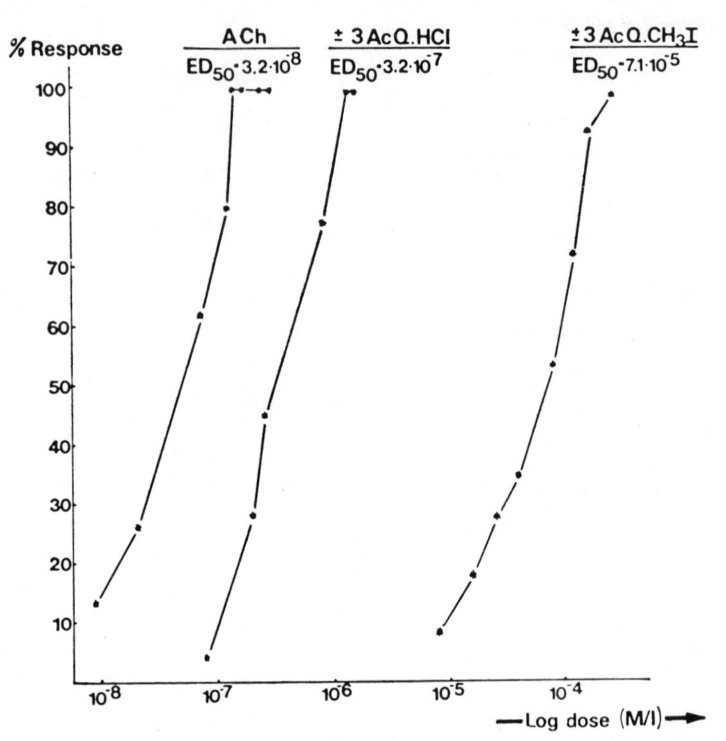

Fig. 2-6 Relative potency of acetylcholine, (±) 3-acetyoxyquinuclidine and its methiodide in perfused isolated guinea-pig ileum. Weinstein et al. (1973).

Table 2-2 Kinetic Data for the Enzymatic Hydrolysis of 3-Acetoxyquinuclidine and its Methiodide

	AChE		BuChE	
Substrate	$K_m(\cdot 10^{-4})$	$V_{max}(\cdot 10^{6})$ (M/min)	$K_m(\cdot 10^{-4})$	$V_{max}(\cdot 10^{6})$ (M/min)
ACh	1.5	3.8	7.5	1.00
(±) 3-AcQ·HCl	5.0	1.0	27.0	0.05
(±) 3-AcQ·CH_3I	5.0	3.3	11.0	0.44
(+) AcQ·CH_3I	Inhibitor		Inhibitor	
BuCh	—	—	3.6	1.60
(±) 3-QBu·CH_3I	—	—	2.3	1.10

ACh = acetylcholine AChE = acetylcholinesterase. BuCh = butyrylcholine. BuChe = butyrylcholinesterase. 3-QBu = 3-butyroxyquinuclidine. Weinstein et al. (1973).

the muscarinic receptor and high activity with enzymes suggests that Ach engages both classes of active sites in very nearly the preferred solution conformation, that is, a gauche conformation with respect to τ_2. This does not, however, imply that all Ach-congeners must be in this conformational state when successfully engaging the receptors. This is amplified elsewhere in this section by the analysis of other congeners.

Lichtenberg et al. (1974) assumed the quantitative validity of the Karplus relationship (Karplus, 1959) and determined the gauche rotamer population about τ_2 to be 91% for Ach; the value of τ_2 was between 65° and 69°, a slight perturbation from true gauche.

By studying the conformations of molecules of the type:

$$R\text{—}\overset{\overset{A}{\|}}{C}B\text{CH}_2\overset{\overset{R''}{|}}{\text{CH}_2}\text{N}R'_2 \qquad \begin{aligned} R &= \text{—C}_6\text{H}_5, \text{—CH}_3 \\ R' &= \text{—CH}_3 \\ R'' &= \text{H, CH}_3 \\ A &= \text{O, S, Se} \\ B &= \text{O, S, Se} \end{aligned}$$

in solution, as well as that of the dimethylsulfonium analog of Ach, the following conclusions have been drawn by Mautner (1974):

$$\text{CH}_3\overset{\overset{\text{O}}{\|}}{\text{C}}\text{OCH}_2\text{CH}_2\overset{+}{\text{S}}(\text{CH}_3)_2$$

dimethylsulfonium analog of Ach

1. The quasicyclic, gauche conformation of Ach and of related β-ammoniumethyl esters is not induced by structured solvent. It is seen not only in D_2O but also in solvents as nonpolar as chloroform.
2. A positively charged group in the β-position (whether trimethylammonium, dimethylammonium, or dimethylsulfonium) tends to interact with the acyloxy oxygen of esters and, to a lesser extent with the nitrogen of amides, and is required for the *gauche* conformation of the —OCCN— grouping.
3. In the absence of a positive charge in the β-position, no specific rotamer is favored. Thus, 2-dimethylaminoethylbenzoate (Bartels, 1965) or 2-dimethylaminoethylacetate (Casy et al., 1971) will not be in the gauche conformation if studied at pHs at which their amino groups are not protonated (Table 2-3).
4. Replacement of the carbonyl oxygens of Ach and related esters with sulfur, or, in the case of local anesthetics, with sulfur with selenium, has only minor conformational effects.
5. Replacement of the acyloxy oxygens of such esters with sulfur or with selenium converts the gauche to the trans conformation.

Table 2-3

Compound	Percent Conformation	
	Gauche	*Trans*
$(CH_3)_2NCH_2CH_2OC(=O)C_6H_5$	67	33
$(CH_3)_2NHCH_2CH_2OC(=O)C_6H_5$	100	0

Mautner (1974).

6. Replacement of the carbonyl oxygen of 2-trimethylammoniumethyl of 2-dimethylammoniumethyl amides with sulfur or with selenium progressively favors the trans conformation. This is presumably due to the increasing contribution of charge-separated forms of the type:

$$-\overset{+}{N}H{=}\underset{A^-}{\underset{|}{C}}- \qquad A = O, S, Se$$

on descending the periodic table (Krackov et al., 1965) (Table 2-4).

7. In all cases studied, the conformation prevailing in the crystal was that prevailing in solution. This appears to be generally true (Casy et al., 1971;

Table 2-4

Compound	Percent Conformation	
	Gauche	*Trans*
$(C_2H_5)_2NHCH_2CH_2NHC(=O)C_6H_5$	74	26
$(CH_2H_5)_2NHCH_2CH_2NHC(=S)C_6H_5$	58	42
$(C_2H_5)_2NHCH_2CH_2NHC(=Se)C_6H_5$	51	49

Mautner (1974).

Partington et al., 1972; Makruyannis et al., 1973; Cushley and Mautner, 1970). The only exception known in this group of compounds is carbamylcholine (Partington et al., 1972; Barrans and Clastre, 1970) and its N-phyenyl analog (Ajo, et al., 1973).

8. There is no simple relationship between conformation and biologic activity. For instance, whereas the cholinergic activity of Ach (gauche) exceeds that of acetylthiolcholine or acetylselenolcholine (trans), the depolarizing activity of methoxycholine (gauche) is much lower than that of its methylthio or methylseleno analogs (trans). Similarly, the local anesthetic activity of 2-dialkylaminoethyl thiolesters and selenolesters (trans) exceeds that of their ester analogs (gauche) (Mautner et al., 1972; Mautner, 1972).

By cleaving the ethylene bridge at C_4 in 3-AcQ and by shortening the ethylene group bonded to the nitrogen to methyl, one obtains the acetate of *N*-methyl-3-piperidinol. This compound and its methiodide were described as muscarinic agonists (Lambrecht, 1971). On thermodynamic grounds, the *N*-methyl-3-acetoxy-piperdine is almost completely in the chair form. An enthalpy difference of 5.5 kcal/mole and a entropy difference of 5 cal/deg·mole have been reported (Lyle et al., 1966; Law, 1961; Eliel, 1966) between the boat and chair forms of six-member alicyclic or heterocylic saturated rings. Consequently, at physiologic temperature (35°C), 0.15% of the *N*-methyl-3-acetoxypiperdine will be in the boat form at thermodynamic equilibrium, that is, one molecule in 650 will be in the boat form.

In Figure 2-7, the possible stereochemical configurations of *N*-methyl-3-acetoxypiperdine are shown. Configurations III and VI correspond to the configurations of the enantiomers of 3-AcQ.

Using the strong muscarinic activity of 3-AcQ for reference, one can postulate with reasonable certainity that the molecules in the boat form in

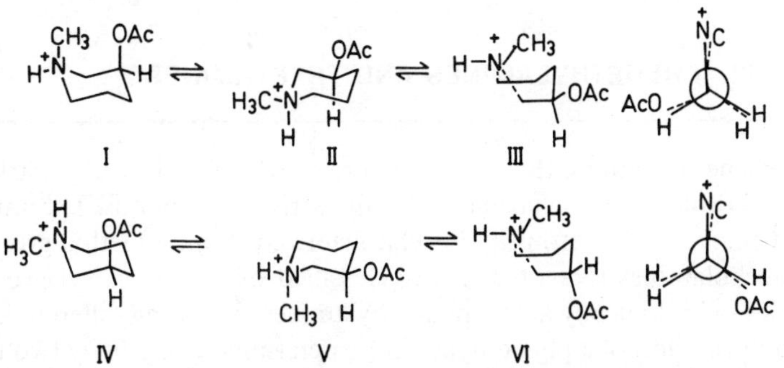

Fig. 2-7 Theoretically possible conformations of *N*-Methyl-3-acetoxypiperidinium ion. One optical enantiomer only is shown. Lambrecht and Mutschler (1974).

the thermodynamic equilibrium of *N*-methyl-3-acetoxypiperdine are responsible for its muscarinic action. Therefore, the difference in the biologic activity of the two drugs should be of the same magnitude as that expected from the proportions of the chair and boat forms of *N*-methyl-3-acetoxypiperdine. Hence comparable pharmacologic investigation of these substances can indicate whether the reaction of muscarinic agents with the muscarinic receptor will produce a sufficient free energy gain to exceed the energy barrier between the boat and the chair forms of six-membered heterocyclic saturated rings. Lambrecht and Mutschler (1974) have compared the muscarinic activity of the two tertiary pairs 3-AcQ and *N*-methyl-3-acetoxypiperdine as well as methyl-*N*-methyl-piperdine-3-carboxylate and methyl quinuclidine-3-carboxylate. If the receptor-agonist interaction induces the energetically unfavorable boat form of the two piperdine derivatives, these two substances should exhibit approximately the same affinity as the quinuclidine compounds. This is, in fact, not the case. This is especially apparent when one compares the percentage of the *N*-methyl-3-acetoxypiperdine in the boat form in the thermodynamic equilibrium and the difference in muscarinic activity between this compound and 3-AcQ. *N*-methyl-3-acetoxypiperdine has 513 times less activity, and, as mentioned above, one in about 650 molecules is in the boat form. Thus one concludes that the interaction of *N*-methyl-3-acetoxypiperdine with the muscarinic receptor does not alter the equilibrium between the boat and chair conformations. Such studies do not resolve whether the *N*-methyl group of the molecules of the piperdine derivatives that are in the boat form is pseudoaxial or pseudoequatorial in the receptor complex. Complementary work by Lambrecht and Mutschler (1974) involving the quaternary salt of 3-AcQ suggests that the pseudoaxial form is the active conformation. However, the important point is that the muscarinic receptor-agonist interaction is not capable of overcoming a conformational energy barrier of 5–6 kcal/mole in the agonist molecule.

III PHENETHYLAMINES AND TRYPTAMINES

Some phenethylamines, the base structure of which is shown as part of the legend of Table 2-5, are pharmacologically active compounds. The amphetamines (Costa and Garattini, 1970), characterized by an α-methyl group, and the catecholamines (Nagotsu, 1973), classified by a β-hydroxyl group, are examples of biologically active phenethylamines exhibiting different actions. The conformation of a phenethylamine is determined largely by two torsion angles, τ_1 and τ_2, also defined in Table 2-5. τ_3, rotation about the C_α—N^+ bond, is of conformational significance when substitutions are made on the

Table 2-5

Conformation of phenethylamines in the solid state and solution. The rotamer population in solution is presented as percent following the conformational code symbol. The general phenethylamine structure is

The definitions of the rotamer states using Newman projections for τ_2 are

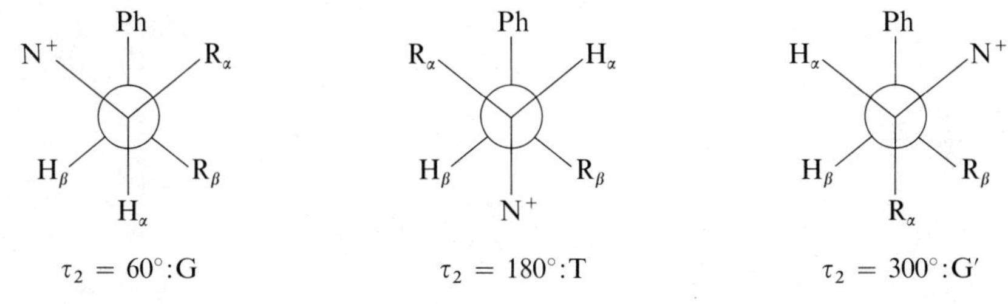

Unless otherwise indicated the solvent is D_2O, and the temperature is approximately 30°C.

The definition of the rotamer states using Newman projections for τ_1 are

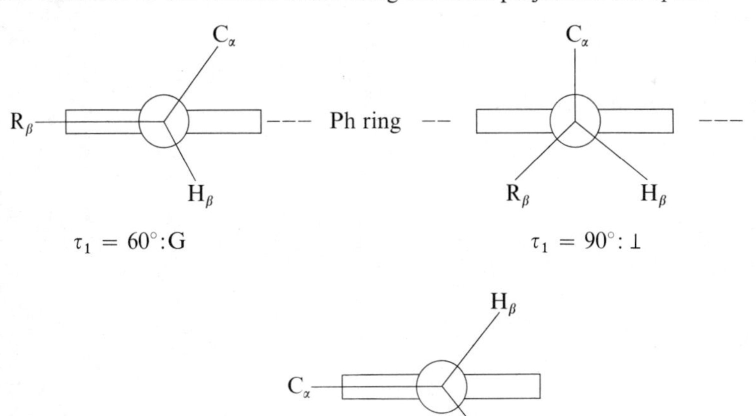

$\tau_1 = 180°$:||

Table 2-5 (*Continued*)

Compound	Solid State τ_1	Solid State τ_2	Ref.	Solution τ_1(%)	Solution τ_2(%)
β-Phenylethylamine·HCl	70°	171°	1	—	G + G′(44)
					T(56)
L(+)ephedrine·H_2PO_4	—	166	23	—	
L(+)ephedrine·HCl	—	169	3	—	G(10)
		164	2	—	T + G′(90)
L(+)ephedrine·HPO_4	—	183	22		
	—	161			
ψ-ephedrine·HCl	—	—	—	—	G + G′(15–17)
				—	T(85–83)
Dopamine·HCl	80	174	4	—	T(43)
				—	G + G′(57)
Noradrenaline·HCl	83	176	5	—	180°
				—	G(14)
				—	T(76)
				—	G′(10)
Amphetamine·H_2SO_4	72	176		—	—
	77	173	6	—	—
	70	172		—	—
	83	166		—	—
Amphetamine·HCl	—	—	—	—	G(45)
				—	T(50)
				—	G′(05)
Amphetamine(free base)	—	—	—	—	G(39)
				—	T(50)
				—	G′(11)
Methamphetamine·HCl	—	—	—	—	G(39)
				—	T(55)
				—	G′(06)
o-methyoxymethamph-etamine·HCl	—	—	—	—	G(36)
				—	T(47)
				—	G′(17)
Benzphetamine·HCl	—	—	—	—	G(35)
				—	T(64)
				—	G′(01)
Mescaline·HCl	89	175	7	—	—
Mescaline·HBr	—	G	8	—	—
Isoprenaline	—	191	21[a]	—	G(11)
	—	178		—	T(83)
				—	G′(06)

[a] Sulfate complex.

Ref.	R_1	R_2	R_3	R_4	R_5	R_6	R_α	R_β	R_7	
15	H	H	H	H	H	H	H	H	H	
16	CH_3	H	H	H	H	H	CH_3	OH	H	[erythro (OH,Me)]
16	CH_3	H	H	H	H	H	CH_3	OH	H	[threo (OH,Me)]
17	H	H	H	OH	OH	H	H	H	H	
18	H	H	OH	OH	H	H	H	H	H	
15	H	H	H	H	H	H	CH_3	H	H	
15	H	H	H	H	H	H	CH_3	H	H	
19	H	H	H	H	H	H	CH_3	H	H	[NH_2 not N^+H_3]
19	CH_3	H	H	H	H	H	CH_3	H	H	
19	CH_3	H	H	OCH_3	H	H	CH_3	H	H	
19	CH_3	H	H	H	H	H	CH_3	H	Ph	
—	H	H	OCH_3	OCH_3	OCH_3	H	H	H	H	
—										
15	iPr	H	OH	OH	H	H	H	OH	H	

Table 2-5 (*Continued*)

Compound	Solid State τ_1	τ_2	Ref.	Solution τ_1(%)	τ_2(%)
Pseudo-ephedrine	75	172		—	—
Cu complex	79	164	9	—	—
	71	179		—	—
2,4,5 trimethoxy-amphetamine·HCl	67	50	10	—	—
3,4,5 trihydroxyphenethylamine	—	T	11	—	T(?)
2,4,5 trihydroxyphenehtylamine	—	T	12	—	—
Adrenaline	⊥	T	13	—	G(17)
				—	T(77)
				—	G′(06)
Adrenalone	⊥	T	14	—	—
β-phenylethanolamine	—	—	—	—	G(10)
				—	T(84)
				—	G′(06)
Phenylephrine	—	—	—	—	G(16)
				—	T(81)
				—	G′(03)
Synephrine	—	—	—	—	G(10)
				—	T(81)
				—	G′(09)
N,N-dimethyl-2-(*o*-bromophenyl)-ethanolamine	—	—	—	—	G(00)
				—	T(96)
				—	G′(04)
Isopropanolamine	—	—	—	—	G(13)
				—	T(79)
				—	G′(08)
Norephedrine	—	—	—	—	G(21)
				—	T + G′(79)
Metaraminol	—	—	—	—	G(21)
				—	T + G′(79)
Butanephrine	—	—	—	—	G(28)
				—	T + G′(72)

Ref.	Structure								
	R_1	R_2	R_3	R_4	R_5	R_6	R_α	R_β	R_7
—	CH_3	H	H	H	H	H	CH_3	OH	H
—									
—									
—	H	OCH_3	H	OCH_3	OCH_3	H	CH_3	H	H
20	H	H	OH	OH	OH	H	H	H	H
—	H	OH	H	OH	OH	H	H	H	H
15	CH_3	H	OH	OH	H	H	H	OH	H
	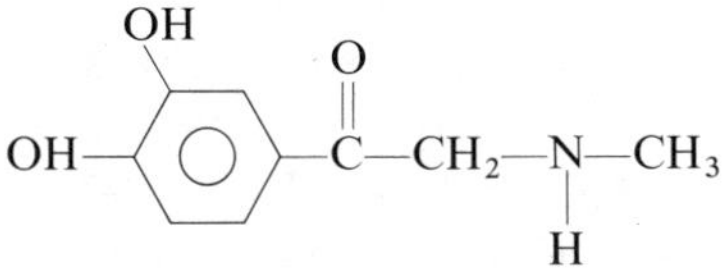 OH, OH—(benzene ring)—C(=O)—CH_2—N(H)—CH_3								
15	H	H	H	H	H	H	H	OH	H
15	CH_3	H	OH	H	H	H	H	OH	H
15	CH_3	H	OH	H	H	H	H	OH	H
15	CH_3	H	Br	H	H	H	H	OH	CH_3
	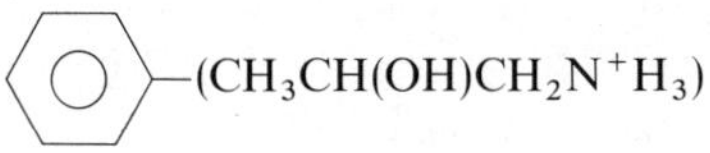 (benzene ring)—$(CH_3CH(OH)CH_2N^+H_3)$								
15	H	H	OH	H	H	H	CH_3	OH	H
15	H	H	H	H	H	H	CH_3	OH	H
15	H	H	OH	OH	H	H	CH_2CH_3	OH	H

Table 2-5 Key

REF.

1. Tsourcaris (1961).
2. Phillips (1954).
3. Bergin (1971).
4. Bergin and Carlström (1968).
5. Carlström and Bergin (1967).
6. Bergin and Carlström (1971).
7. Tsoucaris et al. (1973).
8. Ernst and Cagle (1973).
9. Bailey et al. (1968).
10. Baker et al. (1973).
11. Koldercup et al. (1972).
12. Anderson et al. (1972).
13. Carlström (1973).
14. Bergin (1971).
15. Ison et al. (1973).
16. Porthogese (1967).
17. Bustard and Egan (1971).
18. Gieessner-Prettre and Pullman (1975).
19. Neville, et al. (1971).
20. Pullman et al. (1974).
21. Mathews and Palenik (1971).
22. Hearn et al. (1972).
23. Hearn and Bugg (1972).

(N^+H_3) group, or when the ring contains proton-accepting groups to form hydrogen bonds with the (N^+H_3) protons, as in the histamines.

The torsion angle, τ_1, indicates the orientation of the plane of the side chain with respect to the plane of the ring (e.g., coplanar for $\tau_1 = 0°$ or $180°$, perpendicular for $\tau_1 = 90°$), and the τ_2 defines the orientation of the cationic head with respect to the ring (e.g., gauche for $\tau_2 = 60°$ or $300°$, trans for $\tau_2 = 180°$).

Values of τ_2 have been determined for a number of biologic phenethylamines in the solid state by X-ray crystallography (for a review, see Carlström et al., 1973) and in solution by high-resolution PMR spectroscopy. Table 2-5 contains the conformational properties of a variety of phenethylamines in solution and the solid state. Making the quantitative estimation of the rotamer populations in solution required certain critical assumptions and numerical assignments, which are outlined in general at the beginning of this chapter and discussed in detail specifically for the phenethylamines in Ison et al., 1973; Feeney et al., 1974; and Roberts, 1974.

In the crystal structures, the conformational state about $C_\alpha—C_\beta$ (τ_2) is trans for all compounds investigated except mescaline·HBr and 2,4,5-trimethoxyamphetamine·HCl. However, in aqueous solution, the phenethylamines must be regarded as flexible. A mixture of trans and gauche forms is observed with respect to τ_2. For the catecholamines, the trans form is populated roughly 80% of the time, whereas the other phenethylamines, including the amphetaamines, are in the trans state only about 55% of the time. Even the 80% trans preference in the catecholamines corresponds to a gauche-trans energy difference of about only 0.8 kcal/mole. The added stability in the trans conformation for the catecholamines has been attributed to an interaction between the β-hydroxyl group and the amino group. Other substitutions in the molecule have, in general, relatively small effects on rotamer populations. Ring hydroxyl groups tend to favor the gauche and gauche conformers. These effects may be due to an interaction between the amino group and the ring hydroxyl groups through an intermediatory water molecule (Bustard and Egan, 1971).

Values of τ_1 are known in the solid state. In all molecules studied, the preferred rotamer state possesses a τ_1 value of about 80°. This is a distorted gauche conformation, placing the extended side chain nearly perpendicular to the ring. Only adrenaline (Carlström, 1973) and adrenalone (Bergin, 1971) have different τ_1 values such that, for both molecules, the extended side chain lies nearly in the plane of the ring. A theoretical calculation of some phenethylamines, including adrenaline (Pullman et al., 1972), predicts $\tau_1 \cong 90°$ in all the compounds. Consequently, one might tentatively assign the unusual τ_1 conformations in adrenaline and adrenalone observed in the solid state to crystal packing interactions. No values of τ_1 in solution for this class of compounds have been established, although some theoretical work has been reported (Giessner-Prettre and Pullman, 1975).

Mescaline presents a perplexing problem in conformational preference. In the crystal of its hydrobromide (Ernst and Cagle, 1973), the ethylamine side chain is gauche with respect to the ring, whereas in the crystal of its hydrochloride (Kolderup et al., 1972) the side chain is trans to the ring. This raises the question of which is the intrinsically preferred conformation of the free molecule and what is therefore the role of the crystal packing forces producing the observed distortions. A complementary question concerns the energy difference and barrier between the two forms. The problem is of particular interest because of the divergent proposals that have been put forward in recent years with regard to structural analogies between LSD and mescaline (and also psychotomimetic indolalkylamines). Snyder and Richelson (1968) have suggested that a folded conformation of the ethylamine side chain is the basis of activity; others argue for the extended side chain conformation (Baker et al., 1974; Chothia and Pauling, 1969; Kang and Green, 1970; Johnson, et al., 1973).

The related compound, 2,4,5-trimethoxyamphetamine, also a potent psychotomemetic, exists in the gauche form in the crystal (Baker et al., 1973); two other congeners, 3,4,5- and 2,4,5-trihydroxyphenethylamines exist in the crystal in the trans form (Koldercup et al., 1972; Andersen et al., 1972). Pullman et al. (1974) have carried out conformational energy calculations on mescaline and related compounds in free space. For mescaline, the two different crystal valence-bond geometries lead to different predictions for the preferred conformation that agree in each case with the crystallogrographic findings. This result is another example (like that of acetylcholine discussed in section II of this chapter) of the decisive importance of the valence-bond geometry in determining the preferred conformation. It is apparently through the differences in geometry induced by the crystalline intermolecular force field that the intramolecular forces produce their major effect on preferred conformation.

The tryptamines, whose general backbone structure is shown below, are structurally similar to the phenethylamines. Moreover, many tryptamines have similar types of activities to the phenethylamines. Serotonin (5-hydroxytryptamine) is a central nervous transmitter, whereas *N*, *N*-dimethyltryptamine and psilocybin are both potent hallucinogens. The definitions of conformational states reported in Table 2-5 hold for the conformations of the tryptamines reported in Table 2-6. The side chain is seen to adopt the gauche (τ_1 *and* τ_2) conformer state or the trans (τ_1 *and* τ_2) conformation for four compounds; psilocybin is the exception. It is interesting to note that approximate gauche conformations about τ_1 in the phenethylamines were characterized by a trans conformation about τ_2. 2,4,5 trimethoxyamphetamine HCl is the only exception. In the tryptamines, a gauche τ_1 corresponds to a gauche τ_2 except for psilocybin, which is identical to the phenethylamines. A trans τ_1 in the tryptamines is complemented by a trans τ_2. Pauling (1973) has suggested that $\tau_1 \approx 70°$ and $\tau_2 \approx 180°$ is relevant to biologic activity

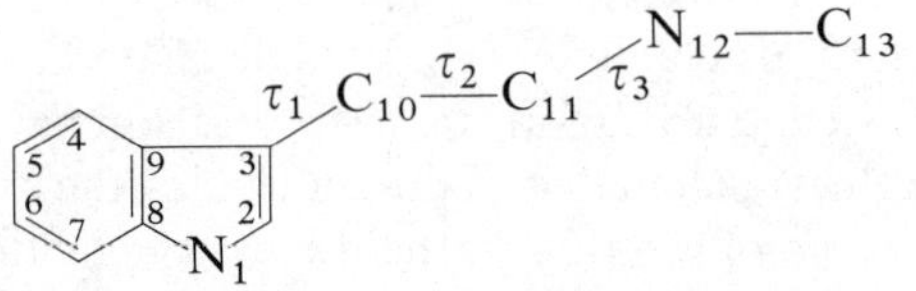

Table 2-6 Conformations of Some Tryptamines

Compound	Method	τ_1[a]	τ_2	Reference
Serotonin picrate	X-ray	66	67	Thewalt & Bugg (1972)
Serotonin Creatinine sulphate	X-ray	12	173	Karle et al., (1965)
Tryptamine HCl	X-ray	69	60	Wakahara et al., (1970)
N,N-Dimethyl-5-methoxytryptamine	X-ray	17	179	Falkenberg & Carlström (1971)
Psilocybin a	X-ray	69	174	Pauling, (1973)
b		103	−165	

[a] τ_1 Refers to $C_9—C_3—C_{10}—C_{11}$.
Pauling (1973).

because of the correlation with the LSD conformation (Baker et al., 1973) as observed in psilocybin.

IV HISTAMINES AND RELATED COMPOUNDS

There is evidence that the physiologic actions of histamine (Black et al., 1972) can be mediated by two types of receptors, the so-called H_1 and H_2 receptors. Kier (1968) made the interesting suggestion, based on his extended Hückel calculations, that the dual activity of histamine might be a consequence of the existence of two preferred conformations of the histamine monocation, namely, the trans and gauche forms illustrated in Figure 2-8. In order to explore further the reasonableness of this hypothesis, the conformation of several histamines, antihistamines, and related compounds have been determined in solution and in the solid state.

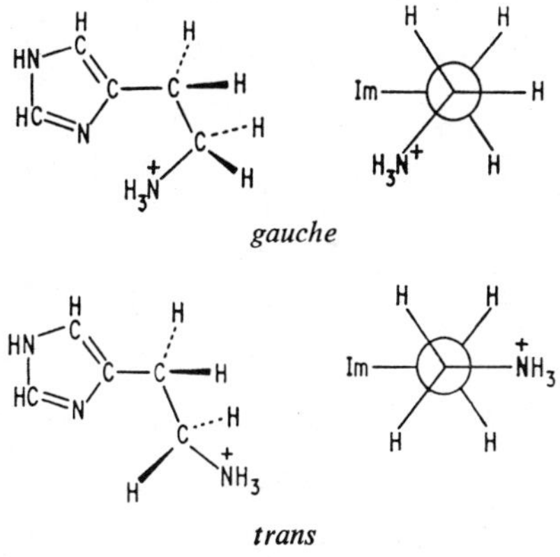

gauche

trans

Fig. 2-8 Gauche and trans conformers of histamine. Ganellin et al. (1973).

Table 2-7 lists the τ_2 rotamer fractions of several histamine-type compounds in solution at various pHs and solution compositions. As a point of reference, in a completely random conformational distribution, the trans form is populated at a one-third fraction and the gauche forms at a combined two-thirds fraction (two gauche forms to one trans form). On this basis, the data in Table 2-7 suggest that the trans conformation is slightly preferred in solution at low pHs, and this trans preference tends to increase with increasing pH. The exceptions to this observation are the α and β methyl-substituted species in which the gauche forms comprise nearly 100% of the

Table 2-7 Solution Conformation Populations of Some Histamines

Part I
Reference Structure

R_9, R_{10} (on N and C of imidazole ring), R_8 (on ring C) — C $\overset{\tau_1}{\circlearrowleft}$ $C_\beta(R_6)(R_7)$ $\overset{\tau_2}{\circlearrowleft}$ $C_\alpha(R_4)(R_5)$ $\overset{\tau_3}{\circlearrowleft}$ $N^+(R_1)(R_3)$—R_2

Compound											Rotamer Fraction About τ_2		
R_1	R_2	R_3	R_4	R_5	R_6	R_7	R_8	R_9	R_{10}	Ref.	N_t	$N_g + N'_g$	Solution
H	H	H	H	H	H	H	H	H	H	1	0.47	0.53	pH 7
										1	0.55	0.45	0.1 *N* D_2SO_4
H	Me	Me	H	H	H	H	H	H	H	1	0.66	0.34	pH 4
										1	0.60	0.40	pH 6.5
										1	0.52	0.48	pH 8.5
										2	0.72	0.28	—
Me	Me	Me	H	H	H	H	H	H	H	1	0.76	0.24	—
H	H	H	H	H	H	H	H	H	Me	2	55[a]	45[a]	—
H	H	H	H	H	H	H	H	Me	H	2	57	43	—
H	H	H	H	H	H	H	Me	H	H	2	57	43	—
H	H	H	H	H	($\frac{1}{2}$Me + $\frac{1}{2}$Me)		H	H	H	2	2[a]	98[a]	—
H	H	H	($\frac{1}{2}$Me + $\frac{1}{2}$Me)		H	H	H	H	H	2	10[a]	90[a]	—

Part II Solution Conformation Populations of Some Histamine-Related Compounds

Compound	Ref.	Rotamer Fraction About τ_2: N_t	$N_g + N_g$	Solution
Betazole (pyrazole ring: H–C=C–H, N–H, N^+H; side chain CH_2–CH_2–N^+H_3)	1	0.52	0.48	pH 7.3
		0.61	0.39	pH1
5-Hydroxytryptamine (HO-indole, NH; side chain CH_2–CH_2–N^+H_3)	1	0.44	0.56	—
Bisnorpheniramine (H–C(phenyl)(pyridyl)–CH_2–CH_2–NH_2)	3	0.53	0.47	D_2O(80°C)
	3	0.92	0.08	D_2O/D_2SO_4(80°C)

(1) Ison (1974).
(2) Ganellin et al. (1973).
(3) Testa (1974).
[a] Estimated from a combination of experimental and theoretical data.

conformational states. This distinct preference is the result of the large steric bulk of the methyl group as compared to that of the proton.

Although no X-ray diffraction analyses have been done on the monocation of histamine, studies of related compounds show that the bimethylene side chain of histamine-diphosphate monohydrate (Veidis and Palenik, 1969; Veidis et al., 1969) tetrachlorocobaltate (Bonnet and Jeannin, 1972), bromide (Decon, 1972), sulfate monohydrate (Yamane et al., 1973), and the free base (Bonnet and Ibers, 1973) is, in the solid state, always in the τ_2-trans conformation. The creatinine sulphate complex, a 5-hydroxy-typtamine analog, is also in the τ_2-trans form in the solid state (Karle et al., 1965), whereas another analog, picrate, is in the τ_2-gauche form in crystals (Bugg and Thewalt, 1970). The solution rotamer populations of the histamines indicates an energy difference between the trans and gauche conformational states of about only 0.3–0.5 kcal/mole for the τ_2 rotation. Rotational pref-

C

erence about C $\circlearrowright$ C^{β}, defined as τ_1, is quite variable in the solid state.

N

The bimethylene side chain of histamine diphosphate bromide lies in the plane of the ring; the tetrachlorocobaltate of histamine, as well as histamine sulfate monohydrate, is perpendicular to the ring, and histamine bromide lies in one of the two planes making an angle of 60° with the plane of the ring. The observed conformation about C^{α} $\circlearrowright$ $N^{+}H_3$, defined as τ_3, is gauche with respect to the ring for the histidine cation, regardless of orthorhombic or monomclinic crystal packing (Donohue and Caron, 1964; Bennett et al., 1970; Candlin and Harding, 1970; Madden et al., 1972 *a,b*). Neutron diffraction studies of the histidine monocation indicates the positions of the hydrogen atoms explicitly and reveals the existence of a bent hydrogen bond N_1 (ring) $\cdots$ H—$N^{+}<$ (Lehmann, et al., 1972). Interestingly, the τ_1 and τ_2 rotational states are also gauche in the ethylamine side chain of histidine.

Because the drugs known as "antihistamines" interact with only one of the two or more types of histamine receptors, they form a group that appears homogeneous in its specific mechanism of action. Their basic structure is a short chain of three or four atoms (with or without a hetero-atom, and possibly part of an alicycle) carrying at one end a basic center, and at the other end two geminal or vicinal aromatic nuclei (which in some cases are part of a tricyclic system).

Like the other chiral antihistaminic drugs, the compounds known as the pheniramines, shown in Figure 2-9 (i.e., pheniramine-I, chlorpheniramine-II, and brompheniramine-III) exhibit a stereoselective activity. The (+)-enantiomers are the active forms and have the S absolute configuration (Shofi'ee and Hite, 1969). Six staggered conformations about the bonds of

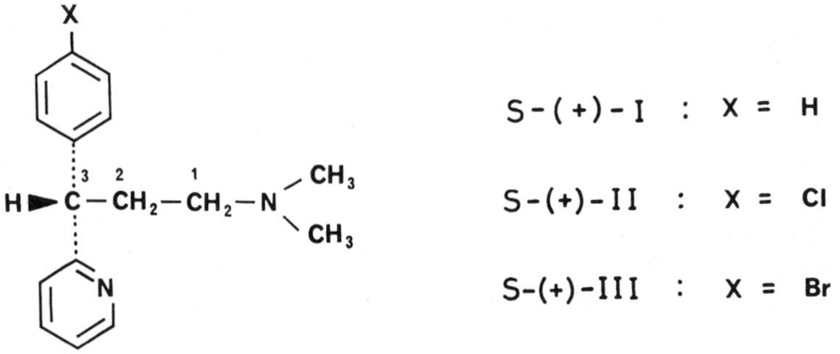

S-(+)-I : X = H

S-(+)-II : X = Cl

S-(+)-III : X = Br

Fig. 2-9 Three pheniramines. Testa (1974).

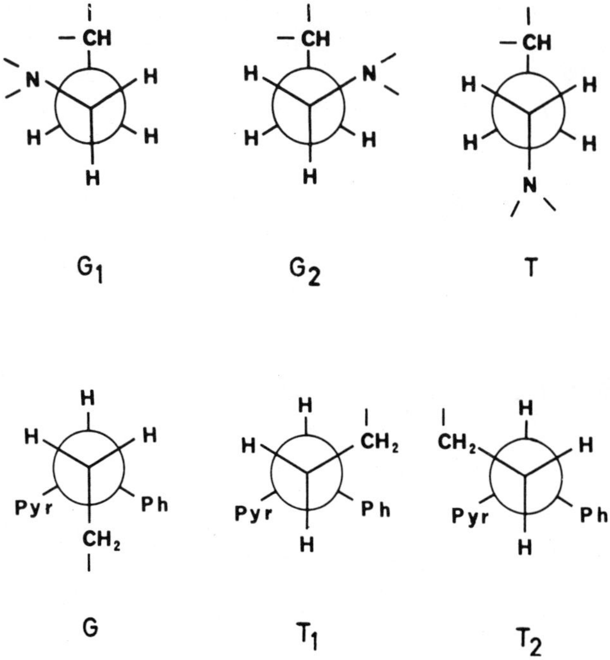

G_1 G_2 T

G T_1 T_2

Fig. 2-10 The staggered conformations about C_1—C_2 and C_2—C_3 bonds of pheniramine. Testa (1974).

the C^3—CH_2^2—CH_2^1—$N(CH_3)_2$ side chain are possible in the pheniramines and are shown in Figure 2-10. In a crystalline form (±)-brompheniramine maleate is observed to be in the fully extended trans conformation (James and Williams, 1971). Testa (1974) has observed marked differences in the CD spectra between the diprotonated, monoprotonated, and nonprotonated species of S-(+)-pheniramine and S-(+)-chlorpheniramine. These spectral differences have been interpreted in terms of the dication existing in a fully extended conformation, and the monocation (the only significant form at physiologic pH) partially folded because of an attractive interaction between ring and side chain nitrogen atoms that is markedly increased in methanol and chloroform. The antihistaminic analog, *bis*-norpheniramine, exists in both folded and extended conformations in D_2O with a significantly increased proportion of the latter in D_2O/D_2SO_4, as shown in Table 2-7. Conformational studies of other antihistamines with flexible side chains indicate that folded and extended conformers coexist in solution (Ham, 1971).

The question of the active conformation of histamine at H_1 receptors has been studied by several authors, and there is general agreement that an extended side chain is essential (Pertiti, 1970; Ganellin, et al., 1973; Ganellin, 1973). Of particular interest is a recent study with conformationally restricted analogs (Shunack, 1973). Among the three positional isomers of piperidyl- and aminocyclohexyl-imidazole, only 4-(3-piperidyl)-imidazole (I) and 4-(2-aminocyclohexyl)-imidazole (II) (seemingly cis + trans) behave as agonists of the H_1 receptor, thus providing additional evidence for an active extended form.

The same conclusions were reached through studies of the structure-activity relationship of H_1 receptor competitive antagonists, that is, the therapeutic class of antihistaminic drugs. Support is provided by conformationally restricted antihistamines such as 1,5-diphenyl-3-dimethylamino-pyrrolidine (III). The trans isomer is found to be a potent and selective long-acting antagonist, whereas the cis isomer is a potent reversible antagonist (Hanna and Ahmed, 1973). This compound is interesting because it suggests

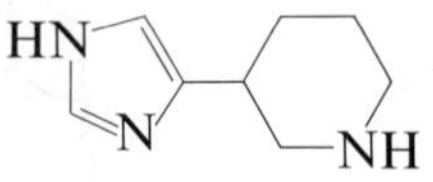

HN N NH

(I)

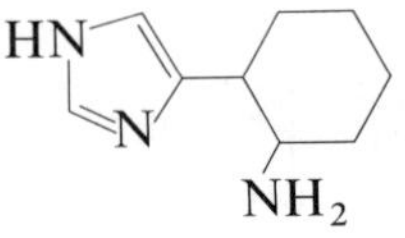

HN N NH_2

(II)

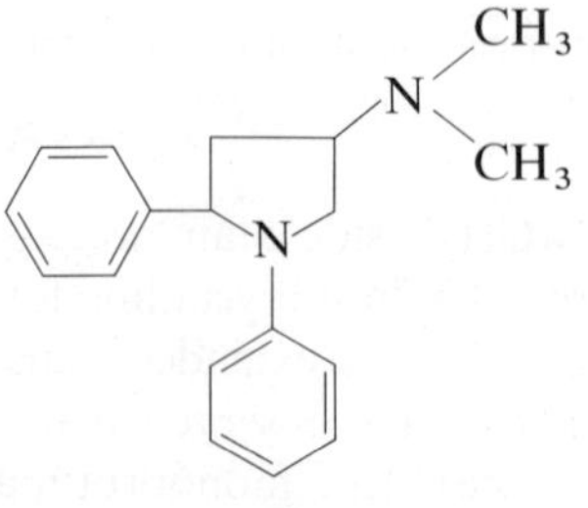

CH_3 N CH_3 N

(III)

that the fully extended form is not a strict requirement for an interaction with the H_1 receptor. Certain variations in the degree of extension of the side chain can be tolerated for activity. An extensional range is also apparent in a series of antihistamines whose conformational freedom is restricted by a double bond (Casy and Ison, 1970; Ison et al., 1973).

V STRUCTURAL ANALYSIS OF α-AMINO ACID RESIDUE UNITS

The α-amino acid residue units, $-\underset{\mathrm{H}}{\underset{|}{\mathrm{N}}} \overset{\phi}{\curvearrowright} \underset{\mathrm{H}}{\underset{|}{\overset{\mathrm{R}}{\overset{|}{\mathrm{C}^{\alpha}}}}} \overset{\psi}{\curvearrowright} \overset{\mathrm{O}}{\overset{|}{\mathrm{C}}}-$, are a set of congeneric compounds whose structural properties have been investigated in several different molecular environments. As such, this class of compounds is well suited to exploration of the influence of external forces on the spatial organization of a set of well-defined structural units. Table 2-8 contains a summary of the conformational behavior of the α-amino acid residue unit when part of (1) the monomeric acid, (2) the "dipeptide", that is, *N*-acetyl-*N′* methyl-(X)-amide, structure, and (3) the homopoly(α-amino acid). Both solution and/or solid-state data are reported as available. Figure 2-11 contains the (ϕ, ψ) contour energy maps for each of the *N*-acetyl-*N′* methyl-(X)-amide compounds as calculated by Pullman and co-workers (Pullman and Pullman, 1974) using the PCILO technique. Superimposed on each appropriate energy map are the (ϕ, ψ) coordinates of the corresponding residues as found in globular proteins whose crystal structures have been determined.

Valence-bond geometry has been considered to be independent of the external medium in the works reported here. The many studies of amino acid geometries suggest that this is a reasonable assumption (Scheraga, 1968; Marsh and Donohue, 1967; Gurskaya, 1968; Kennard and Watson, 1970; Hamilton et al., 1972; Kitano et al., 1973; Kitano and Kuchitus, 1973; Katz and Post, 1960; Senti and Harker, 1940). However, it has been suggested (Pullman et al., 1974) that the external force field modifies the valence-bond geometry, which, in turn, modifies the intramolecular force field that ultimately controls the resultant molecular conformation.

The monomeric α-amino acids may be generated by adding a proton to the *N*-terminus and a OH group (for the neutral form) to the C-terminal end of the α-amino residue groups. X-ray diffraction analysis of the amino acid crystals does not allow determination of the location of protons. Consequently, ϕ has assumed to be $-180°$, corresponding to the energy minimum

Table 2-8 Conformation of α-amino Residue Groups, —N(H)—ϕ—C(R)(H)—ψ—C$^{\alpha}$(=O)—, in Different Environments

Side chain Unit R	Source Structure	Solution[a,b]				Solid State[b,c]		
		Confor-mation	ϕ	ψ	Ref.	Confor-mation	ϕ	ψ
$-H$	GLYCINE					$\tilde{\beta}$	−180	161.3
	N-acetyl-*N*′-methylglycylamide	C_7[d]	−75	50	(1–5)			
		C_5[e]	−180	−180				
	Acetyl *N*-methylamine[c]						−60	−180
							−60	0
	Polyglycine					β[f]		
						3_1[g]		
$-CH_3$	L-ALANINE					$\tilde{\beta}$	−180	163.8
						$\tilde{\beta}$	−180	160.4
	N-acetyl-*N*′methyl alaylamide	C_7	−75	50	(1, 3, 4, 6, 7, 8)			
		β/C_5*	−160	[160, 180]				
	Poly(L-alanine)	α_R[h†]				α_R		
		α_R/R[i]				β		
		β						
		R						
		α'[j]						

—$CH(CH_3)_2$	L-VALINE HBr					β	−180	166.8
	HCl					β	−180	171.7
	$HCl.H_2O$					β	−180	175.3
	N-acetyl-*N*′methylvalylamide	C_7	−75	50	(1)			
		β/C_5	−152	[150, 180]				
	Poly(L-valine)	β/R				β		
—$CH_2CH(CH_3)_2$	L-LEUCINE HBr					~β	−180	166.4
	N-acetyl-*N*′methyl leucylamide	C_7	−75	50	(1)			
		β/C_5	−154	[155, 180]				
	Poly(L-leucine)	α_R/R				α_R		
—$CH(CH_3)$— (C_2H_5)	L-ISOLEUCINE HBr					~β	−180	162.5
	HCl					~β	−180	176.1
	Poly(L-isoleucine)	β/R				β		
—CH_2OH	L-SERINE					~β	−180	177.4
	N-acetyl-*N*′methylserylamide	C_7	−75	50				
		β/C_5	−160	[160, 180]	(1)			
	Poly(L-serine)	β				β		
		R						
—$CH(CH_3)OH$	L-THREONINE					~β	−180	165.0
	Poly(L-threonine)					β		
—CH_2COOH	L-ASPARTIC ACID HCl					~β	−180	147.7
	Poly(L-aspartic acid)	α_R/β/R(?)				α_R		
—CH_2CH_2COOH	L-GLUTAMIC ACID					~β	−180	145.8
	HCl					~β	−180	162.2
	Poly(L-glutamic acid)	α_R				α_R		
		CC^k						
		R						

Table 2-8 (*Continued*)

Side Chain Unit R	Source Structure	Solution[a,b] Confor-mation	ϕ	ψ	Ref.	Solid State[b,c] Confor-mation	ϕ	ψ
$—(CH_2)_4NH_2$	L-LYSINE HCl,H_2O					~β	−180	161.1
	Poly(L-lysine)	α_R CC β R				α_R		
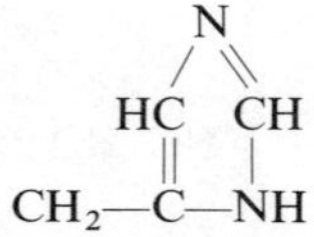 $CH_2—C(=CH—N=CH—NH)$ (imidazole)	L-HISTIDINE HCl,H_2O					~β	−180	180.0
	Poly(L-histidine)	α_R R				α_R		
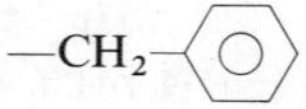 $—CH_2—C_6H_5$	L-PHENYLALANINE HCl					~β	−180	177.8
	N-acetyl-*N'* methyl phenylalayl-amide	C_5 C_7^{eq}			(1)			
	Poly(L-phenylalanine)	α_R				α_R		
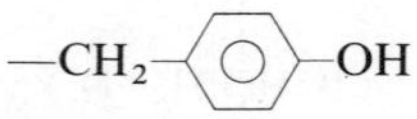 $—CH_2—C_6H_4—OH$	L-TYROSINE HBr					~β	−180	148.4
	HCl					~β	−180	146.8
	Poly(L-tyrosine)	α_R				α_R		
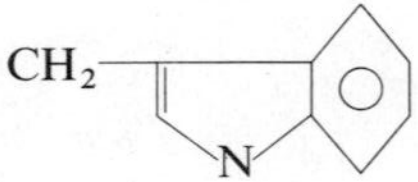 CH_2— (indole)	L-TRYPTOPHAN HBr					~β	−180	−172.8
	Poly(L-tryptophan)	α_R				α_R		
$—CH_2—SH$	L-CYSTEINE HCl					~β	−180	172.0
	N-acetyl-*N'*methylcysteinylamide	C_7	−75	50	(1)			
	Poly(L-cysteine)					β(?)		

—$(CH_2)_2SCH_3$	L-METHIONINE					~β	−180	148.0
	N-acetyl-N′methylmethionylamide	C_7	−75	50	(1)			
	Poly(L-methionine)	α_R/R				α_R		
—N—C^α—, CH_2 CH_2, CH_2 (ring)	L-PROLINE cis						−120	
	trans						−120	
	trans-N-acetyl-L-proline	C_7			(9)			
	trans-Acetyl-L-proline-N-methylamide						−76.3	−15.8 (10)
	Poly(L-proline) cis	10_3R.H.				10_3R.H.		
	trans	3_1L.H.				3_1L.H.		
—$(CH_2)_3$—(NH)— −$C(NH_2)_2$	L-ARGININE $2H_2O$					~β	−180	168.6
	2HI					~β	−180	161.0
	HBr,H_2O(molecule 1)					~β	−180	175.4
	HBr,H_2O(molecule 2)					~β	−180	153.7
	HCl,H_2O(molecule 1)					~β	−180	174.0
	HCl,H_2O(molecule 2)					~β	−180	154.1
	HCl(molecule 1)					~β	−180	126.7
	HCl(molecule 2)					~β	−180	138.8
	Poly(L-arginine)					α_R		
						β		
—(CH_2)—C(=O)—NH_2	L-GLUTAMINE					~β	−180	164.6
	N-acetyl-N′ methyl		30	60				
	glutaminyl-amide		−75	50	(1, 11)			
	Poly(L-glutamine)							

[a] All observed conformations are reported.
[b] The conformations of the homopolypeptides are taken from Walton and Blackwell (1973).
[c] The amino acid conformations are taken from Lakshminarayanan et al. (1967).

Table 2-8 (*Continued*)

[d] This conformation is defined in Fig. 2–12a.
[e] This conformation is defined in Fig. 2–12b.
[f] β—beta conformation.
[g] 3_1—three residues per turn of the helix in either direction.
[h] α_R—right-handed alpha helix.
[i] R—disordered, or random, chain conformation.
[j] α'—distorted alpha helix according to Parrish and Blout (1972).
[k] CC—charged coil, see Walton and Blackwell (1973).
* A slash(/) between conformational designations indicates that the individual polypeptide chains are thought to possess both (all) of these states at one time, that is, there is a conformational distribution along the chain.
† The conformations listed below one another represent the different conformations observed in different solvents, at different temperatures, etc.

(1) Marraud (1971).
(2) Avignon (1972).
(3) Marraud et al. (1970).
(4) Avignon et al. (1969).
(5) Cung et al. (1973).
(6) Mizushima (1957).
(7) Bystrov et al. (1969).
(8) Avignon and Huong (1970).
(9) Schellman and Neilsen (1967).
(10) Matsuzaki and Iitaka (1971).
(11) Marraud and Neel (1973).

A structural analysis of hydrated peptide crystals known as of 1963 in addition to hydrated crystals of many other organic compounds is given by Clark (1964).

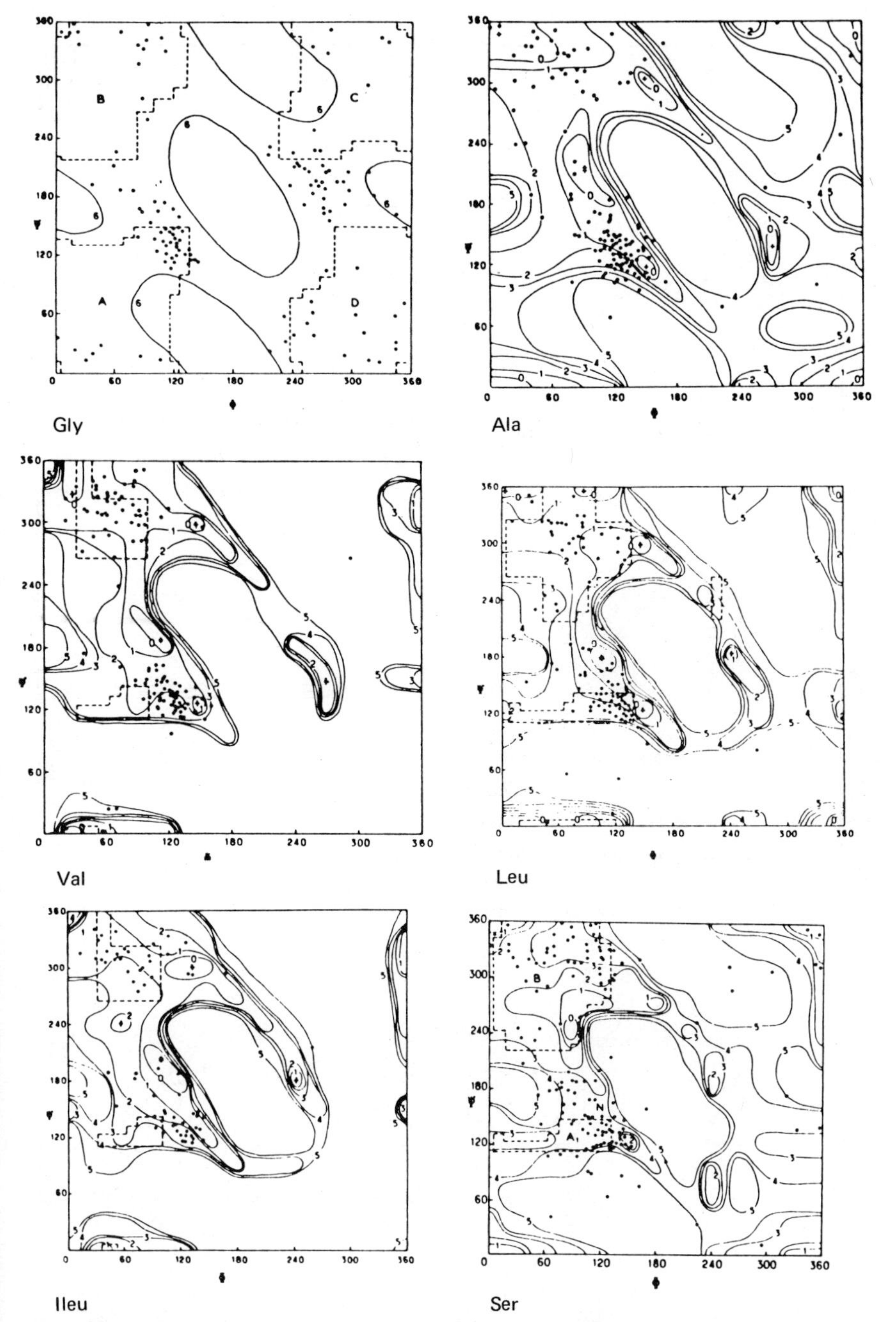

Fig. 2-11 (ϕ, ψ) energy contour maps of the amino acid "dipeptides," that is, *N*-acetyl-*N'*-methyl-(X)-amides as determined by Pullman, Pullman, and co-workers. The energy contours are in kcal/mole relative to the global minimum. Superimposed on the maps are the (ϕ, ψ) coordinates of residues found in globular proteins. See Pullman and Pullman (1974).

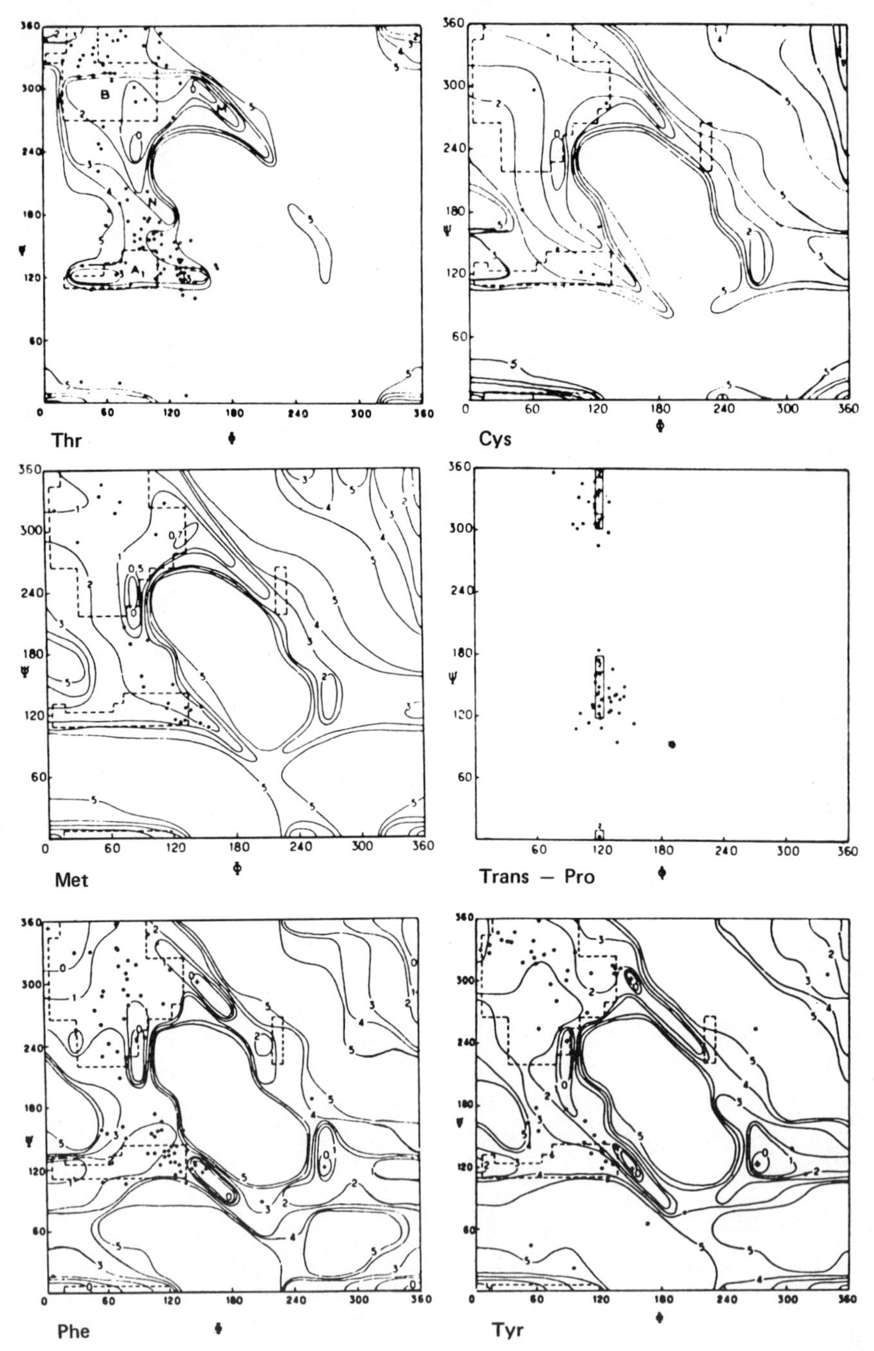

Fig. 2-11 (*Continued*)

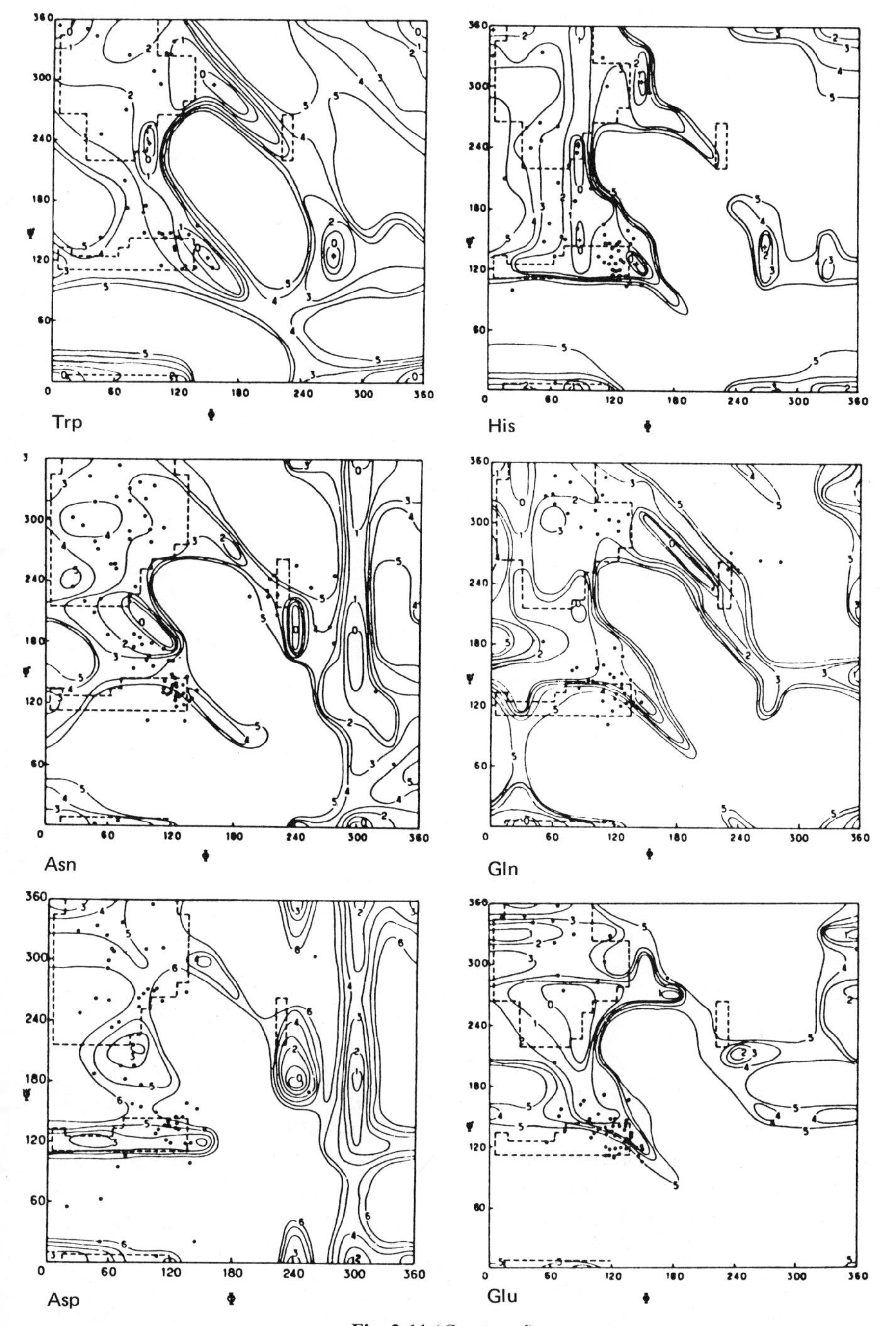

Fig. 2-11 (*Continued*)

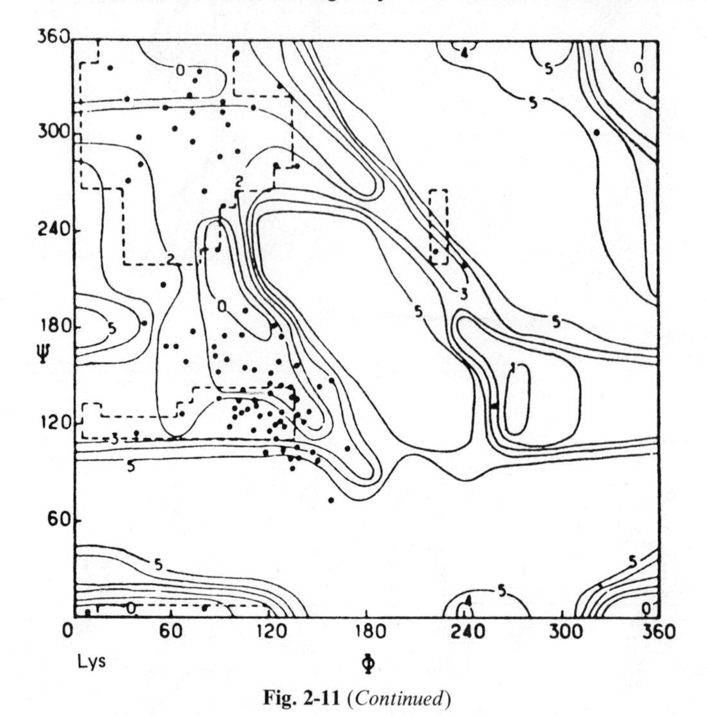

Fig. 2-11 (*Continued*)

of the torsional energy barrier proposed for $H_2N \overset{\phi}{\curvearrowright} CH_3$ (Hershbach, 1962) in the solid-state data in Table 2-8. The corresponding ψ reported for the amino acids in the solid state represents an average based on the two dihedral angles computed for the N—C$^{\alpha}$ $\curvearrowright$ C—O group using both carboxyl oxygens. It is quite clear from Table 2-8 that the amino acids all prefer a near β, that is, all trans, conformation, which corresponds to the minimum intramolecular torsional potential energy. In the case of glycine, deviations from $-180°$, the trans value for ψ, can be attributed to the intermolecular packing forces. In the other amino acids, the deviations from trans planarity can be assigned to packing plus backbone-side chain interactions. The magnitude of the packing energies can be estimated by assuming (1) that the torsional potential for ψ in the amino acids is identical to that for $CH_3 \curvearrowright COOH$ (Hershbach, 1962) and (2) that the conformation of the amino acid in a crystal represents an equilibrium state established through the balance of intramolecular torsional forces about ψ with intermolecular packing forces. The estimated packing energies, based on these approximations, are listed as part of Table 2-9.

Table 2-9 Interaction Energies Involving α-Amino Acid Residue Units

			Fractional Number of Protein Residues[b]					
			Intraprotein Residue Energies, kcal/mole					
Residue	Crystal Packing Energies[a] kcal/mole	Total No. Residues	0–1	1–2	2–3	3–4	4–5	≥5
gly[c]	0.96	—	—	—	—	—	—	—
ala	1.07	110	0.46	0.23	0.17	0.04	0.08	0.02
val	1.26	92	0.37	0.26	0.29	0.02	0.03	0.03
leu	1.18	90	0.32	0.38	0.19	0.07	0.00	0.04
ileu	1.30	55	0.44	0.29	0.24	0.02	0.00	0.01
ser	1.24	106	0.08	0.05	0.23	0.40	0.10	0.14
thr	1.13	77	0.05	0.05	0.10	0.57	0.08	0.14
cys	1.27	7	0.14	0.43	0.43	0.00	0.00	0.00
met	0.79	22	0.18	0.68	0.14	0.00	0.00	0.00
phe	1.22	47	0.11	0.51	0.30	0.04	0.04	0.00
tyr	0.73	45	0.07	0.47	0.20	0.16	0.04	0.06
trp	1.19	25	0.16	0.20	0.44	0.16	0.00	0.04
his	0.65	40	0.10	0.58	0.10	0.13	0.05	0.04
asn	—	69	0.04	0.10	0.15	0.28	0.20	0.23
gln	1.06	58	0.26	0.24	0.22	0.03	0.07	0.17
asp	0.73	37	0.00	0.00	0.00	0.19	0.27	0.54
glu	0.98	49	0.02	0.16	0.35	0.10	0.08	0.29
lys	1.02	84	0.37	0.23	0.27	0.05	0.03	0.05

[a] Using the torsional potential functions: $V(\phi) = 0.99\,(1 + \cos 3\,(\phi - 60°))$
$V(\psi) = 0.65\,(1 + \cos 3\,(\psi - 60°))$
See the text for a description of the computation. These are average energies for residues having multiple crystal structures reported.
[b] This is the fractional number of the total number of residues observed in globular proteins lying between x and $x + 1$ kcal/mole of the calculated "dipeptide" global energy minimum shown in Fig. 2-11.
[c] Energy contours are not sufficiently resolved to allow an estimation.

The *N*-acetyl-*N′* methyl-(X)-amides can be constructed from the α-amino residue groups by addition of a $(CH_3)\text{—}\overset{O}{\overset{\|}{C}}\text{—}$ group to the *N*-terminal end and an $NH(CH_3)$ group to the C-terminus. These compounds have often been thought of as first nearest-neighbor models for polypeptide structure (Scheraga, 1968, 1971; Lewis et al., 1973). The preferred solution conformation, with CCl_4 or, in few instances, water, as the solvent, for these molecules is a seven-membered ring structure, denoted C_7, completed through the formation of a hydrogen bond-like interaction between the amide hydrogen on the *N*-terminal side of the molecule and the carbonyl oxygen of the C-terminus. At least four different C_7 conformations have been proposed. *N*-acetyl-*N'* methylglycylamide adopts the C_7 conformation shown in Figure 2-12*a* when in CCl_4 and a similar conformation shown in Figure 2-12*f* when in water. Note in Figure 2-12*f* that the structure has been postulated to be partly stabilized through a water molecule–hydrogen bond bridge. Nonglycyl compounds adopt the C_7 conformations shown in Figure 2-12*d* and *e*. These two structures differ only in whether the C^{α}—C^{β} bond is axial or equatorial with respect to the ring.

A second conformation in solution has been found to have a significant statistical weight. This is the all trans, β-like conformation shown in Figure 2-12*b*. Within the context of ring structures, this conformation can be considered a five-membered ring, denoted C_5, in which the ring is completed by a pseudo-hydrogen bond between the amide proton and carbonyl oxygen of the α-amino residue group. Theoretical calculations (Pullman and Pullman, 1974) have suggested that the conformation shown in Figure 2-12*c* is a stable, deep energy minimum for some *N*-acetyl-*N'*-methyl-(X)-amides. This conformation has not been observed experimentally. Appropriate references for the discussion of *N*-acetyl-*N'* methyl-(X)-amides are given as part of Table 2-8.

Perhaps the most significant structural finding from an analysis of the amino acid and *N*-acetyl-*N'* methyl-(X)-amides compounds is that the preferred conformations are largely independent of the side chain. Also, the equilibrium distribution of C_7 and C_5 conformational states, to a good approximation, is governed by the minimization of the sum of the torsional energy expended in going from the all trans C_5 conformation to the C_7 conformation and that gained through hydrogen bond formation when the C_7 structure is formed.

There is excellent agreement between the PCILO energy minima in Figure 2-11 and the preferred solution conformations listed in Table 2-8. Thus it is reasonable to assume that the energy surfaces computed by the PCILO calculations are realistic descriptions of the conformational energetics of the *N*-acetyl-*N'* methyl-(X)-amides. In this regard it is possible to estimate the minimum intraprotein energies required to maintain the observed protein residue conformations. The populations reported in Table 2-9 are considered to be meaningful to within one energy contour because of the uncertainty in the precision of the (ϕ, ψ) determinations coupled with the reliability of PCILO calculations. Nevertheless, at least 30% of the residues found in globular proteins are in conformational states at least 2 kcal/mole or more above the preferred conformations in the corresponding *N*-acetyl-*N'*methyl-(X)-amides. Moreover, for those residues capable of specific side chain interactions—serine, threonine, tyrosine, asparagine, glutamine, and histidine—the percent of globular protein residues within 2 kcal/mole of the *N*-acetyl-

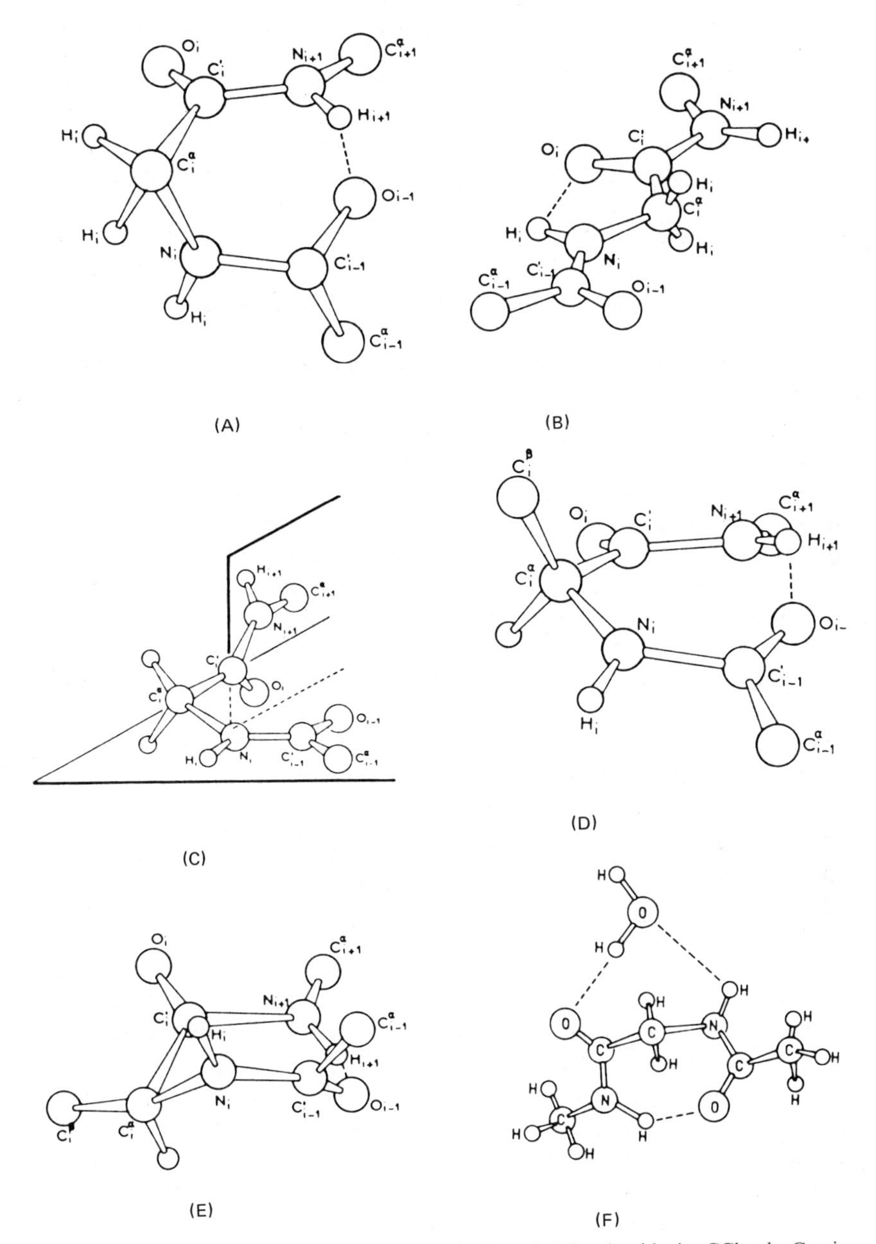

Fig. 2-12 *a.* C_7 ring conformation of *N*-acetyl-*N′* methylglycylamide in CCl_4. *b.* C_5 ring conformation of *N*-acetyl-*N′* methylglycylamide in CCl_4. *c.* The secondary minimum energy conformation, denoted by M, found from PCILO calculations. Pullman and Pullman (1974). *d.* The axial C_7 ring conformation determined for nonglycyl amides. *e.* The equatorial C_7 ring conformation determined for nonglycyl amides. *f.* Hydrogen-bonded association of the C_7 equatorial ring conformation of a glycyl (or alanyl) dipeptide stabilized by interaction with a water molecule, following Avignon (1972). Pullman and Pullman (1974).

*N'*methyl-(X)-amide global energy minimum is even higher: 86%, 88%, 46%, 86%, 49%, and 32%, respectively. Interestingly, the fractional number of globular protein residues within 2 kcal/mole of the *N*-acetyl-*N'*methyl-(X)-amide minima, approximately 60–70%, is about the highest accuracy with which protein secondary structure can be predicted with any of the available empirical methods (see Fasman and Chou, 1974, and references cited therein). Hence, while it is clear from an inspection of Figure 2-11 that each specific globular protein residue conforms to the general stereochemistry defined by the appropriate *N*-acetyl-*N'*methyl-(X)-amide, there are frequently significant conformational energy excursions from the first nearest neighbor energy minima that are probably due to the formation of stable intraprotein interactions involving groups on residues farther displaced than nearest neighbors in the primary structure.

Homopoly(α-amino acids) in the solid state seem to be torn between adopting the β conformation most acceptable to the isolated individual monomeric units and the α helical conformation. Roughly, the preferential state results from the lower free energy due to intrachain hydrogen bond formation and low-to-medium chain packing at the expense of some intramolecular torsional energy (the α helix), or medium-to-good chain packing, possibly through interchain hydrogen bonds with little expenditure of intramolecular torsional energy (β conformation). In solution, things become more complicated mainly through configurational entropy. There is now the possibility of a distribution of conformational states in the chain. The disordered, or random, chain conformation becomes important through entropic contributions to the total free energy.

Scheraga and co-workers have developed a method of evaluating helix-coil stability constants for the naturally occurring amino acids in water. The helix-coil stability constants are to be used in the prediction of protein tertiary structure. To date, about half of the natural amino acids have been analyzed. In this technique, random copolymers are synthesized in a "host-guest" fashion in which the amino acid in question is the guest in a polymer composed of either N^5-(3-hydroxybutyl)-L-glutamine or N^5-(4-hydroxybutyl)-L-glutamine "host" residues. The results of these studies are reported in Table 2-10. Included in this table are the standard-state thermodynamic quantities for 20°C; ΔG^0 (the free energy), ΔH^o (the enthalpy), and ΔS^0 (the entropy) for the conversion of a coil residue of an amino acid to a helical one at the end of a long helical sequence. Also listed in Table 2-11 are the Zimm-Bragg parameters (1959) σ and s analyzed according to the Lifson-Allegra-Poland-Scheraga theory (VonDreele et al., 1971*a,b*; Ananthanarajanan et al., 1971). In Figure 2-13 are plots of s versus temperature for the various residues. For reference, s = 1 defines an equal disposition for helix formation or destruction. The curves quantitatively

Table 2-10 Helix-Coil Stability Constants for Amino Acids in Water (Von Dreele et al., 1971a,b)

Residue	ΔH° (cal/mole)	ΔS° eu	ΔG°_{20} (cal/mole)	σ_{20}	s_{20}[a]	References
glycine	625 ± 100	1.0 ± 0.3	~ 327	1×10^{-5}	0.591	Ananthanarayanan et al. (1971)
L-alanine	-242 ± 21	-0.703 ± 0.067	-40 ± 6.5	0.0008 ± 0.0002	1.068	Platzer et al. (1972)
L-serine	-101 ± 95	-0.9 ± 0.3	158 ± 7	7.5×10^{-5}	0.757	Hughes et al. (1972)
L-leucine	-149.2	-0.267	-76.0	33×10^{-4}	1.14	Alter et al. (1972)
	-23.8	$+0.189$	-81.5	12×10^{-4}	1.15	
L-phenylalanine	-170 ± 80	-0.46 ± 0.28	-44 ± 12	0.0018	1.07	Van Wart et al. (1973)
L-glutamic acid						
a) pH = 2.8	-1070 ± 330	-3.1 ± 1.1	-174 ± 49	1.0×10^{-2}	1.35	Maxfield et al. (1975)
b) pH = 8 in 0.1 *N* KCl	-190 ± 110	-0.72 ± 0.37	17 ± 19	6.0×10^{-4}	0.97	
L-valine	640 ± 140	2.05 ± 0.47	40.4	1.0×10^{-4}	0.93	Alter et al. (1973)
L-tyrosine	-930 ± 100	-3.1 ± 100	-14 ± 27	66×10^{-4}	—	Scheule et al. (1976)

[a] Used as primary indicator of helix-breaking potential.

Table 2-11 Bond Torsion Angles in Proline Diketopiperazines

Angles	cyclo (Pro-Leu) Crystal Structure	cyclo(Pro-Pro)		cyclo(Pro-D-Pro)	
		Calc'd	Expt'l	Calc'd	Expt'l
χ_1	−32	−33	−30	−37	−40
χ_2	36	34	42	36	29
χ_3	−25	−23	−21	−22	−12
χ_4	4	2		−1	
ϕ	−42	−16		−6	
ψ	34	26		5	
ω	6	−10		−14	

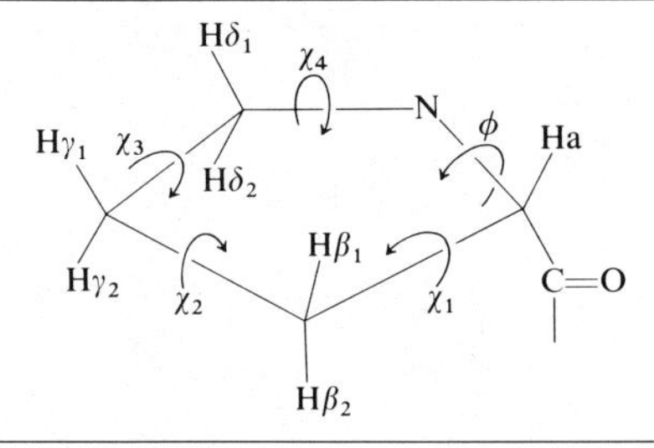

Young et al. (1973)

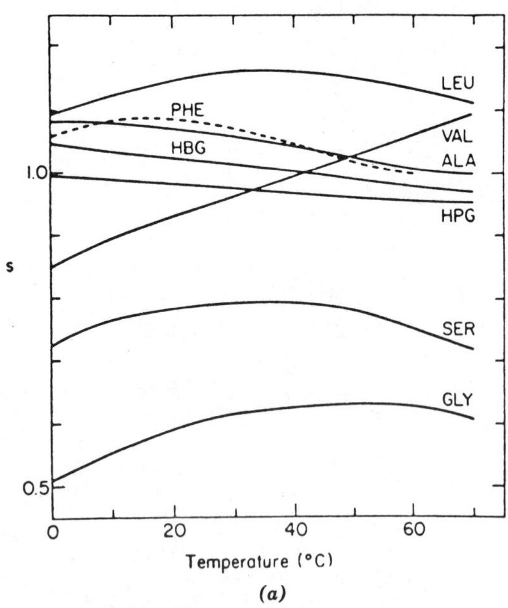

(a)

Fig. 2-13*a* Temperature (°C). Temperature dependence of *s* for various amino acid residues in water. The sources of the data are given in the text. Alter et al. (1973).

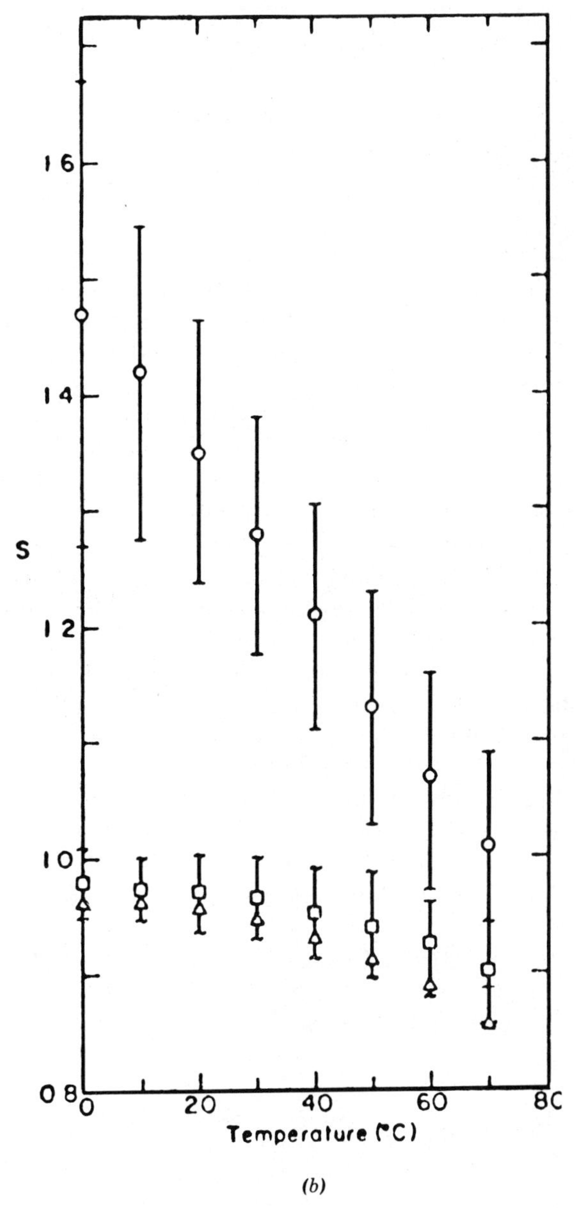

(b)

Fig. 2-13*b* A plot of s vs. T for poly (L-glutamic acid) at pH 2.3 (○), at pH 8 in 0.1 N KCI (□), and at pH 8 in water (△). The error symbols for the values at pH 2.3 (I), and at pH 8 in 0.1 N KCl (I), are described in Maxfield et al. (1975).

reflect the empirical helix-breaking properties derived for amino acid residues (Fasman, 1974 and references therein).

VI D_2O SOLUTION CONFORMATIONS OF SOME AMINO ACIDS

Although crystal structures have been determined for many amino acids (see Table 2-8) and related molecules, the solid-state structures are often characterized by a number of hydrogen bonds. Hence the solid-state conformation may not be representative of the situation in solution, and, consequently, it may be misleading to base, for example, structure-activity relationships, on the solid-state conformation.

In the solid state, when there is a strong intramolecular force stabilizing a particular structure, the solid and solution results should agree. This is found, for example, in the peptide group, in which the double-bond character of the C—N bond ensures a planar group. But with single bonds, as, for example, with the 1,2-distributed ethane derivatives, $XCH_2—CH_2Y$, there is usually no well-defined intramolecular force, except those which tend to stagger single bonds. Consequently, rotamers I, II, and III, shown in Figure 2-14, are possible. It is here that a crystal structure may present a C—C dihedral angle, which is a compromise between the intermolecular forces, such as crystal packing, hydrogen-bonding possibilities, and steric effects, and any intramolecular stabilizing forces.

Ham (1974) has studied the distribution of conformational states of five physiologically active amino acids in D_2O using high-resolution NMR. Each of these molecules can be considered a 1,2-ethane derivative, and Ham has characterized torsional rotations about the C—C bonds by measuring the vicinal coupling constants and applying the appropriate Karplus-type relationship.

The results are presented in Figure 2-15 in terms of characteristic interaction distances involving the N and C-terminal atoms. β-alanine, $^+NH_3CH_2 \cdot CH_2COO^-$, is a minor amino-acid in the brain, but it is one that inhibits the

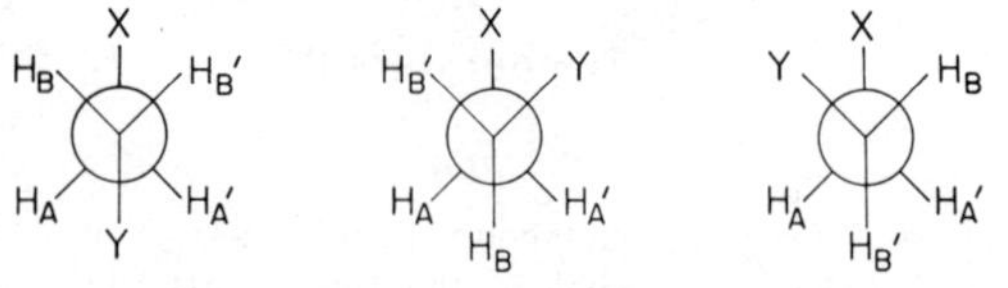

X
H_B H_B'
H_A H_A'
Y

X
H_B' Y
H_A H_A'
H_B

X
Y H_B
H_A H_A'
H_B'

I II III

Fig. 2-14 Ham (1974).

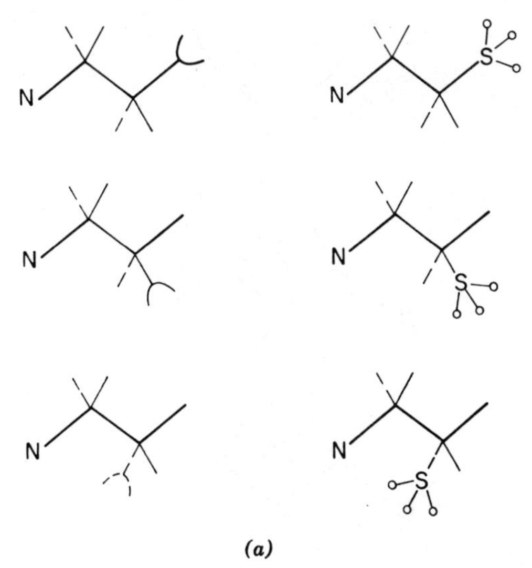

(a)

Fig. 2-15*a* Solution conformations of the zwitterions of β-alanine $+NH_3CH_2COO—$ and taurine, $+NH_3CH_2CH_2SO_3—$. The top conformation in each case corresponds to the trans rotamer with the NCCX skeleton having maximum extension. In D_2O, the pmr spectra indicate that the proportion of trans rotamer is 33% for β-alanine and 25% for taurine. In this rotamer, the N-S distance is about 4 Å in taurine and the corresponding distance to carbon in β-alanine 3.8 Å. Ham (1974).

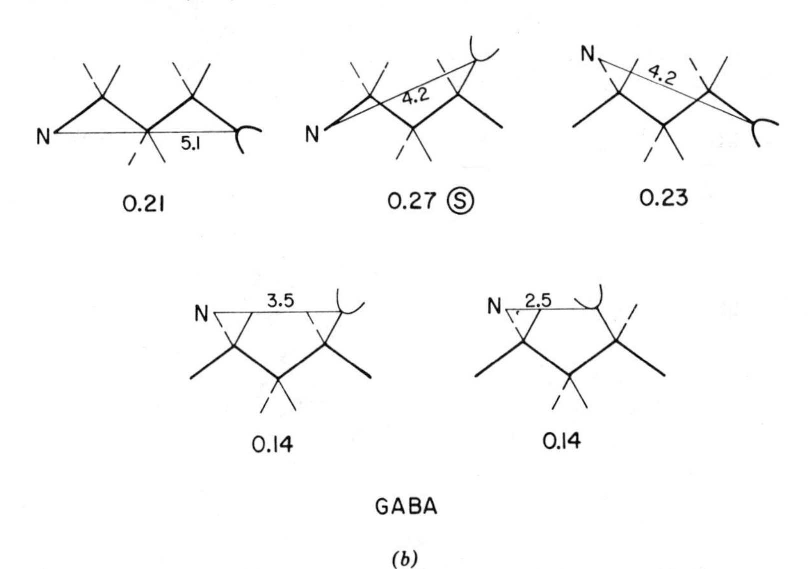

(b)

Fig. 2–15*b* Solution conformations of the zwitterion of γ-aminobutyrate $+ NH_3CH_2CH_2$ CH_2-COO—in D_2O. The population appropriate to the indicated + N—C distance (in Å) are beneath each rotamer. Apart from the fully extended rotamer in the upper left, there is a paired rotamer in which either the +N or COO—group is in a different but equivalent position. The conformation found in the solid is indicated by S. Ham (1974).

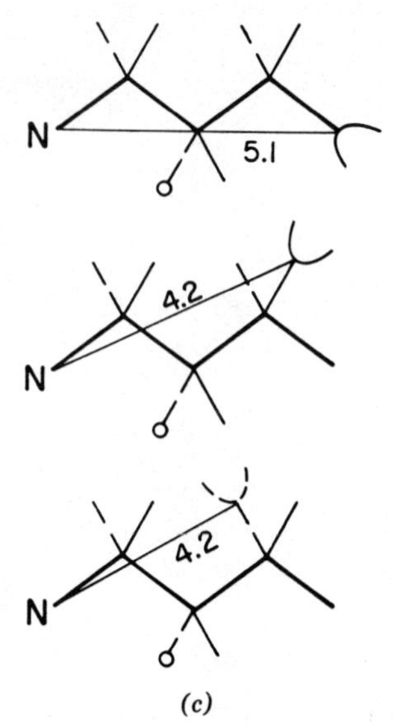

Fig. 2-15*c* The most populated solution conformations of β-OHγ-aminobutyrate in D_2O. These total about 70%; the other N, OH gauche conformation comprises nearly 30%, and there is a small population with N, OH trans. The $+N$—C distance indicated is in Å. Ham (1974).

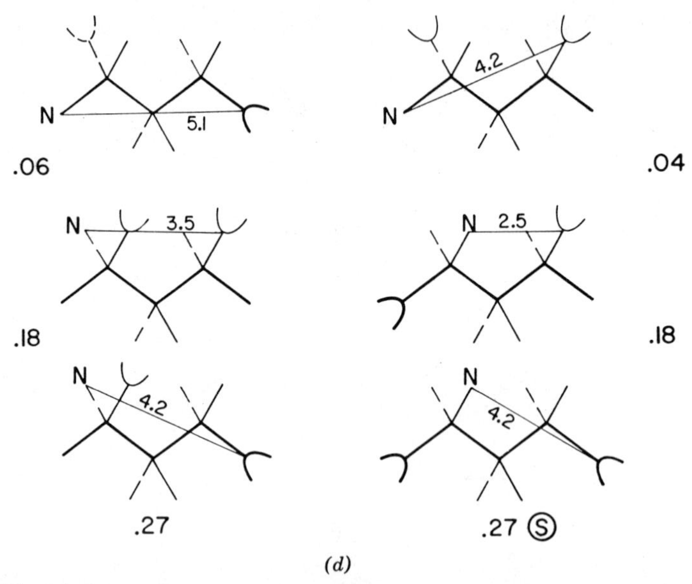

Fig. 2-15*d* Solution conformations for glutamate. The populations appropriate to the indicated ^+N—C distances (in Å) are beneath each rotamer. The conformations in the upper left and in the bottom row are unique. Each of the other three has a paired conformation with the same $+N$—C distance but a different distance between the carboxylic groups. The conformation most closely related to that found in a neutron diffraction study of L-glutamic acid hydrochloride is indicated by S. Ham (1974).

firing of spinal neurons (McIlwain and Bachelard, 1971). The PMR spectrum shows that there is no conformational preference around the central C—C bond. In the solid state, there is a crystal structure of a Ni-complex in which the $^{+}NCCCOO^{-}$ is in a gauche arrangement, but here the chelating effect may be a considerable influence in stabilizing the gauche conformation (Jose et al., 1964).

Taurine, $^{+}NH_3CH_2CH_2SO_3^{-}$, is the sulphonic analog of β-alanine and, after the dicarboxylic acids, is the most abundant of the amino acids in the brain. It seems to be released in the central nervous system on stimulation of certain neurons, and there are indications of an uptake mechanism for its deactivation or termination of transmitter action. In the solid, the zwitterion has a gauche conformation in the NCCS chain (Okaya, 1966), and the PMR results show about 25% of the trans conformation to be present in solution.

γ-aminobutyric acid, $^{+}NH_3CH_2CH_2CH_2COO^{-}$, a homolog of β-alanine, is an inhibitory amino acid in the central nervous system and is formed by enzmatic decarboxylation of glutamate. With two CH_2—CH_2 single bonds, there are trans and gauche possibilities about each. In solution, the zwitterion gives a spectrum that enables the population of states to be estimated. The numbers below the structures in Figure 2-15*b* indicate the fractional distribution of conformational states.

The most populated conformations in solution are those with N to C distances near 4.2 Å. In the solid state, the zwitterion has a trans conformation about the $^{+}NCCC$ end and a gauche arrangement at the $CCCCOO^{-}$ end (Steward et al., 1973). This is one of the more populated solution conformations: the other has the reverse arrangement—gauche then trans.

The PMR spectrum of β-hydroxy γ-aminobutyric acid indicates that there is considerable preference around the $^{+}NCH_2$—CHOH bond and, in analogy with other systems, the N, OH are largely in a gauche arrangement. Choline, for example, has about 10% of a trans NCCO form, with the other 90% divided equally between the two gauche forms (Partington et al., 1972). Interpretation of the coupling constants of the $^{+}NCH_2$—CH system indicates that there is nearly 70% of the gauche form with the CH_2COO^{-} between the CH_2 protons (see Figure 2-15*c*). There is roughly 30% of the other gauche form with the H_β protons between the CH_2 protons and a small percentage of the conformer with N, OH trans. For the most populated conformations, the ^{+}N—COO^{-} distance can be in the range of 4.2 to 5.1 Å. The CH_2COO^{-} part of the PMR spectrum is consistent with no preference around the HOC—CH_2COO^{-} bond, although this observation is not conclusive.

The amino acid glutamate, $^{-}OOC\ CH(NH_3^{+})CH_2CH_2COO^{-}$, is the immediate precursor of γ-amino-butyrate, yet in the brain the dicarboxylic acid has completely opposite neuronal activity. Glutamate is an excitant, and it increases the rate of cell firing in both the brain and spinal cord (Krnjeck, 1970; McIlwain and Bachelard, 1971). The CH—CH_2CH_2 system with two single bonds again allows the possibility of trans and gauche forms around each bond. The rotameric states may be grouped by the $^+N \cdots COO^-$ distance, although pairs with the same N—C distance will differ in the distance between carboxylic groups.

The PMR spectrum can be analyzed in terms of the conformational states and populations shown in Figure 2-15*d*. The fully extended form is unique and very slightly populated. There are four conformations, making up a total of 36%, with the shorter N—C distances of 3.5 Å and 2.5 Å. Conformations with the 4.2 Å separation make up the remainder and comprise nearly 60% of the solution conformations. The zwitterion conformation closest to the atomic arrangement in solid L-glutamic acid hydrochloride (see Table 2-8) is indicated by the symbol S (Sequeria, 1972).

For the amino acids reported, a range of conformational states are available to the molecules in solution. No single conformation is exclusively favored. Molecular flexibility is considerable, and it may be necessary so that these molecules can respond to the perturbations presented by the receptor during their mutual interaction. It is interesting to note that, for each molecule studied, the solid-state conformation is found to be one of the more probable solution conformations. However, the solid-state conformation is not necessarily the *most* probable solution conformation.

VII CYCLIC PEPTIDES

A Background

Cyclic peptides are more restricted in conformational freedom than their corresponding linear analogs. Consequently, it is reasonable to anticipate that the solution and solid-state conformations are more likely to agree than are linear oligopeptides. Cyclization, however, introduces additional types of molecular flexibility through ring conformers. Deber et al. (1976) discuss the advantages of using cyclic peptides to unravel the nature of peptide conformational behavior.

B Diketopiperazines

Information on the simplest cyclic peptides, the diketopiperazines, is now quite extensive. NMR data is available on over 20 diketopiperazines (Hruby,

1974) in addition to a wealth of X-ray structural data. Diketopiperazine itself, or cyclo(Gly-Gly), which is the prototype molecule in this series, maintains a planar ring conformation in the crystal (Corey, 1938; Degeilh and Marsh, 1959). Kopple and Marr (1967), using PMR and a Karplus-type relationship, concluded that cyclo-(Gly-Gly) has a planar conformation in Me_2SO, whereas, in trifluoroacetic acid, the coupling is extremely small and clearly inconsistent with the planar ring conformer. Moreover, the coupling is not that expected from the most reasonable alternate to the planar conformation. This is a rapid interconversion between boat conformers:

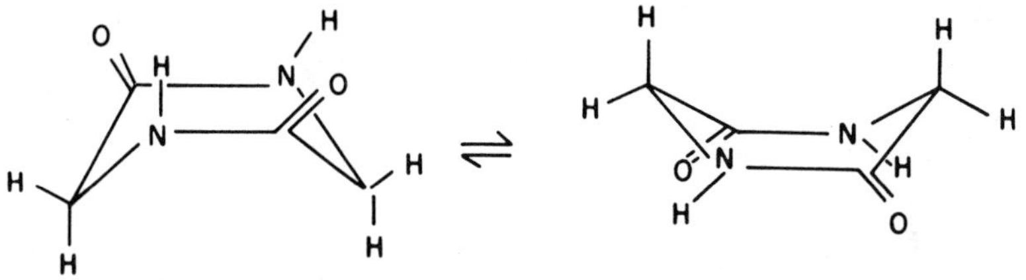

Cyclo(Ala-Ala), that is, L-*cis*-3,6-dimethyl-2,5 piperazinedione, has the equatorial boat conformation shown in Figure 2-16*a*, with planar amide groups inclined at about 30° to each other in crystalline state (Benedetti, et al., 1969; Sletten, 1970). The DL or trans isomer has a planar ring, as in Figure 2-16*b*. In solution both compounds exhibit very small values of the coupling constant, $J_{N\alpha}$, probably not exceeding 1 Hz (Bovey, 1974). These data are consistent with the X-ray findings for the cis compound, but how such a cis ⇌ trans interconversion might occur, if at all, is not clear. Consequently, NMR results do not appear to provide a clear picture of the behavior of these compounds in solution. Cyclo(Gly-Ala) also possesses $J_{N\alpha}$ coupling constants near zero in Me_2SO, which is not consistent with any proposed conformation.

The Gly-X diketopiperazines, in which X is an aromatic amino acid, possess rather interesting conformational properties. The glycine protons of cyclo(Gly-Tyr), or monobenzyl diketopiperazine, show a differentiation in chemical shift of the order of one part per million, in contrast to a negligible difference when X is nonaromatic (Kopple and Marr, 1967; Kopple and Ohnishi, 1969; Bovey, 1969). This feature has been found in a number of similar compounds (Kopple and Ohnishi, 1969) and strongly suggests a folded conformation with the aromatic ring positioned over the glycine methylene group, as shown in Figure 2-16*c*. It is very likely that the aromatic ring is positioned above the diketopiperazine ring as a result of attractive π—π interactions between rings. Webb and Lin (1971) find this structure for cyclo(Gly-Tyr) in the solid state. The coupling constant, $J_{\alpha\beta}$, is in agreement with this conformation, whereas the $J_{N\alpha}$ values for the glycine residues are

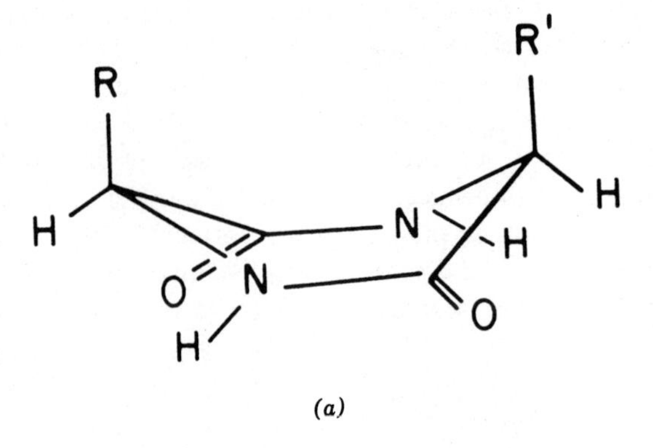

(a)

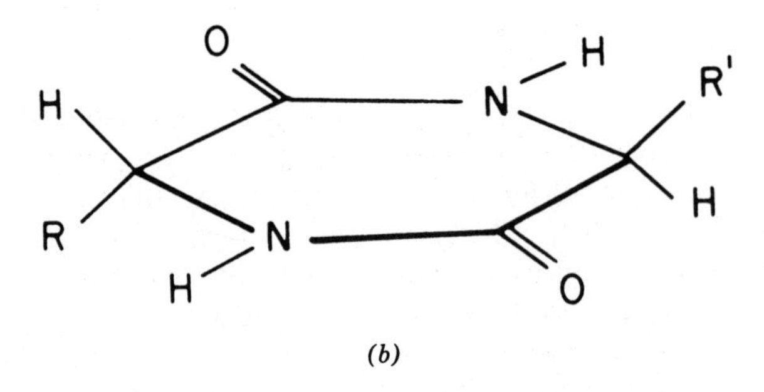

(b)

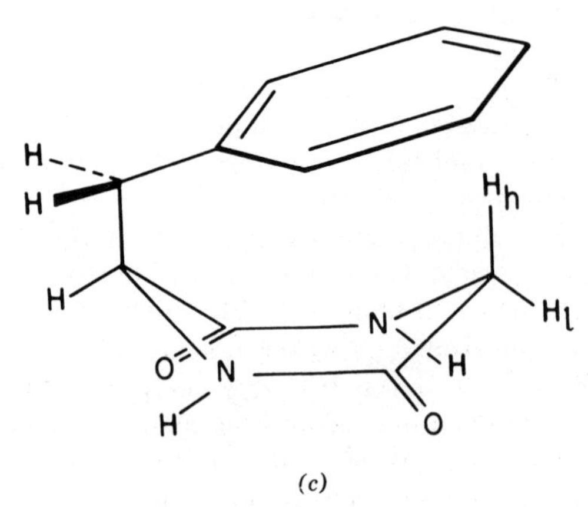

(c)

Fig. 2-16 *a*. Cyclo(Ala-Ala). *b*. Cyclo(Ala-Ala) *trans* isomer. *c*. Cyclo(Gly-Aromatic amino acid).

not entirely consistent. Thus, although NMR and X-ray data agree in general for the conformations of the cyclo(gly-X); X aromatic, compounds, $J_{N\alpha}$, one of the major parameters in the NMR study of larger peptides, is not entirely consistent among these compounds.

The conformations of a pair of isomeric diketopiperazines cyclo(Pro-Pro) and cyclo(Pro-D-Pro) have been determined by Young, Madison and Blout (1973) using the vicinal proton couplings in the proline ring along with the europium complex Eu(fod)$_3$ to sort out the ring syn and anti protons. In addition, energy-minimization calculations were employed as part of the analysis. Both NMR and energy calculations indicate that the LL isomer is a boat with the β carbons in the equatorial positions (Figure 2-16*a*), whereas, in the DL isomer, the ring is near planar. In Table 2-11 (p. 54), the NMR and energy-calculation parameters are compared to those for cyclo(Pro-Leu) in the solid state (Karle, 1972). The agreement is excellent except for a larger value of ϕ in cyclo(Pro-Pro), indicating a more tightly folded main ring for the LL compound than for cyclo(Pro-Leu). Unlike most of the other cyclic peptides discussed in this section, the cyclo(Pro-X) compounds are highly rigid, especially when X = proline. Thus distortions that may occur through intermolecular hydrogen bounding and other interactions peculiar to the crystal, or perhaps even the choice of solvent, should have minimal control on molecular conformation. In turn, the X-ray and NMR structures should agree well for these rigid compounds.

This inverse relationship between structural rigidity and intermolecular modifications on molecular conformation is also exhibited by some of the tripeptide analogs to cyclo(Pro-Pro). Two particularly interesting molecules are the tripeptides cyclo(tri-L-proline) and cyclo(Pro-Pro-Hyp). Cyclo(tri-L-proline) was first synthesized by Rothe (1965). Deber et al. (1971) used vicinal proton couplings to determine the proline ring conformations in the same way as applied to the proline diketopiperazines. NMR analysis suggests a molecule with three equivalent rings, each with the nitrogen atom fairly strongly puckered exo to the carbonyl group. Three such units can be assembled only into a ring with a cis conformation, $\omega \cong 0°$, at the peptide bonds that must be very nearly planar. The structure is shown in stereo in Figure 2-17. The conformations of cyclo (Pro-Pro-Hyp) and its acetate and benzoate esters are essentially identical.

Kartha et al. (1974) have determined the crystal structure of cyclo(tri-L-proline). In general, NMR and X-ray structural findings are in good agreement. A major difference is that the Hyp residue has a torsional angle, $\omega = 18°$, in the crystal. The deviation of the Hyp unit from the NMR conformation is probably a result of the involvement of the proline ring OH group in a strong hydrogen bond (2.7 Å) to a carbonyl group of a neighboring molecule.

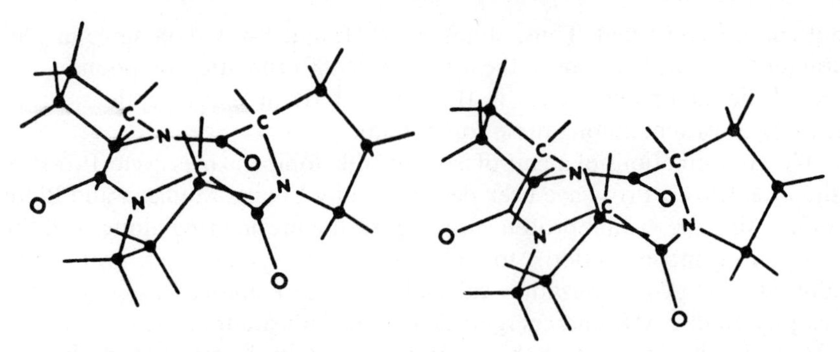

Fig. 2-17 A steroview of cyclo(tri-L-proline). [Reprinted with permission from Deber et al., *J. Am. Chem. Soc.*, **93**, 4893 (1971). Copyright by the American Chemical Society.]

C Cyclic Hexapeptides

This class of cyclic peptides has received the most attention because these compounds can be synthesized with relative ease and because many bind cations and/or have antibiotic activity. Solution conformations, based mainly on NMR, have been suggested for at least 31 compounds in this class (Hruby, 1974). The conformation that has been most frequently proposed, and in some cases convincingly demonstrated, is the antiparallel β conformation, normally with β turns at the ends and two cross-ring hydrogen bonds as first suggested by Schwyzer (Schwyzer et al., 1958; Schwyzer, 1959; Schwyzer et al., 1964).

Despite extensive solution studies, few cyclic hexapeptides are available for a detailed comparison of solution and solid-state structures. One compound in which such a comparison is possible is cyclo(Gly-Gly-DAla-DAla-Gly-Gly):

```
  6    1    2
DAla-Gly-Gly
 |         |
DAla-Gly-Gly
  5    4    3
```

The proton spectrum is relatively complex (Tonelli and Brewster, 1972) because of the lack of symmetry; there are spectra for each of the six residues. The absence of proline or bulky side chains suggest that there may be a number of low-energy conformations between which equilibration will be rapid. It is observed that although all of the glycine residues exchange NH protons rapidly, one of them has a temperature coefficient about one-third as great as the others, suggesting an internal hydrogen bond. The crystal conformation shown in Figure 2-18 shows the expected two intramolecular

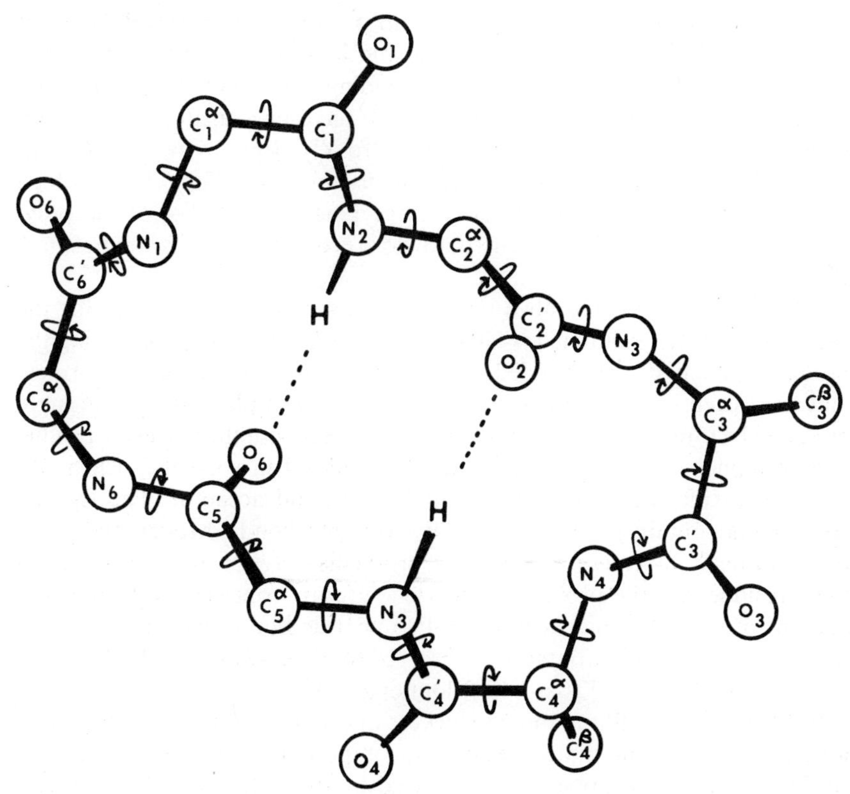

Fig. 2-18 The X-ray conformation of cyclo(Gly-Gly-DAla-DAla-Gly-Gly). [Reprinted with permission form Karle et al., *J. Am. Chem. Soc.*, **92,** 3755 (1970). Copyright by the American Chemical Society.]

hydrogen bonds with all trans peptide bonds, the two D-alanine residues forming one of the turns. If the two methyl side chains are disregarded, the conformation has an approximate center of symmetry. This is a strained antiparallel β structure, and, with regard to intramolecular forces alone, it is of high energy. In the crystal, however, this strain energy is more than counterbalanced by the presence of nine external hydrogen bonds to three H_2O molecules of crystallization per peptide molecule and one intermolecular NH···O=C hydrogen bond, in addition to the two Gly_1-Gly_4 intramolecular hydrogen bonds.

In contrast, energy calculations indicate that, in solution, there are 25 low-energy conformations consistent with the observed values of $J_{N\alpha}$, all of which have energies about 15 $\pm$ 1 kcal *less* than that of the crystal conformation. Eight of these are calculated to have one $(NH)_{Gly_3}\cdots(O{=}C)_{Ala_2}$

hydrogen bond that is different from both of those in the crystal. Hence it appears that this molecule is fairly flexible in solution, with none of the probable solution conformations corresponding to the crystal conformation.

Blout and co-workers (Blout et al., 1974) have investigated the solution structures of congeneric sets of cyclo(X-Pro-Y)$_2$ peptides by NMR techniques. These cyclic hexapeptides serve as useful models for β turns, which have been recognized to comprise a large portion (27–33%) of the polypeptide chains of globular proteins (Crawford et al., 1973; Chow and Fasman, 1974).

In general, the favored all-trans peptide bond conformer found for the cyclo(X-Pro-Gly)$_2$ examined is C_2-symmetric with the residue preceding the proline that is hydrogen bonded. It is clear, however, that other conformers may exist even in these simple compounds. For example, if the prolines are preceded by L-residues, it is found that a conformer with the two cis X-proline peptide bonds exists. This conformer is stabilized in solvents such as dimethylsulfoxide, which is a strong hydrogen-bond acceptor. The two cis conformers contain no intramolecular hydrogen bonds. Model studies on the all-trans conformers reveal steric problems between the side chain of the L residue and the proline S methylene group. Consistent with this, the amount of the conformer with cis X-Pro peptide bonds, which appears to relieve the unfavorable contacts, is observed to increase as the side chain size increases, as shown in Table 2-12 (Blout et al., 1974).

Investigations by Blout and co-workers of cyclo(Gly-Pro-X)$_2$ peptides, in which X is an L residue, indicate that these compounds exist in a C_2-symmetric conformation that is apparently stabilized by two intramolecular Gly-Gly

Table 2-12 Distribution[a] of all-trans and one cis conformers for cyclo(Gly-Pro-X)$_2$ hexapeptides

Peptide	Solvent	All trans (%)[b]	One cis (%)[b]
cyclo(Gly-Pro-Gly)$_2$	D_2O	~100	~0
cyclo(Gly-Pro-Ala)$_2$	D_2O	76	24
cyclo(Gly-Pro-Ser)$_2$	D_2O	77	23
cyclo(Gly-Pro-Phe)$_2$[c]	Me_2SO-d_6	~100	~0
cyclo(Gly-Pro-Val)$_2$	D_2O	58	42
	Me_2SO-d_6	23	77

[a] Measured by ^{13}C NMR.
[b] Uncertainty = ±10%.
[c] Synthesized by P. E. Young.
Blout et al. (1974)

hydrogen bonds. In addition, several cyclic hexapeptides of this type display an asymmetric conformation that has been shown to contain one cis X-Pro bond. In dimethylsulfoxide solutions, when the X-residue is valine, the asymmetric conformer is favored over the symmetric. In contrast, in the phenylalanine cyclic hexapeptide, only an all-trans conformer is observed. These results suggest that the predominant conformation is determined not only by the size of the X residue, but also by subleties of its side chain. Table 2-13 summarizes data on the amounts of cis conformers for these cyclic hexapeptides as a function of varying side chains.

It is clear from Tables 2-12 and 2-13 that multiple conformations are found in the cyclo(X-Pro-Y)$_2$ hexapeptides depending on solvent and side chain. Since these hexapeptides are models for β bends in globular proteins, it is reasonable to speculate that similar relationships between conformation and solvent and/or primary structure may exist in globular proteins.

Table 2-13 Distribution[a] of all-trans **and two** cis **conformers for** cyclo**(X-Pro-Gly)$_2$ hexapeptides**

Peptide	Solvent	All trans (%)[b]	Two cis (%)[b]
cyclo(Gly-Pro-Gly)$_2$	D_2O	~100	~0
	Me_2SO-d_6	~100	~0
cyclo(Ala-Pro-Gly)$_2$	D_2O	88	12
cyclo(Ser-Pro-Gly)$_2$[c]	D_2O	75	25
	Me_2SO-d_6	20	80
cyclo(Phe-Pro-Gly)$_2$	Me_2SO-d_6	20	80
cyclo(Val-Pro-Gly)$_2$	D_2O	13	87
	Me_2SO-d_6	<20	>80

[a] Measured by ^{13}C NMR.
[b] Uncertainty = $\pm 5\%$.
[c] Measured by ^{1}H NMR.
Blout et al. (1974)

D Gramicidin S

The most extensively studied naturally occurring cyclic peptide is gramicidin S, cyclo(Val-Orn-Leu-DPhe-Pro)$_2$. Proposed solution conformations for gramicidin S have resulted from the work of Schwyzer and Ludescher (1968, 1969), Stern et al. (1968), and Ovchinnikov et al. (1970). The suggested structures, all derived from NMR analyses, differ in some respects, but establish that this cyclic peptide exists in a C_2-symmetric conformation, with four intramolecular hydrogen bonds in an antiparallel β-sheet conformation with all peptide bonds trans. This general type of structure with C_2 symmetry was also suggested by the early X-ray study by Hodgkin and Oughton (1957) and has been confirmed by the partially complete X-ray study by De Santis et al. (1973).

E Antamanide

This cyclic decapeptide occurs in the poisonous mushroom *amanita phalloides* and acts as an antidote to its poisons, amanitin and phalloiden, by preventing their accumulation in the liver. It is of particular interest because it forms strong complexes with Li^+, Na^+, K^+, and Ca^{2+}. The strongest complex is formed with Na^+, and it is probably in this form that it is biologically active (Wieland, 1972).

Antamanide has an unusually high proportion of proline and phenylalanine and is without symmetry:

```
8   9   10  1   2
Pro-Phe-Phe-Val-Pro
|               |
Pro-Phe-Phe-Ala-Pro
7   6   5   4   3
```

Depending on the solvent, several different conformations can be realized in solution. In weak hydrogen-bond-accepting solvents such as acetic acid and acetonitrile, the most probable structure is now believed to be an antiparallel β structure with four cross-ring hydrogen bonds, as depicted in Figure 2-19 (Patel, 1973). The antiparallel β structure conformation requires that the Val_1-Pro_2 and Phe_6-Pro_7 bonds be trans and that the Pro_2-Pro_3 and Pro_7-Pro_8 bonds be cis.

When the sodium complex is formed, there are major conformational changes evident in the NMR data (Patel, 1973), although there is no change in the conformations about the peptide bonds. The proposed structure of the sodium complex, based on NMR data, is a saddle-shaped structure with a twofold pseudo-axis, with the carbonyl groups turned inward to line the cavity in which the sodium ion is held. There are two hydrogen bonds: Phe_6NH to $Pro_3C{=}O$ and Val_1NH to $Pro_8C{=}O$. The carbonyl carbon spectrum shows that four carbonyl groups form ligands to the metal (Patel, 1973; Bystrov et al., 1972).

Figure 2-20 shows the X-ray structure of the lithium complex of antamanide reported by Karle et al. (1973). It is believed that the sodium complex does not differ significantly. There is very close similarity of the X-ray

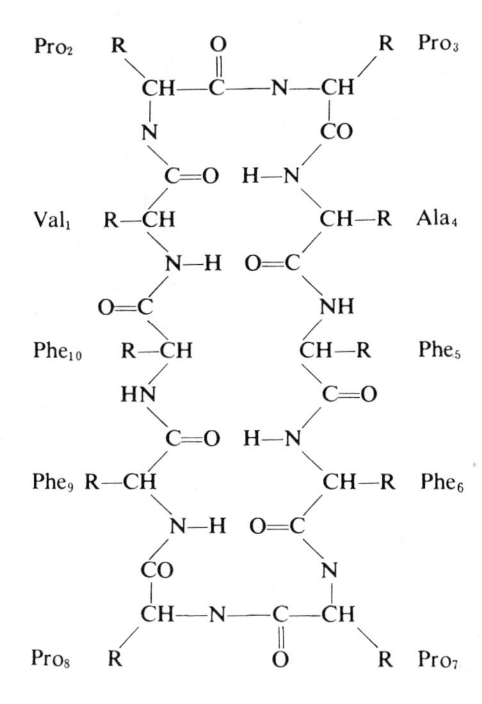

Fig. 2-19 The antiparallel β conformation of antamanide. [Reprinted with permission from Patel, *Biochemistry*, **12**, 667 (1973). Copyright by the American Chemical Society.]

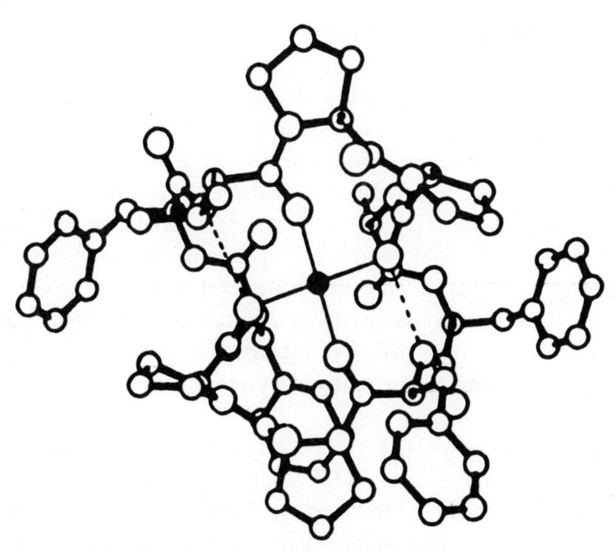

Fig. 2-20 One view of the X-ray structure of the lithium complex of antamanide Karle et al. (1973).

structure and that deduced by NMR in solution. There are some torsion angular differences, and the X-ray structure is of high energy according to conformational energy calculations. In the NMR structure, the torsion angles move to lower-energy positions. This movement probably corresponds to a relaxation in solution on elimination of stabilizing forces present only in the crystal.

The coordination of the metal in the crystal is unusual in that the four carbonyl ligands are arranged pyramidally to the metal, with a molecule of acetonitrile (crystallizing medium) forming a fifth ligand at the apex of the pyramid.

F Valinomycin

Valinomycin is a cyclic depsipeptide antibiotic that has the structure:

```
        L              D               D              L
 ( NH—CH—CO—O—CH—CO—NH—CH—CO—O—CH—CO )
       |             |              |             |
       CH            CH             CH            CH3
      / \           / \            / \                  3
   CH3 CH3       CH3 CH3        CH3 CH3

 |______________| |______________| |______________| |______________|
      L-Val           D-HyIv           D-Val            L-Lac
```

It is capable of forming a potassium complex, and both the complexed and uncomplexed forms have been the subject of a number of NMR investigations (Ivanov et al., 1969; Ohnishi and Urry, 1969; Urry and Ohnishi, 1970; Patel and Tonelli, 1973; Patel, 1973). Like antamanide, valinomycin undergoes conformational isomerization with change of temperature and solvent, but always exhibits a spectrum consistent with threefold symmetry on the NMR time scale. This, however, does not rule out a pseudo-rotating structure that may be unsymmetric at any one instant. In a predominantly hydrocarbon environment, the "pore", or "bracelet", structure is dominant. In this conformation, all the valine NH protons are internally hydrogen bonded. In ether solvents at low temperatures, propeller-like conformations are formed in which the DVal hydrogen bonds are retained, but the LVal NH protons are turned outward to bond with the solvent, causing the "pore" to close up. Other conformations may exist under other conditions (Patel and Tonelli, 1973).

The NMR structure of the potassium complex (Ivanov et al., 1969; Ohnishi and Urry, 1969; Urry and Ohnishi, 1970; Patel, 1973) resembles the pore, or bracelet, structure. However, it differs in that all six of the ester carbonyls, which are turned outward in the bracelet, turn *inward* to form ligands to K^+ (Patel, 1973; Ohnishi et al., 1972).

The crystal structures of the uncomplexed form determined by Duax et al. (1972) and the K^+ complex determined by Pinkerton et al. (1973) are shown in Figures 2-21*a* and *b*, respectively. The structure of the complex is threefold symmetric and is in good agreement with the NMR structure. In contrast, the X-ray structure of the uncomplexed form possesses a pseudocenter of symmetry, but does not possess threefold symmetry. Hence this structure differs from any of the solution structures. Eight of the twelve residues are in conformations closely resembling those of the complex, but the other four (one each of LVal, DVal, LLac, and DHyIv) are different. It is possible that this structure is one of several pseudorotamers, the average of which is seen in solution, and that the instantaneous solution structure is not symmetric. However, Ivanov (1969) measured the dipole moment of the uncomplexed valinomycin in carbon tetrachloride and found a value of 3.5D. Patel and Tonelli (1973) found a dipole moment of 3.4D in benzene.

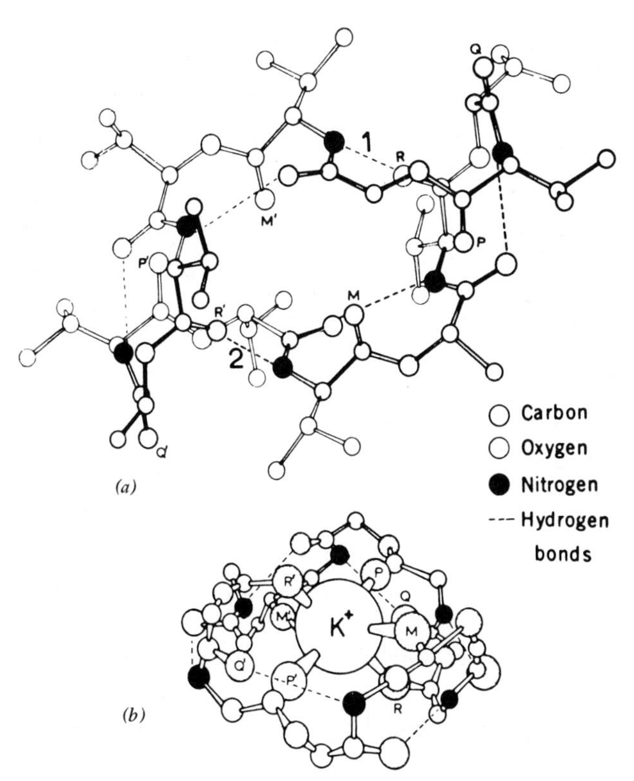

Fig. 2-21 X-ray conformations of valinomycin (top) and its K^+ complex (bottom). [Reprinted with permission for Duax et al., *Science*, **176, 911** (1972). Copyright 1972 by the American Association for the Advancement of Science.]

These values agree with the calculated value (Patel and Tonelli, 1973) of 3.9D for the proposed "bracelet" structure. The calculated dipole moment for the Duax crystal structure is 6.9D. Thus it appears that the solution conformation is probably truly threefold symmetric, or nearly so, and really differs from the crystal structure. Further, a recent Raman spectroscopic study (Rothschild et al., 1973) strongly indicates that the crystal conformation is not unique, but that the structure, and particularly the number of internal hydrogen bonds, depends on the solvent from which the cyclic peptide is crystallized.

G Stendomycin

Pitner and Urry (1972) have carried out the 220-MHz PMR studies of the antifungal tetradecapeptide antibiotic stendomycin in trifluoroethanol and methanol. Using three techniques, deuteron proton-exchange rates, chemical shift temperature dependence, and trifluoroethanol solvent mixtures, they concluded that all peptide protons were shielded, to varying degrees, from the solvent. These findings are in agreement with the work of Bodansky and Bodansky (1968), who found that stendomycin did not react with the modified Rydon reagent, suggesting that all peptide protons were "inside" the molecule. Pitner and Urry propose a folding of the lactone ring using β and β-like turns, and a left-handed α-helical segment for the series of D-amino acids in the linear segment.

REFERENCES

Abraham, R. J. and G. Gatti (1969). *J. Chem. Soc.* (B), 961.

Ajo, D., A. Damiani, R. Fidenzi, A. Lapiccierella, and N. Russo (1973). *Biochem. Biophys. Res. Commun.*, **52**, 807.

Alter, J. E., R. H. Andreatta, G. T. Taylor, and H. A. Scheraga (1973). *Macromolecules*, **6**, 564.

Alter, J. E., G. T. Taylor, and H. A. Scheraga (1972). *Macromolecules*, **5**, 739.

Ananthanarajanan, V. S., R. H. Andreatta, D. Poland, and H. A. Scheraga (1971). *Macromolecules*, **4**, 417.

Andersen, A. M., A. Mostad, and Ch. Romming (1972). *Acta Chem. Scand.*, **26**, 2670.

Avignon, M. (1972). Thése, Univ. de Bordeaux.

Avignon, M., C. Garrigou-Lagrange, P. Bothorel (1972). *J. Chim. Phys. Physicochim. Biol.*, **69**, 62.

Avignon, M. and P. V. Huong (1970). *Biopolymers*, **9**, 427.

Avignon, M., P. V. Huong, J. Lascombe, M. Marraud, and J. Néel (1969). *Biopolymers*, **8**, 69.

Bailey, N. A., P. M. Harrison, and R. Mason (1968). *Chem. Commun.*, 559.

Baker, R. W., C. H. Chothia, P. J. Pauling, and T. J. Petcher (1971). *Nature (New Biol.)* **234**, 439.

Baker, R. W., C. H. Chothia, P. J. Pauling, and H. P. Weber (1973). *Mol. Pharmacol.*, **9**, 23.

Balashova, T. A. and Yu. A. Ovchinnikov (1973). *Tetrahedron*, **29**, 873.

Barfield, M. and H. L. Gearhart (1973). *J. Am. Chem. Soc.*, **95**, 641.

Barrans, Y. and J. Clastre (1970). *Compt. Rend. Acad. Sci.*, **270c**, 306.

Bartels, E. (1965). *Biochim. Biophys. Acta*, **109**, 194.

Behr, J. P. and J. M. Lehn (1972). *Biochem. Biophys. Res. Commun.*, **49**, 1573.

Belleau, B. (1970). In *Fundamental Concepts in Drug Receptor Interactions* (J. F. Danielli, Ed.). Academic Press, New York, p. 121.

Bennett, I., A. G. H. Davidson, M. M. Harding, and A. Hoy (1970). *Acta Cryst.*, **B26**, 1722.

Benedetti, E., P. Corradini, M. Goodman, and C. Pedone (1969). *Proc. Natl. Acad. Sci. USA*, **62**, 650.

Bergin, R. (1971). *Acta Cryst.*, **B27**, 381.

Bergin, R. (1971). *Acta Cryst.*, **B27**, 2139.

Bergin, R. and D. Carlström (1968). *Acta Cryst.*, **B24**, 1506.

Bergin, R. and D. Carlström (1971). *Crystallography*, **B27**, 2146.

Black, J. W., W. A. M. Duncan, G. J. Durant, C. R. Ganellin, and M. E. Parsons (1972). *Nature*, **236**, 385.

Blout, E. R., C. M. Deber, and L. G. Pease (1974). In *Peptides Polypeptides and Proteins*, (E. R. Blout, F. A. Bovey, M. Goodman, and N. Lotan, Eds.). Wiley-Interscience, New York, p. 266.

Bodansky, M. and A. Bodansky (1968). *Nature*, **220**, 73.

Bonnet, J. J. and J. A. Ibers (1973). *J. Am. Chem. Soc.*, **95**, 4829.

Bovey, F. A. (1969). *Nuclear Magnetic Resonance Spectroscopy*. Academic Press, New York, p. 67.

Bovey, F. A. (1974a). In *Macromolecular Reviews*. Wiley Interscience, New York, vol. 9, p. 1.

Bovey, F. A. (1974b). In *Peptides, Polypeptides, and Proteins* (E. R. Blout, F. A. Bovey, M. Goodman and N. Lotan, Eds.). Wiley-Interscience, New York, p. 248.

Brimblecome, R. W. (1974). *Drug Actions on Cholinergic Systems*. University Park Press, Baltimore.

Bugg, C. E. and V. Thewalt (1970). *Science*, **170**, 852.

Bustard, T. M. and R. S. Egan (1971). *Tetrahedron*, **27**, 4457.

Bystrov, V. F., V. T. Ivanov, S. A. Kozmin, I. I. Mikhaleva, K. Kh. Khalilulina, Y. A. Ovchinnikov, E. Fedin, and P. V. Petrovski (1972). *FEBS Lett.*, **21**, 34.

Bystrov, V. F., S. L. Portnova, V. I. Tsetlin, V. T. Ivanon, and Y. A. Ovchinnikov (1969). *Tetrahedron*, **25**, 493.

Candlin, R. and M. M. Harding (1970). *J. Chem. Soc.*, A384.

Carlström, C. (1973). *Acta Cryst.*, **B29**, 262.

Carlström, D. and R. Bergin (1967). *Acta Cryst.*, **23**, 313.

Carlström, D. and R. Bergin (1967). *Acta Cryst.*, **23**, 313.

Carlström, C. R. Bergin, and G. Falkenberg (1973). *Quart. Rev. Bioph.*, **6**, 257.

Casy, A. F., M. M. A. Hassan, and E. C. Wu (1971). *J. Pharm. Sci.*, **60**, 67.

Casy, A. F. and R. R. Ison (1970). *J. Pharm. Pharmacol.*, **22**, 270.

Chothia, C. and P. Pauling (1969). *Proc. Natl. Acad. Sci., USA*, **63**, 1069.

Chow, P. Y. and G. D. Fasman (1974). *Biochemistry*, **13**, 222.

Chynoweth, K. R., B. Ternai, L. S. Simeral, and G. E. Maciel (1973). *Mol. Pharmacol.*, **9**, 144.

Clark, J. R. (1964). *Rev. Pure Appl. Chem.*, **18**, 50.

Corey, R. B. (1938). *J. Am. Chem. Soc.*, **60**, 1598.

Costa, E. and S. Garattini (1970). *Internat. Symp. on Amphetamines and Related Compounds.* Raven Press, New York.

Crawford, J. L., W. N. Lipscomb, and C. G. Schellman (1973). *Proc. Natl. Acad. Sci. USA*, **70**, 538

Culvenor, C. C. J. and N. S. Ham (1966). *Chem. Commun.*, 537.

Cushley, R. J. and H. G. Mautner (1970). *Tetrahedron*, **26**, 2151.

Cung, M. T., M. Marraud, and J. Néel (1973). In *Conformation of Biological Molecules and Polymers* (E. D. Bergmann and B. Pullman, Eds.). Academic Press, New York, p. 69.

Dangoumaw, J., Y. Barrans, and R. Gay (1969). *Therapie*, **24**, 479.

Deber, C. M., D. A. Torchia, and E. R. Blout (1971). *J. Am. Chem. Soc.*, **93**, 4893.

Deber, C. M., V. Madison, and E. R. Blout (1976). *Accts. Chem. Res.*, **9**, 106.

Degeilh, B. and R. E. Marsh (1959). *Acta Cryst.*, **12**, 1007.

DeSantis, P. (1973). In *Proceedings of the International Symposium on the Conformation of Biological Molecules and Polymers*, (E. Bergmann and B. Pullman, Eds.). Academic Press, New York, p. 493.

Dewar, M. J. S. and R. C. Fahey (1963). *J. Am. Chem.*, **85**, 2704.

Donohue, J. and A. Caron (1964). *Acta Cryst.*, **17**, 1178.

Duax, W. L., H. Hauptman, C. M. Weeks, and D. A. Norton (1972). *Science*, **176**, 911.

Eliel, E. L. (1966). In *Stereochemie der Kohlenstoffverbindungen.* Verlag Chemie, Weinheim, p. 254.

Ernst, S. R. and F. W. Cagle Jr. (1973). *Acta Cryst.*, **B29**, 1543.

Falkenberg, G. and D. Carlström (1971). *Acta Cryst.*, **B27**, 411.

Fasman, G. D. and P. Y. Chou (1974). In *Peptides, Polypeptides and Proteins* (E. R. Blout, F. A. Bovey, M. Goodman, and N. Lotan, Eds.). Wiley-Interscience, New York, p. 114.

Feeney, J., P. E. Hansen, and G. C. K. Roberts (1974). *Chem. Commun.*, 503.

Ganellin, C. R. (1973). *J. Med. Chem.*, **16**, 620.

Ganellin, C. R., G. N. J. Port, and W. G. Richards (1973). In *Conformation of Biological Molecules and Polymers* (E. D. Bergmann and B. Pullman, Eds.). Academic Press, New York, p. 579.

Ganellin, C. R., G. N. J. Port, and W. G. Richards (1973). *J. Med. Chem.*, **16**, 616.

Genson, D. W. and R. F. Christoffersen (1973). *J. Am. Chem. Soc.*, **95**, 362.

Giessner-Prettre, C. and B. Pullman (1975). *J. Mag. Res.*, **18**, 564.

Gurskaya, G. V. (1968). *The Molecular Structure of Amino Acids.* Consultants Bureau, New York.

Ham, N. S. (1971). *J. Pharm. Sci.*, **60**, 1764.

Ham, N. S. (1974). In *Molecular and Quantum Pharmacology* (E. D. Bergmann and B. Pullman, Eds.). Reidel, Dordrecht-Holland, p. 261.

Hamilton, W. C., M. N. Frey, L. Golic, T. F. Koetzle, M. S. Lehmann, and J. J. Verbist (1972). *Mater. Res. Bull.*, **7**, 1225.

Hanna, P. E. and A. E. Ahmed (1973). *J. Med. Chem.*, **16**, 963.

Hearn, R. A. and C. E. Bugg (1972). *Acta Cryst.* (B), **28**, 3662.

Hearn, R. A., G. R. Freeman, and C. E. Bugg (1973). *J. Am. Chem. Soc.*, **95**, 7150.

Hershbach, D. R. (1962). *Bibliography for Hindered Internal Rotation and Microwave Spectroscopy*. Lawrence Radiation Laboratory, Univ. of Calif.-Berkeley.

Hruby, V. J. (1974). In *Chemistry and Biochemistry of Amino Acids, Peptides and Proteins*, Vol. 3 (B. Weinstein, Ed.). Dekker, New York.

Hughes, L. J., R. H. Andreatta, and H. A. Scheraga (1972). *Macromolecules*, **5**, 187.

Ison, R. R. (1974). In *Molecular and Quantum Pharmacology* (E. Bergmann and B. Pullman, Eds.). Reidel-Dordrecht, Holland, p. 55.

Ison, R. R., F. M. Franks, and K. S. Soh, (1973). *J. Pharm. Pharmacol.*, **25**, 887.

Ison, R. R., P. Partington, and G. C. K. Roberts (1973). *Mol. Pharmacol.*, **9**, 756.

Ivanov, V. T., I. A. Laine, N. D. Abdullaev, L. B. Senyavina, E. M. Popov, Y. A. Ovchinnikov, and M. M. Shemyakin (1969). *Biochem. Biophys. Res. Commun.*, **34**, 803.

James, M. N. G. and G. J. B. Williams (1971). *J. Med. Chem.*, **14**, 670.

Johnson, C. L., S. Kang, J. P. Green (1973). In *Conformation of Biological Molecules and Polymers* (E. D. Bergmann and B. Pullman, Eds.). Academic Press, New York, p. 517.

Jose, P., L. M. Pant, and A. B. Biswas (1964). *Acta Cryst.*, **17**, 24.

Kang, S. and J. P. Green (1970). *Proc. Natl. Acad. Sci. USA*, **67**, 62.

Karle, I. L. (1972). *J. Am. Chem. Soc.*, **94**, 81.

Karle, I. L., K. S. Dragonette, and S. A. Brenner (1965). *Acta Cryst.*, **19**, 713.

Karle, I. L., J. W. Gibson, and J. Karle (1970). *J. Am. Chem. Soc.*, **92**, 3755.

Karle, I. L., J. Karle, Th. Wieland, W. Burgermeister, H. Faulstich, and B. Witkop (1973). *Proc. Natl. Acad. Sci. USA*, **70**, 1836.

Karplus, M. (1959). *J. Chem. Phys.*, **30**, 11.

Karplus, M. (1963). *J. Am. Chem. Soc.*, **85**, 2870.

Karplus, S. and M. Karplus (1972). *Proc. Natl. Acad. Sci. USA*, **69**, 3204.

Kartha, G., G. Ambody, and P. V. Shankar (1974). *Nature*, **247**, 204.

Katz, B. (1966). *Nerve Muscle Synapse*. McGraw-Hill, New York.

Katz, J. L. and B. Post (1960). *Acta Cryst.*, **13**, 624.

Kennard, O. and D. G. Watson, Eds. (1970). *Molecular Structures and Dimensions*, **1**, 343.

Kier, L. B. (1967). *Mol. Pharmacol.*, **3**, 487.

Kier, L. B. (1968). *J. Med. Chem.*, **11**, 441.

Kitano, M., T. Fukuyama, and K. Kuchitsu (1973). *Bull. Chem. Soc. Japan*, **46**, 384.

Kitano, M. and K. Kuchitsu (1973). *Bull. Chem. Soc. Japan*, **46**, 3048.

Koldercup, M., A. Mostad, and Ch. Romming (1972). *Acta Chem. Scand.*, **26**, 483.

Kopple, K. D. and D. H. Marr (1967). *J. Am. Chem. Soc.*, **89**, 6193.

Kopple, K. D. and M. Ohnishi (1969). *J. Am. Chem. Soc.*, **91**, 962.

Kopple, K. D., G. R. Wiley, and R. Tauke (1973). *Biopolymers*, **12**, 627.

Krackov, M. H., C. M. Lee, and H. G. Mautner (1965). *J. Am. Chem. Soc.*, **87**, 892.

Krnjevic, K. (1970). *Nature*, **228**, 119.

Lakshminarayanan, A. V., V. Sasisekharan, and G. N. Ramachandran (1967). In *Conformation of Biopolymers* (G. N. Ramachandran, Ed.). Academic Press, New York, Vol. 1, p. 61.

Lambrecht, G. (1971). In *Cyclische Acetylcholinanaloga*. H. & P. Lang, Bern-Grankfurt.

Lambrecht, G. and E. Mutschler (1974). In *Molecular and Quantum Pharmacology* (E. Bergmann and B. Pullman, eds.). Reidel-Dordrecht, Holland, p. 179.

Law, H. H. (1961). *Angew. Chem.*, **73**, 425.

Lehmann, M. S., T. F. Koetzle, and W. C. Hamilton (1972). *Int. J. Peptide Protein Res.*, **4**, 229.

Lewis, P. N., F. A. Momany, and H. A. Scheraga (1973). *Israel J. Chem.*, **11**, 121.

Levy, G. C. (1973). *Acc. Chem. Res.*, **6**, 161.

Lichtenberg, D., P. A. Kroon, and S. I. Chan (1974). *J. Am. Chem. Soc.*, **96**, 5934.

Lichter, R. L. and J. D. Roberts (1970). *J. Org. Chem.*, **35**, 2806.

Liquori, A. M., A. Damiani, and J. L. DeCoen (1968). *J. Mol. Biol.*, **33**, 445.

Lyle, R. E., D. J. Mac Mahon, W. E. Krueger, and C. K. Spicer (1966). *J. Org. Chem.*, **31**, 4164.

Madden, J. J., E. L. McGandy, and N. C. Seeman (1972*a*). *Acta Cryst.*, **B28**, 2377.

Madden, J. J., E. L. McGandy, N. C. Seeman, M. M. Harding, and A. Hoy (1972*b*). *Acta Cryst.*, **B28**, 2382.

Madison, V. S. (1974). In *Peptides, Polypeptides and Proteins* (E. R. Blout, F. A. Bovey, M. Goodman, and N. Lotan, Eds.). Wiley-Interscience, New York, p. 89.

Makriyannis, A. (1974). *Pharmacologist*, **16**, 285.

Makriyannis, A., R. F. Sullivan, and H. G. Mautner (1973). *Proc. Natl. Acad. Sci. USA*, **69**, 3416.

Marraud, M. (1971). Thése, Univ. de Nancy.

Marraud, M. and J. Néel (1973). *J. Chim. Phys.*, **70**, 947.

Marraud, M., J. Néel, M. Avignon, and P. V. Huong (1970). *J. Chim. Phys.*, **67**, 959.

Marsh, R. E. and J. Donohue (1967). *Adv. Protein Chem.*, **22**, 235.

Martin-Smith, M., G. A. Smail, and J. B. Stenlake (1967). *J. Pharm. Pharmacol.*, **19**, 561.

Mashkovsky, M. D. (1963). In *Proc. of the 1st Internat. and Pharmac. Meeting*, Stockholm (K. J. Burnings and P. Lindgren, Eds.). Pergamon Press, London, Vol. 1, p. 359.

Mathews, M. and G. J. Palenik (1971). *J. Am. Chem. Soc.*, **93**, 497.

Matsuzaki, T. and Y. Iitaka (1971). *Acta Cryst.*, Sect. B, **27**, 507.

Mautner, H. G. (1972). *Ann. N.Y. Acad. Sci.*, **192**, 167.

Mautner, H. G. (1974). In *Molecular and Quantum Pharmacology* (E. Bergmann and B. Pullman, Eds.). Reidel-Dordrecht, Holland, p. 119.

Mautner, H. G., D. D. Dexter, and B. W. Low (1972). *Nature*, **238**, 87.

Maxfield, F. R., J. E. Alter, G. T. Taylor, and H. A. Scheraga (1975). *Macromolecules*, **8**, 479.

McConnell, H. M. (1957). *J. Mol. Spectrosc.*, **1**, 11.

McIlwain, H. and H. S. Bachelard (1971). In *Biochemistry and the Central Nervous System*, 4th ed., Chaps. 8 and 14. Churchill & Livingstone, Edinburgh.

Mizushima, S., T. Shimanouchi, M. Tsuboi, and T. Ozakana (1957). *J. Am. Chem. Soc.*, **79**, 5357.

Nagatsu, T. (1973). *Biochemistry of Catecholamines*. University Park Press, Baltimore.

Neville, G. A., R. Deslauriers, B. J. Blackburn, and I. C. Smith (1971). *J. Med. Chem.*, **14**, 717.

Ohnishi, M., M.-C. Fedarko, J. D. Baldeschwieler, and L. F. Johnson (1972). *Biochem. Biophys. Res. Commun.*, **46**, 312.

Ohnishi, M. and D. W. Urry (1969). *Biochem. Biophys. Res. Commun.*, **36**, 194.

Okaya, Y. (1966). *Acta Cryst.*, **21**, 726.

Ovchinnikov, Y. A., V. T. Ivanov, V. F. Bystrov, A. I. Miroshnikov, E. N. Shepel, N. D. Abdullaev, E. S. Efremov, and L. B. Senyavina (1970). *Biochem. Biophys. Res. Commun.*, **39**, 217.

Parrish, J. R. and E. R. Blout (1972). *Biopolymers*, **11**, 1001.

Partington, P., J. Feeney, and A. S. V. Burgen (1972). *Mol. Pharmacol.*, **8**, 269.

Patel, D. J. (1973*a*). *Biochemistry*, **12**, 667.

Patel, D. J. (1973*b*). *Biochemistry*, **12**, 496.

Patel, D. J. and A. E. Tonelli (1973). *Biochemistry*, **12**, 486.

Pauling, P. J. (1973). In *Conformation of Biological Molecules and Polymers* (E. Bergamann and B. Pullman, Eds.). Academic Press, New York, p. 505.

Periti, P. R. (1970). *Pharmacol. Res. Commun.*, **2**, 309.

Phillips, D. C. (1954). *Acta Cryst.*, **7**, 159.

Pinkerton, M., L. D. Steinrauf, and P. Dawkins (1969). *Biochem. Biophys. Res. Commun.*, **35**, 512.

Pitner, T. P. and D. W. Urry (1972). *Biochemistry*, **11**, 4132.

Platzer, K. E. B., V. S. Ananthanarayanan, R. H. Andreatta, and H. A. Scheraga (1972). *Macromolecules*, **5**, 177.

Port, G. N. J. and A. Pullman (1973). *J. Am. Chem. Soc.*, **95**, 4059.

Porthogese, P. S. (1967). *J. Med. Chem.*, **10**, 1057.

Pullman, B., H. Berthod, and A. Pullman (1974). *An. Quim.*, **70**, 1204.

Pullman, B., J. L. Coubeils, P. Courrière, and J. P. Gervois (1972). *J. Med. Chem.*, **15**, 17.

Pullman, B. and P. Courrière (1972). *Mol. Pharmacol.*, **8**, 371.

Pullman, B. and A. Pullman (1974). In *Advances in Protein Chemistry* (C. B. Anfinsen, J. T. Edsall and F. M. Richards, Eds.). Academic Press, New York, Vol. 28, p. 347.

Ramachandran, G. N., R. Chandrasekaran, and K. D. Kopple (1971). *Biopolymers*, **10**, 2113.

Roberts, G. C. K. (1974). In *Molecular and Quantum Pharmacology* (E. Bergmann and B. Pullman, Eds.). Reidel-Dordrecht, Holland, p. 77.

Rothe, M., K. D. Steffen, and I. Rothe (1965). *Angew. Chem., Int. Ed.* (English), **4**, 356.

Rothschild, K. J., I. M. Asher, E. Anastassakis, and H. E. Stanley (1973). *Science*, **182**, 384.

Schellman, J. A. and E. B. Neilsen (1967). In *Conformation of Biopolymers* (G. N. Ramachandran, Ed.). Academic Press, London, Vol. 1, p. 109.

Scheraga, H. A. (1968). *Adv. Phys. Org. Chem.*, **6**, 103.

Scheraga, H. A. (1971). *Chem. Rev.*, **71**, 195.

Scheule, R. K., F. Cardinaux, G. T. Taylor, and H. A. Scheraga (1976). *Macromolecules*, **9**, 23.

Schunack, W. (1973). *Arch. Pharmaz.*, **306**, 934.

Schwyzer, R. (1969). *Rec. Chem. Progr.*, **20**, 147.

Schwyzer, R. (1971). Cited as a private communication in R. J. Weinkam and E. C. Jorgenson. *J. Am. Chem. Soc.*, **93**, 7038.

Schwyzer, R., J. D. Carrion, B. Gorup, H. Nolting, and A. Tun-Kyi (1964). *Helv. Chim. Acta*, **47**, 441.

Schwyzer, R. and U. Ludescher (1968). *Biochemistry*, **7**, 2514.

Schwyzer, R. and U. Ludescher (1969). *Helv. Chim. Acta*, **52**, 2033.

Schwyzer, R., P. Sieber, and B. Gorup (1958). *Chimica*, **12**, 90.

Senti, F. and D. Harker (1940). *J. Am. Chem. Soc.*, **62**, 2008.

Sequeira, A., H. Rajagopal, and R. Chidambaram (1972). *Acta Cryst.*, **B28**, 2514.

Shafi'ee, A. and G. Hite (1969). *J. Med. Chem.*, **12**, 266.

Shefter, E. (1971). *Cholinergic Ligand Interactions* (D. J. Triggle, Ed.). Academic Press, New York, p. 83.

Sletten, E. (1970). *J. Am. Chem. Soc.*, **92**, 172.

Snyder, S. H. and E. Richelson (1968). *Proc. Natl. Acad. Sci. USA*, **60**, 206.

Sogn, J. A., W. A. Gibbons, and E. W. Randall (1973). *Biochemistry*, **12**, 2100.

Stern, A., W. A. Gibbons, and L. C. Craig (1968). *Proc. Natl. Acad. Sci. USA*, **61**, 734.

Steward, E. G., R. B. Player, and D. Warner (1973). *Acta Cryst.*, **B29**, 2038.

Stothers, J. B. (1972). *Carbon-13 NMR Spectroscopy*. Academic Press, New York.

Suwalsky, M. and W. Traub (1972). *Biopolymers*, **11**, 623.

Testa, B. (1974). In *Molecular and Quantum Pharmacology* (E. Bergmann and B. Pullman, Eds.). Reidel-Dordrecht, Holland, p. 241.

Thewalt, B. and C. E. Bugg (1972). *Acta Cryst.*, **B28**, 82.

Thong, C. M., D. Canet, P. Granger, M. Marraund, and J. Néel (1969). *C. R. Acad. Sci.*, C, **269**, 580.

Tonelli, A. E. and A. I. Brewster (1972). *J. Am. Chem. Soc.*, **94**, 2851.

Tsoucaris, G. (1961). *Acta Cryst.*, **14**, 909.

Tsoucaris, D., C. de Rango, G. Tsoucaris, C. Zelwer, R. Parthasarathy, and F. E. Cole (1973). *Cryst. Struct. Comm.*, **20**, 193.

Urry, D. W. and M. Ohnishi (1970). In *Spectroscopic Approaches to Biomolecular Conformation* (D. W. Urry, Ed.). American Medical Association, Chicago, p. 263.

vonDreele, P. H., N. Lotan, V. S. Ananthanarayanan, R. H. Andreatta, D. Poland, and H. A. Scheraga (1971*b*). *Macromolecules*, **4**, 408.

vonDreele, P. H., D. Poland, and H. A. Scheraga (1971*a*). *Macromolecules*, **4**, 396.

Veidis, M. V. and G. J. Palenik (1969). *Chem. Commun.*, 196.

Veidis, M. V., G. J. Palenik, R. Schaffrin, and J. Trotter (1969). *J. Chem. Soc.* (A), 2659.

Wakahara, A., T. Fujiwara, and K. Tomita (1970). *Tetrahedron Lett.*, 4999.

Walton, A. G. and J. Blackwell (1973). *Biopolymers*. Academic Press, New York.

Webb, L. E. and C.-F. Lin (1971). *J. Am. Chem. Soc.*, **93**, 3818.

Wieland, Th. (1972). In *Chemistry and Biology of Peptides* (J. Meinhofer, Ed.). Ann Arbor Science Publishers, Ann Harbor, Mich., p. 377. See also, *Angew. Chem.*, (1968), **80**, 209.

Weinstein, H., B. Z. Apfelderfer, S. Cohen, S. Maayani, and M. Sokolovsky (1973). In *Conformation of Biological Molecules and Polymers* (E. D. Bergmann and B. Pullman, Eds.). Academic Press, New York, p. 531.

Yamane, T., T. Ashida, and M. Kakudo (1973). *Acta Cryst.*, **B29**, 2884.

Young, P. E., V. Madison, and E. R. Blout (1973). *J. Am. Chem. Soc.*, **95**, 6142.

Zimm, B. H. and J. K. Bragg (1959). *J. Chem. Phys.*, **31**, 526.

CHAPTER 3

Intermolecular Interactions and Drug Action

This chapter discusses the extent to which we can use our present understanding of intermolecular interactions to explain and ultimately predict biologic activity in congeneric sets of drug compounds. Two extreme approaches are employed. Linear and additive free-energy parameters are applied in the one class of analyses, whereas molecular orbital indices are used in the alternate method. A cooperative program encompassing both of these techniques mediated by recent developments in empirical structural mechanics has been suggested as the best way to proceed in molecular design. Applications from many of the recent and sophisticated studies of intermolecular interactions reported in this book are generally too difficult or simply not practical to apply in large-scale molecular design studies.

I CONCEPTS AND THEORY OF HANSCH ANALYSIS

Perhaps the most definitive demonstration of the interactive role played by the molecular environment on biologic activity is seen in the action of drugs. It was Hansch (Hansch and Fujita, 1964) who first proposed and subsequently verified that biologic activity very often depends *parabolically* on the lipid-aqueous phase partition coefficient of the drug. Hansch and Fujita have defined a hydrophobicity factor, π, which measures logarithmic changes in the partition coefficient. By applying measured π constants Hansch has been able to obtain startling correlations of molecular structure with observed activity in a wide variety of in vitro and in vivo biologic tests. Hansch Analysis, as this technique has come to be known, along with other procedures proposed to establish quantitative structure activity relationships (QSAR), have been the subject of several reviews (Hansch, 1966; Singer and Purcell, 1967; Burger, 1970; Ariëns, 1971–1976; Tute, 1972; Purcell et al., 1973; Redl et al., 1974; Harper and Simmonds, 1966).

The basic Hansch equation, a starting point for many later correlations, was derived (Hansch and Fujita, 1964) by considering the general case of a drug applied to any biologic system. In any biologic test, only two quantities can be measured: the amount of compound given (the dose) and the biologic activity (response) obtained. The response is determined by the structure, that is, by the physicochemical properties of the compound, and, within a closely related or so-called congeneric series of compounds, changes in structure can be related to changes in biologic activity. It has long been recognized that drug absorption from the applied phase, and ensuing transport to a sensitive site, is governed largely by a lipophilic-hydrophilic balance that can be expressed in terms of the corresponding partition coefficient. The work of Collander (1954*a*,*b*) is often quoted, since it was he who first demonstrated that the rate of transport of many organic compounds through the cellular material of *Nitella* cells is proportional to the logarithm of their partition coefficients between an organic solvent and water. Collander's work on log P (logarithm of the partition coefficient) and transport was reexamined by Milborrow and Williams (1968), who confirmed the original findings. Brodie (1964) has reviewed this general topic along with the role of ionization in the absorption and excretion of drugs.

Thus, although many factors are clearly involved in biologic action (for some recent general discussions, see Callingham, 1974; Davies and Prichard, 1973), Hansch began by assuming that, for any congeneric series of compounds, one particular "reaction" could be critical and rate determining. If K_X is an equilibrium or rate constant for this rate-determining reaction, which is possibly, but not necessarily, at the site of action of the drug, C is the applied concentration, that is, dose, and A represents the probability of a drug molecule reaching this critical site in a given time interval, then the expression for the rate of biologic response is

$$\frac{d(\text{response})}{dt} = ACK_X \tag{3-1}$$

Hansch was aware of Collander's work and determined to relate A to log P and changes in A, as one compared molecules in a congeneric series, to changes in log P. As mentioned in the introductory paragraph, these changes can be expressed in the form of substituent constants, which Hansch et al. (1963) have termed π-constants and rigorously defined as

$$\begin{aligned} \pi &\equiv \log(P_X/P_H) \\ \pi &\equiv \log P_X - \log P_H \end{aligned} \tag{3-2}$$

where P_X and P_H are the partition coefficients of substituted and parent molecules, respectively. A necessary consequence of the definition of π is that if any molecule is divided into n arbitrary parts, each of whose π value is known, then the log P of the entire molecule is simply

$$\log P = \sum_{i=1}^{n} \pi_i \tag{3-3}$$

For a reference standard, partition, and consequently π, coefficients have all been measured in the 1-octanol water system. Clearly this system has been chosen as a model for biologic lipid and aqueous phases. A negative π value thus indicates a change toward greater affinity for the aqueous phase, and a positive value indicates greater affinity for the lipid phase. In essence, π expresses the relative *free-energy change* on moving a derivative from one phase to another.

Hansch next chose to assume as a working hypothesis that the probability A is related to log P for the complete molecule, or to π for changes in congeneric series, through a Gaussian distribution function

$$A = a \exp[-(\pi - \pi_0)^2/b] \tag{3-4}$$

where a and b are constants, and π_0 is the π value corresponding to the maximum in the distribution. The choice for this functional relationship is based on the fact that, in many series of compounds tested in biologic systems, as the relative lipophilicity was increased, activity rose to a maximum, fell off, and eventually reached zero. Indeed, in vivo it is generally to be expected that, for highly water-soluble drugs (low or negative log P), the probability of reaching some distant receptor site will be low because of rapid excretion.

Hansch pointed out that approximately linear relationships between biologic activity with log P can be expected when compounds within a limited range of log P values, either all higher or lower than the optimum, are chosen for study. This neatly accounts for the results of Collander.

The substitution of the expression for A given in equation 3-4 into the fundamental rate equation 3-1 gives

$$\frac{d(\text{response})}{dt} = a \exp[-(\pi - \pi_0)^2/b]CK_X \tag{3-5}$$

If it is stipulated that C is measured as the compound concentration necessary to produce a particular constant response (LD_{50}, LD_{95}, ED_{50}, percentage inhibition, etc.) in a fixed time interval, then $d(\text{response})/dt$ can be considered to be constant, and equation 3-5 becomes

$$\log(1/C) = -k_1\pi^2 + k_2\pi\pi_0 - k_3\pi_0^2 + \log K_X + k_4 \tag{3-6}$$

Since π_0 is a constant, being the value of π that yields a log P which, in turn, yields an optimum activity, equation 3-6 is an expression relating biologic activity as measured through $\log(1/C)$ to a measurable free-energy difference parameter π and to the unknown rate constant K_X for the critical reaction.

The final critical assumption inherent to the complete derivation of the structure-function equation is, as first suggested by Hansen (1962) that K_X is a function of the electron release or withdrawal of the substituents, the hydrophobicity of the drug, and the sterochemical geometry. Thus log K_X can be expressed as

$$\log K_X = k_5\pi + k_6\sigma + k_7E_S \tag{3-7}$$

where σ is the Hammett constant (Hammett, 1940) that expresses changes in electronic properties of substituents, and E_S is a steric parameter as suggested by Taft (Taft, 1956; Taft and Lewis, 1959). Substitution of equation 3-7 into 3-6, noting that $(1/C)$ is a measure of biologic response, BR, leads to the working equation

$$\log BR = -k_1\pi^2 + k_2\pi + k_3\sigma + k_4E_S + k_5 \tag{3-8}$$

where $k_2 = C_1(\pi_0 + C_2)$ and $k_5 = C_3 - C_4\pi_0^2$.

II CONFIRMATION OF THE PARABOLIC RELATIONSHIP BETWEEN ACTIVITY AND LOG *P*

The empirically predicted parabolic relationship between log BR, or log $(1/C)$ and π, or log P has received experimental justification and theoretical support.

The way in which log P determines the concentration of a drug at the site of action was demonstrated in an early analysis by Hansch et al. (1965) of the results of Soloway et al. (1960) for a set of antitumor compounds. Soloway injected each of a series of 25 substituted benzene-boronic acids (I) into tumor-bearing mice and measured the actual concentration of boron in brain (C_b) and tumor (C_t) tissue after a 15-minute interval. Thus C_b and C_t

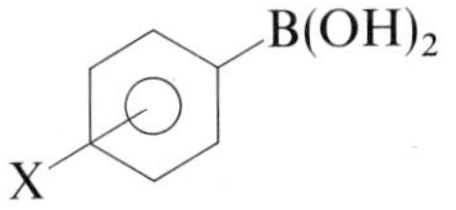

$B(OH)_2$

X

I

represent rates of accumulation of compound. Hansch listed the compounds with their corresponding values of log C_b, log C_t, σ and π values, and applied multiple regression analysis to determine the equations of best fit. The data for brain tissue was best described by equation 3-9, in which the π^2 term was highly significant; that is, the relationship

$$\log C_b = 0.765\pi - 0.540\pi^2 + 1.505 \qquad (3\text{-}9)$$

was parabolic. An equation in π alone did not explain very much of the variation in C_b, and addition of a σ term did not improve the correlation.

The data for tumor tissue was, in contrast, best described by π^2 and σ terms alone, as in equation 3-10:

$$\log C_t = -0.130\pi^2 - 0.405\sigma + 1.341 \qquad (3\text{-}10)$$

The optimum partition coefficient, log P_0, for penetration of brain tissue was found by determining the extremum of equation 3-9, which yielded $\pi_0 = 0.7$. Subsequently, the partition coefficient of the unsubstituted parent compound (I, $X = H$) was measured and found to be log $P_H = 1.58$. Since log P and π are additive,

$$\log P_0 = \log P_H + \pi_0 = 2.3 \qquad (3\text{-}11)$$

Subtraction of equation 3-9 from 3-10 gives equation 3-12, which expresses a "therapeutic index" constructed in terms of substituent constants:

$$\log(C_t/C_b) = -0.765\pi + 0.410\pi^2 - 0.405\sigma - 0.164 \qquad (3\text{-}12)$$

Hansch was able to suggest on the basis of these equations that substituents with π values in the range of -1.0 to -2.0 would be worthy of investigation, since these would still allow a therapeutic concentration in tumor tissue (equation 3-10), while giving high selectivity (equation 3-11).

The concept of an optimum log P value has been theoretically justified by consideration of a simple kinetic model (Penniston et al., 1969) to describe the movement of molecules through a series of aqueous compartments separated by lipid barriers. A set of differential equations giving the concentration of molecules in particular compartments after chosen time intervals was solved for arbitrary values of time, number of barriers, and initial drug concentrations. The results gave a parabolic relationship between the logarithm of the concentration in any chosen compartment and log P.

More convincing than either the direct measurements of accumulation of compound or results from the kinetic model is the practical demonstration of parabolic dependence of activity on lipophilicity in many systems, as indicated by the high significance of a squared term in π or log P. In those cases in which a squared term is not necessary, the reason is usually that only compounds with suboptimal or supraoptimal values of log P have been included in the study. Consequently, all data points fall either on the solely increasing or decreasing sides of the parabola.

III LIPOPHILIC/HYDROPHILIC INTERACTIONS AND DRUG DESIGN

A necessary consequence of the Hansch model is that structurally different sets of compounds, acting by the same mechanism on the same receptor sites, should have the same value for log P_0. The validity of this has certain far reaching implications on the prediction of compounds that will exhibit a particular activity, that is, drug design. In addition to being able to predict activity within a congeneric series of compounds on the basis of a hydrophilic-lipophilic balance, it should be possible to evaluate the potential of new "lead" compounds, that is, nonmembers of a known congeneric series, in terms of this balance.

This principle has been illustrated in a study (Hansch et al., 1968) on the hypnotic activity of various sets of drugs including barbiturates of types 2–4 and nonbarbiturates of types 5–10 given in Table 3-1. Data were gathered from a variety of sources covered in an extensive literature survey, and different methods of measuring a standard hypnotic response in a variety of animals were involved.

In each series, electronic and steric effects were neglected, and log P_0 was calculated from a least-squares fit of the data using equation 3-8 appropriately modified. The results are presented as part of Table 3-1.

The value of log P_0 was found to be relatively constant in spite of the diverse sources of data and simplified correlation equation. Employing only those sets of data for which the 90% confidence limits were satisfied in the calculation of log P_0, an average log $P_0 = 1.8$ was computed for the barbiturates. The nonbarbiturates had a similar log P_0 value, and attention was drawn to the fact that the well-known central nervous system depressants chloroform (1.97), chloretone (2.03), and glutethimide (1.90) also have log $P_0 \approx 2.0$.

Table 3-1 Hypnotics and Log P_0

Series	Number in Set	Test[a]	Log P_0	r^b
(2) 5,5-disubstituted barbituric acid (R_1, R_2)	13	AD_{50} (mice)	1.80	.969
	11	MED (rabbits)	1.66	.896
	9	MED (rabbits)	2.25	.744
	17	MAD (rats)	2.08	.531
	15	MED (rats)	1.65	.855
	13	AD_{50} (mice)	2.69	.702
(3) spiro-cyclohexane barbiturate (R)	10	ND_{50} (mice)	2.69	.915
(4) 5-R-5-(C(R″)=CHR′) barbiturate	14	AD_{50} (mice)	2.71	.937
(5) thiazine-dione (R, R′)	6	HD_{50} (mice)	1.97	.858
(6) RR′C(OH)C≡CH	8	HD_{50} (mice)	1.79	.965
	8	HD_{50} (mice)	2.09	.944
(7) RR′C(OCONH$_2$)C≡CH	8	HD_{50} (mice)	1.56	.947
(8) RR′COH(R′CHC≡CH)	11	MED (rabbits)	2.21	.826
	13	ED_{50} (guinea pigs)	1.92	.805
(9) $Me_2C(SR)CONH_2$	6	HD_{50} (mice)	1.59	.913
(10) RCONHCONHCOR′	14	MED_{50} (mice)	1.69	.918

Reproduced, with permission, from Hansch et al. (1968). [a] Biological test parameters are doses causing a standard "hypnotic" (H) or "narcotic" (N) response. Thus AD_{50} is the active dose giving the standard response in 50% of the animals, MED represents minimum effective dose and MAD represents minimum active dose. [b] Multiple correlation coefficient.

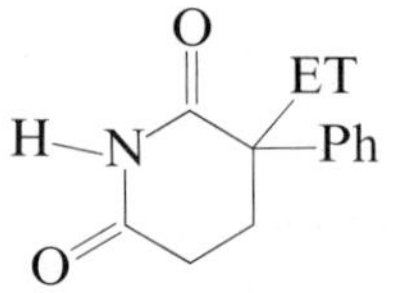

Glutethimide

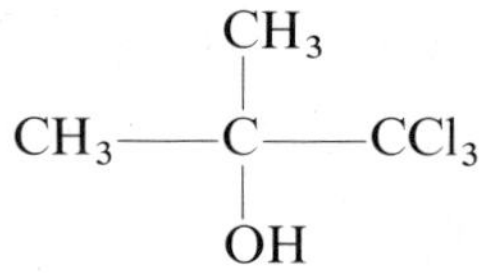

Chloretone

Thus it was suggested that almost any organic compound with log $P \approx 2.0$, provided it was not rapidly metabolized or eliminated, might have hypnotic properties. It does indeed appear that a value of log $P \approx 2.0$ encourages rapid transport to the central nervous system.

IV ALTERNATE METHODS OF DRUG DESIGN

In some examples of QSAR studies, it is inconvenient or impossible to obtain physiochemical parameters for the series of drug molecules. A model that is independent of these parameters was developed by Free and Wilson (1964). It is assumed that the various substituent groups contribute additively and linearly to the biologic activity. For example, a series of molecules represented generally by

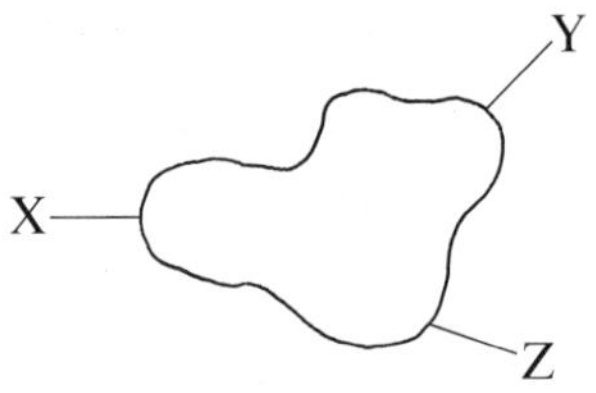

generates a set of equations

$$\log(1/C_i) = a[X_i] + b[Y_i] + c[Z_i] + \mu$$

where $[X_i]$ is the contribution to the activity for the substituent at the X position of the ith congener, and $[Y_i]$ and $[Z_i]$ have analagous definitions for substituents Y and Z, respectively. The term μ is the contribution of the parent part of the molecule. The constants a, b, and c are usually determined through multiple linear regression analyses over the set of congeners after certain restrictions are made (Purcell et al., 1973; Free and Wilson, 1964).

Various molecular orbital indices have been used in QSAR investigations. The Pullmans pioneered and brought to fruition the application of quantum chemistry to problems in biochemistry (Pullman and Pullman, 1963; Bergmann and Pullman, 1973, 1974). Among others, Kier (1971), Rogers and

Cammarata (1969), Genson and Christoffersen (1973), and Liu (1974) have described methods of using molecular orbital concepts in drug design.

New types of substituent constants have joined the ranks of σ, π, and E_S in Hansch-type QSAR studies. The molar refractivity, MR, of substituents is finding its way increasingly more often into QSAR. Pauling was one of the first to emphasize the importance of MR in the interactions of biomacromolecules (Pauling and Pressman, 1945). Agin et al. (1965) developed an expression to relate MR to $\log(1/c)$ for biochemical structure-activity relationships (QSAR). Indicator variables (Draper and Smith, 1966) that take on discrete values of 0 or 1 are being used to account for structural features not parametrized by the continuous and general QSAR variables, for example, σ, π, E_S, etc. An alternate set of π-like constants called f-hydrophobic fragmental constants, has been proposed by Rekker and Nys (1974). A concise summary of methods in drug design concurrent in date to this book is that of Cramer (1976).

V PROSPECTS IN DRUG DESIGN

Rather meaningless debates have taken place between various proponents of the different QSAR methods. These discussions usually result in pointless competition about the merits of one method over another. The point that should be made, however, is that those working in QSAR have the same objective: to gain a better understanding of drug mechanisms and to be able to select target molecules from among the staggeringly large number of possibilities. Within this framework of extracting the strength from the different methods and presenting a composite approach for selecting a target molecule for synthesis and evaluation, Purcell et al. (1973) have suggested the procedure shown in Figure 3-1. In essence three major steps are required:

1. Analyze existing data by constructing a table of molecular structures and biologic activities. Apply Free-Wilson, Hansch, and pattern recognition (Kowalski and Bender, 1972) if feasible.
2. Based on the results in (1), select parameters to be measured and/or calculated. Repeat (1).
3. Continue to recycle as new information becomes available. Use MO calculations to provide "fine tuning" input to QSAR models. When a statistically significant model is found, use this to select molecules for synthesis and evaluation.

Perhaps the most disturbing aspect in most QSAR studies is that in vivo biologic actions are probably multistep processes. The action of the input

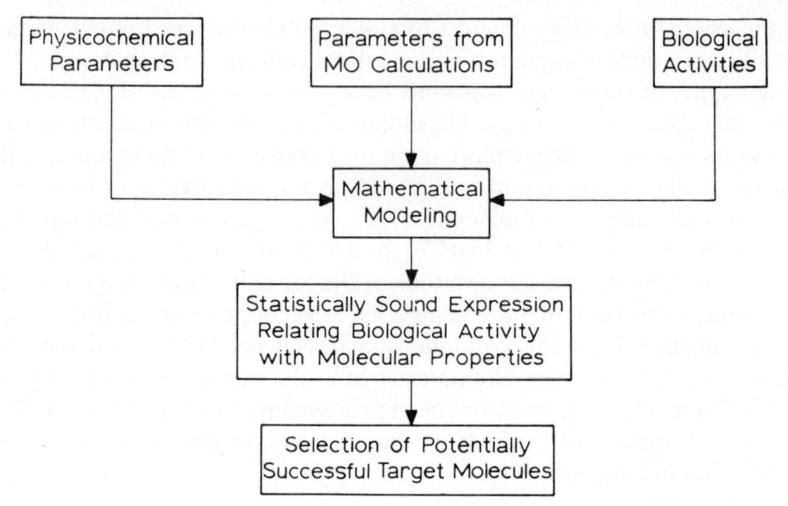

Fig. 3–1 Combined approach using QSAR in drug design. Purcell (1974).

drug molecule may be only the first in a series of reactions necessary to produce the observed biologic response. Intermediate reactions can be thought of as enhancing or diminishing the biologic response relative to the strength of the initial drug-receptor interaction. Thus an association of the properties of a drug molecule to the observed biologic response may be as dangerous as extrapolating from the shape of the drop to the shape of the keyhole in the "multistep" process of opening a door depicted in Figure 3-2!

Baker et al. (see Silipo and Hansch, 1976, for a set of references) recognized this limitation and demonstrated vividly that starting at the enzyme level,

Fig. 3–2 An alternate way to open a locked door: What about drug action? Mautner (1974).

where the inhibitory activity could be *directly* measured, rather than with whole animals, where multiple steps can distort QSAR, constitutes a powerful technique for meaningful QSAR. Baker et al. (see Silipo and Hansch, 1976, for references) synthesized a large number of variations of II in the search for a selective drug that has a strong inhibiting effect on dihydrofolate reductase isolated from Walker 256 and L1210 leukemia tumors, but does not alter the function of dihydrofolate reductase of healthy tissue.

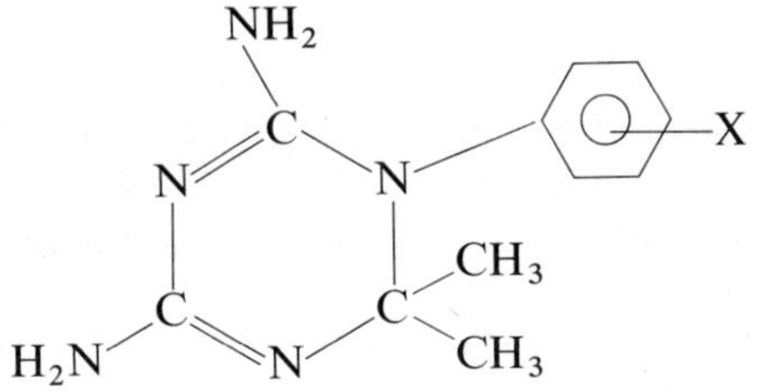

II

Before Baker's death in 1971, he and his students had progressed in their work to synthesizing III, a drug now in clinical trials against cancer. This

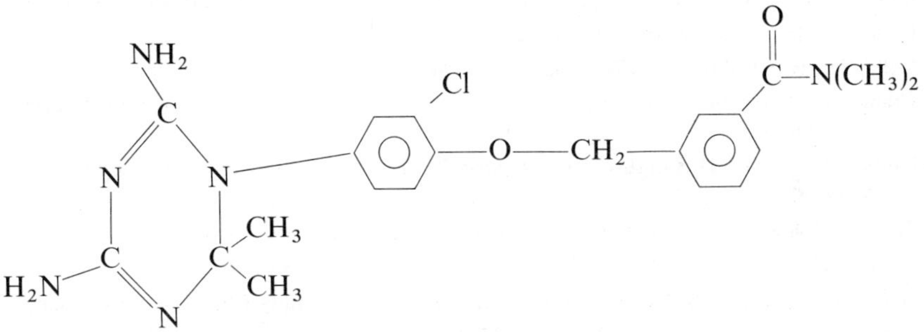

III

study appears to constitute the largest published set of biochemical congeners whose activity has been measured quantitatively in one laboratory. In all, Baker's group synthesized about 260 variations of II and measured their activities. As such, it presents a great challenge to those interested in QSAR. It may in fact become a classic standard to use in the evaluation of the merits and faults of new QSAR methods. Hansch and Silipo (1974, 1976) have carried out the first QSAR analyses of the 4,6-diamino-1,2-dihydro-2,2-dimethyl-1-(x-phenyl)-s-triazines (II). In their latest study (1976), 76% of the variance is correlated by indicator variables and 85% by indicator variables plus hydrophobic and molar refractivity parameters. Hansch has discussed the potential of direct enzyme studies as a source of strategy in drug design (Hansch, 1975).

The approach to drug design using *molecular activity* data has also been exploited by Hitchings and Burchall (Hitchings, 1973; Burchall, 1973) in the development of allopurinal for gout and the new antibacterial agent, trimethoprim. The potential design of anticancerous drugs based on selective DNA-intercalation on the molecular level is discussed in Chapter 8.

REFERENCES

Agin, D., L. Hersh, and D. Holtzman (1965). *Proc. Natl. Acad. Sci. USA*, **53**, 952.

Ariëns, E. J. (1971–1976). *Drug Design*. Academic Press, New York, Vols. 1–6.

Bergmann, E. D. and B. Pullman (1973). *Conformation of Biological Molecules and Polymers*. Academic Press, New York.

Bergmann, E. D. and B. Pullman (1974). *Molecular and Quantum Pharmacology*. Reidel-Dordrecht, Holland.

Brodie, B. B. (1964). In *Absorption and Distribution of Drugs* (T. Fujita, Ed.) E. & S. Livingstone, Edinburgh, p. 16.

Burchall, J. J. (1973). *J. Infec. Dis.*, **128**, s437.

Burger, A. (1960). *Medicinal Chemistry*. Wiley-Interscience, New York, p. 970.

Callingham, B. A. (1974). *Drugs and Transport Processes*. University Park Press, Baltimore.

Collander, R. (1954*a*). *Physiol. Plant.*, **7**, 420.

Collander, R. (1954*b*). *Acta Chem. Scand.*, **5**, 774.

Cramer, R. D., III (1976). In *Annual Reports in Medical Chemistry* (F. H. Clarke, Ed.). Academic Press (in press).

Draper, N. R. and H. Smith (1966). *Applied Regression Analysis*. Wiley-Interscience, New York, p. 134.

Free, S. M. and J. W. Wilson (1964). *J. Med. Chem.*, **7**, 395.

Genson, D. W. and R. E. Christoffersen (1973). *J. Am. Chem. Soc.*, **95**, 362.

Hammett, L. P. (1940). In *Physical Organic Chemistry* McGraw-Hill, New York, chap. 7.

Hansch, C. (1966). In *Annual Reports in Medicinal Chemistry* (C. K. Cain, Ed.). Academic Press, New York, Vol. 34, p. 347.

Hansch, C. and T. Fujita (1964). *J. Am. Chem. Soc.*, **86**, 1616.

Hansch, C., R. M. Muir, T. Fujita, P. P. Maloney, F. Geiger, and M. Streich (1963). *J. Am. Chem. Soc.*, **84**, 2817.

Hansch, C., A. R. Steward, S. M. Anderson, and D. L. Bentley (1968). *J. Med. Chem.*, **11**, 1.

Hansch, C., A. R. Steward, and J. Iwasa (1965). *Mol. Pharmacol.*, **1**, 87.

Hansch, C. and C. Silipo (1974). *J. Med. Chem.*, **17**, 661.

Hansch, C. (1975). In *Advances in Pharmacology and Chemotherapy* (S. Garattini, F. Hawking, A. Goldin, and I. J. Kopin, Eds.). Academic Press, New York.

Hansen, O.R. (1962). *Acta Chem. Scand.*, **16**, 1593.

Harper, N. J. and A. B. Simmonds (1966). *Advances in Drug Research*. Academic Press, New York, Vol. 3.

Hitchings, G. H. (1973). *J. Infec. Dis.*, **128**, s433.

Kier, L. B. (1971). *Molecular Orbital Theory in Drug Research*. Academic Press, New York.

Kowalski, B. R. and C. F. Bender (1972). *J. Am. Chem. Soc.*, **94**, 5632.

Liu, T. K. (1974). *J. Med. Chem.*, **17**, 151.

Mautner, H. G. (1974). *Molecular and Quantum Pharmacology* (E. D. Bergmann and B. Pullman, Eds.). Reidel-Dordrecht, Holland, p. 119.

Milborrow, B. V. and D. A. Williams (1968). *Physiol. Plant.*, **21**, 902.

Pauling, L. and D. Pressman (1945). *J. Am. Chem. Soc.*, **67**, 1003.

Penniston, J. T., L. Beckett, D. L. Bentley, and C. Hansch, (1969). *Mol. Pharmacol.*, **5**, 333.

Pullman, B. and A. Pullman (1963). *Quantum Biochemistry*, Wiley-Interscience, New York.

Purcell, W. P. (1974). *Molecular and Quantum Pharmacology* (E. D. Bergmann and B. Pullman, Eds.). Reidel-Dordrecht, Holland, p. 37.

Purcell, W. P., G. E. Bass, and J. M. Clayton (1973). *Strategy of Drug Design*. John Wiley & Sons, New York.

Rang, H. P. (1973). *Drug Receptors*. University Park Press, Baltimore.

Redl, G., R. D. Cramer III, and C. E. Berkoff, (1974). *Chem. Soc. Rev.*, **3**, 273.

Rekker, R. F. and G. G. Nys (1974). *Molecular and Quantum Pharmacology* (E. D. Bergmann and B. Pullman, Eds.) Reidel-Dordrecht, Holland, p. 457.

Rogers, K. S. and A. Cammarata (1969). *Biochim. Biophys. Acta*, **193**, 22.

Silipo, C. and C. Hansch (1976). *J. Med. Chem.*, in press.

Singer, J. A. and W. P. Purcell (1967). *J. Med. Chem.*, **10**, 1000.

Soloway, A. H., B. Whitman, and J. R. Messer (1960). *J. Pharmacol. Exp. Ther.*, **129**, 310.

Taft, R. W. (1956). *Steric Effects in Organic Chemistry*. John Wiley, London.

Taft, R. W. and I. C. Lewis (1959). *J. Am. Chem. Soc.*, **81**, 5343.

Tute, M. S. (1971). In *Advances in Drug Research* (N. J. Harper and A. B. Simmons, Eds.). Academic Press, New York, Vol. 6, p. 1.

CHAPTER 4

Interaction of Solvent Molecules with Polypeptides

With a high diversity of chemical groups, the polypeptides have the potential to undergo highly specific interactions with a wide variety of solvents. Also, polypeptides can adopt several different ordered chain conformations so that the structural implications of the interactions with the solvent molecules can be monitored. These ordered conformations are usually stabilized by highly specific complementary intra- and interpolypeptide interactions. Thus the polypeptides can additionally be used to calibrate the magnitude of interactions with solvent. This has been done only to a limited degree. The most definitive studies of polypeptide-solvent interactions have been done using highly solvated crystals, and form the basis for this chapter.

I INTRODUCTION AND BACKGROUND

Different solvents have generally been chosen for polypeptide solutions to regulate the equilibrium conformations of the polypeptide chain without much regard for the nature of the associated solvent-polypeptide interactions. The early experiments on solvent-induced conformational transitions in synthetic homopolypeptides (Steinberg et al., 1960; Urnes and Doty, 1961; Singer, 1962) are examples of such applications. Because of the many interaction possibilities and wide range of solvent mixture properties, little progress has been made in understanding and systematizing the solvent-solvent and solvent-peptide interactions. Still, detailed knowledge of polypeptide conformations rests almost entirely on X-ray diffraction studies of materials containing little or no solvent. It is ultimately an assumption to assign a conformation(s) observed by X-ray analysis to a molecule exhibiting certain spectroscopic and/or thermodynamic properties in solution.

II SOLVATED POLYPEPTIDE FILMS

Some workers have attempted to incorporate solvent molecules into liquid-crystalline phases containing polypeptides in the hope of obtaining sufficient regular molecular packing so that oriented X-ray diffraction patterns could be obtained (Robinson et al., 1958; Luzzati et al., 1961, 1962; Saludjian et al., 1963*a,b*; Sasisekharan, 1960; Parry and Elliott, 1965). Traub and co-workers (Traub and Schmueli, 1964; Schmueli and Traub, 1965*a,b*; Traub and Yonath, 1966; Suwalsky and Yonath, 1966) have determined the structures of several partially crystalline polypeptide-solvent systems and, by comparing these with the structures of the dry polypeptides, have investigated the effects of solvent incorporation on the conformation and packing of the polypeptide chains. The detailed structural information concerning the organization of the solvent molecules about the macromolecule found in these types of studies merits some discussion.

Table 4-1 lists the structural formulae of some of the polypeptides and solvents studied by Traub and co-workers, along with the estimated molecular weights of the polypeptides. Table 4-2 contains a detailed summary of the structural organization of the dry and solvated polypeptide materials. Solvated poly-L-proline I crystalline materials demonstrate what might be considered a general rule associated with the interaction of polypeptides and solvent. The unit cell dimensions followed by the mode of chain packing are more susceptible to change as a consequence of increasing solvation than the individual chain conformation. This is schematically illustrated in

Table 4–1 Polypeptide-solvent Systems Investigated

Polypeptide	Molecular Weight	Solvent
Poly-L-proline I 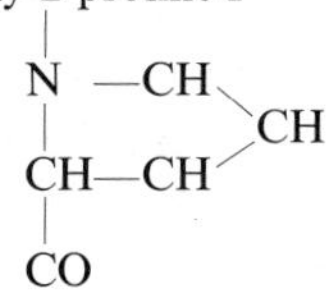 $-N-CH-CH_2-CH_2-CH_2$ (ring: N—CH, CH—CH, CH) CO	10,000–15,000	Propionic acid CH_3CH_2COOH Acetic acid CH_3COOH Formic acid $HCOOH$
Poly-(L-prolyl-glycyl-glycine) 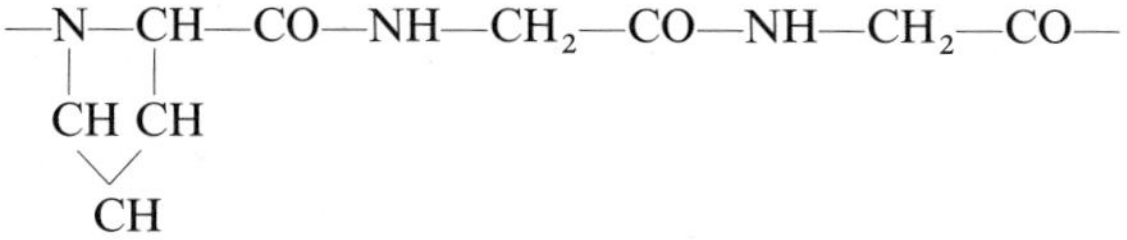 $—N—CH—CO—NH—CH_2—CO—NH—CH_2—CO—$ CH CH CH	3,500	Formic acid $HCOOH$
Poly-(L-proly-glycyl-L-proline) 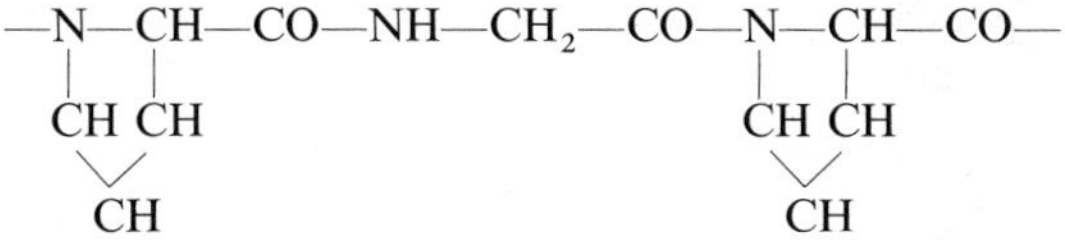 $—N—CH—CO—NH—CH_2—CO—N—CH—CO—$ CH CH CH CH CH CH	6,000	Water H_2O
Poly-γ-benzyl-L-glutamate 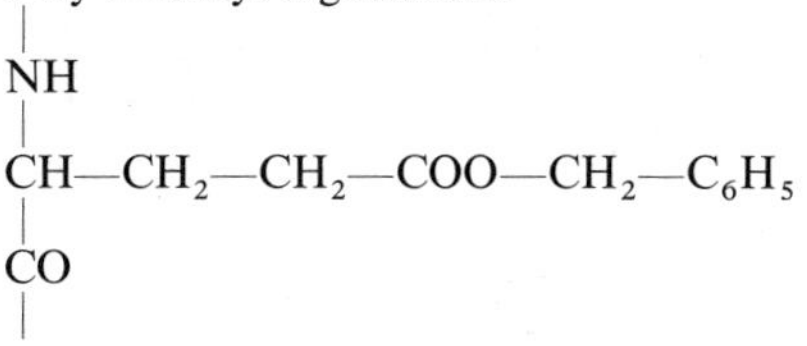 NH $CH—CH_2—CH_2—COO—CH_2—C_6H_5$ CO	200,000	*m*-Cresol $CH_3C_6H_4OH$ (1,3)
Poly-L-lysine hydrochloride 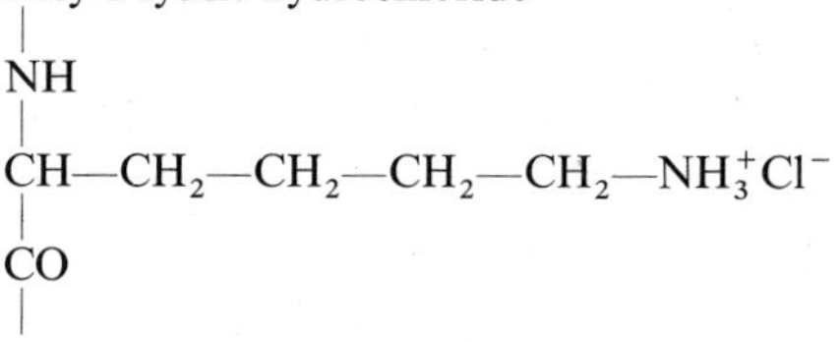 NH $CH—CH_2—CH_2—CH_2—CH_2—NH_3^+Cl^-$ CO	50,000	Water H_2O
Poly-L-arginine hydrochloride 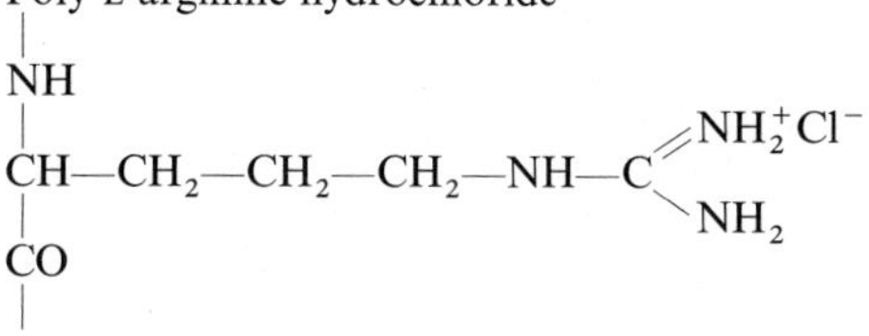 NH $CH—CH_2—CH_2—CH_2—NH—C(=NH_2^+Cl^-)NH_2$ CO	28,000	Water H_2O

From Traub et al., (1967).

Table 4-2 The Structure of Several Partially Crystalline Polypeptide-Solvent Materials Taken From the Work of Traub and Co-workers

Polypeptide	Solvent	Unit Cell: Chain Packing	Dimensions: a(Å)	b(Å)	c(Å)[a]	Angle	Chain Conformation	Number Solvent Molecules per Residue
Poly-L-proline I	Dry	Hexagonal	9.05	9.05	19.0	—	RH 10_3 helix	0
	Propionic Acid							
	(i) low solvation	Tetragonal center	9.13	9.13	19.0	—	RH 10_3 helix	—
	(ii) high solvation	Orthogonal	9.00	25.1	19.0	—	RH 10_3 helix	—
	Acetic acid	Tetragonal	9.13	9.13	19.0	—	RH 10_3 helix	—
	Formic acid	Tetragonal	8.92	8.92	19.0	—	RH 10_3 helix	—
Poly(L-pro-gly-gly)	Dry	Monoclinic $P2_1$	12.2	4.9	9.3	$\beta = 90°$	L.H.3_1 helix	0
	Formic Acid	Monoclinic $P2_1$	13.5	4.9	9.3	$\beta = 94°$	L.H.3_1 helix	—
Poly(L-prolyl-glycyl-L-proline)	0–52% relative humidity H_2O	Hexagonal	12.5	12.5	28.5	—	Collagen-like triple helix	At least 1/3
	>52% relative[b] humidity H_2O	Hexagonal	13.6	13.6	2.85–2.87	—	Collagen-like triple helix	2–4
Poly(γ-benzyl-L-glutamate)	Dry[c]	Moderate chain packing	—	—	—	—	α helix	—
	m-cresol	Poor chain packing	—	—	—	—	α helix	—

Poly(L-lysine)hydrochloride	Dry	Orthogonal	4.62	15.2	6.66	—	β-2_1 helix	0
	Up to 84% relative humidity H_2O	Orthogonal	4.78	17.0	6.66	—	β-2_1 helix	3–5
	≥84% relative humidity H_2O	Hexagonal	16.0	16.0	—	—	α helix	4
		Hexagonal	16.8	16.8	5.53	—	α helix	5
			17.6	17.6	—	—	α helix	6
			18.4	18.4	—	—	α helix	7
			19.0	19.0	—	—	α helix	9
		Little crystallinity	—	—	—	—	Partial α helix or random	15–20
Poly(L-arginine)hydrochloride	Dry	Hexagonal	14.4	14.4	5.4	—	α helix	0
	Low to moderate H_2O content	Hexagonal	15.8	15.8	5.4	—	α helix	1–5
	High H_2O content	Monoclinic	9.3	22.1	6.8	$\alpha = 109°$	β 2_1 helix[e]	5–20

[a] This is the chain axis.
[b] X-ray patterns are much sharper at these higher humidities.
[c] Based on the work of Elliott et al. (1965), Parsons and Martius (1965).
[d] Chain axis repeat not given.
[e] Packing is composed of adjacent antiparallel chains.

Figure 4-1 where a view, parallel to the *c*-axis, of the chain packing as a function of the solvation with propionic acid is shown.

The relatively large number of water molecules that can complex with poly(L-lysine) hydrochloride and poly(L-arginine) hydrochloride as compared to the nonionizable polypeptides is indicative of the large solvation capacity of the ionic groups. If, for example, the proposed packing structure for poly(L-lysine) hydrochloride at 0% relative humidity shown in Figure 4-2 is essentially correct, preferential solvation of NH_3^+ and Cl^- ions at the ends of the side chains would account for the observed increase in the *a* and *b* lattice dimensions. Further, the experimental $\beta \rightarrow \alpha$ transition can be predicted at high degrees of solvation because the Cl^- ions become partially mobile, leading to highly repulsive $NH_3^+ \cdots NH_3^+$ interactions in the β structure. These repulsions can be reduced through the observed packing-conformational transition.

Darke and Finer (1975) carried out an analysis of poly(L-lysine) hydrobromide in water and some organic solvents. In general, the behavior of this polymer in solution parallels the hydrochloride analog, although there are some significant deviations. Mixtures of poly(L-lysine) hydrobromide with water form a solid phase up to about five molecules of water per residue. Since cooling from the liquid imparts more anistropy to the water motions than warming from the solid, it seems likely that the solid does not contain the same type of ordered chain structure as is found in higher water-content states. By comparison with the results of Schnueli and Traub (1965*a*), it is likely that the solid contains β sheets. It should be noted that, if the liquid is cooled very slowly, an extremely high-ordered gel phase can be obtained.

The gel phase exists up to about 13 water molecules per residue at room temperature, and a liquid forms at higher water contents. The water content associated with the gel-liquid transition is highly temperature dependent. The gel phase contains packed α helices for all water contents. The individual residues have reduced translational motion. The gel phase also contains residues in a disordered state. The fraction of residues in the disordered state decreases with decreasing temperature and water content. Approximately one-third to one-fourth of the lysine residues are in the disordered state in the gel at room temperature. Darke and Finer (1975) have determined that approximately seven water molecules are bound (in the sense that their motions are rendered anisotropic) to each helical residue of poly(L-lysine) hydrobromide in the gel state. This may be compared with the four determined from absorption isotherm measurements (Breuer and Kennerley, 1971) and 5 ± 1 from NMR measurements of infreezable water (see Chapter 5). It is thought (Breuer and Kennerley, 1971) that three water molecules bind to N^+H_3 groups and that there is direct hydration of peptide

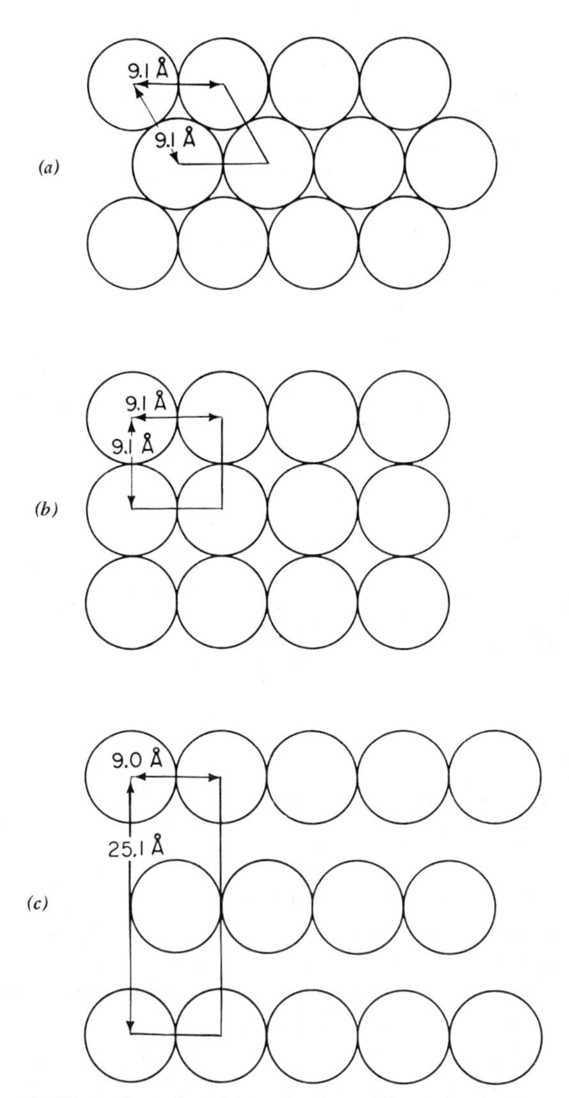

Fig. 4-1 Schematic illustration of packing of polypeptide chains in (*a*) dry polyproline I, (*b*) less highly solvated polyproline I-propionic acid complex, (*c*) more highly solvated polyproline I-propionic acid complex. All three structures are viewed along the polypeptide chains, that is, parallel to the *c*-axis. Traub et al. (1967).

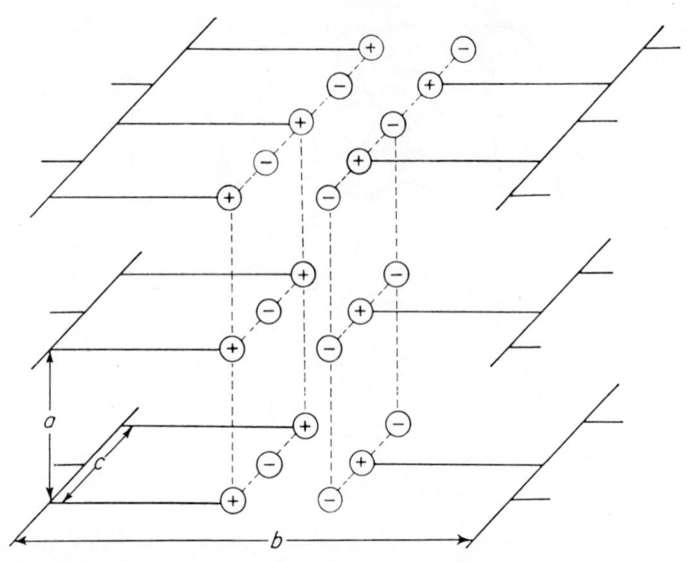

Fig. 4-2 Schematic illustration of a structure proposed for polylysine hydrochloride at 0% relative humidity. Polypeptide chains lie with their lengths along the *c*-axis and are hydrogen bonded together to form pleated sheets parallel to the *ac* plane. The lysine side chains are roughly perpendicular to this plane, and there is a regular alternation of NH_3^+ and Cl^- ions. Traub et al. (1967).

bonds (Breuer and Kennerley, 1971; Baddiel et al., 1972). The extra two or three bound waters observed by Darke and Finer may be associated with the anion and/or, loosely, to the peptide backbone. Yet why are an additional six water molecules (a total of 13) needed to sustain the gel phase? It would seem that the addition of water molecules above seven progressively lowers the free energy of the disordered form until it equals that of the helical form, and a gel transition occurs. This is an entropy-driven process that might be due to an increase in the segmental motion of the lysine side chains; the extra water increases the molecular spacing. The water molecules associated with the helical regions of poly(L-lysine) hydrobromide in the gel have an average activation energy of 3.1 $\pm$ 0.3 kcal per water molecule. Darke and Finer (1975) believe that individual poly(L-lysine) chains are composed of a distribution of α-helical and disordered regions in the gel phase. It appears that there is residual structure in the peptide backbone region of the "disordered" state in the liquid phase that is not present to the same extent in trifluoroacetic solution.

Poly(L-glutamic acid) has been precipitated in the form of small chain-folded single crystals with the β structure (Keith et al., 1964*a*). Three distinct forms of the crystals have been grown: (1) crystals of the calcium, strontium, or barium salts of poly(L-glutamic acid) in which the side groups of the polypeptide are ionized and compensated electrostatically by alkaline earth cations; (2) crystals of free acid obtained from crystals of these salts by hydrogen ion exchange in the solid state, and (3) crystals of the free acid precipitated directly from aqueous solutions of the sodium salt by gradual lowering of pH. Keith et al. (1969*b*) have carried out X-ray structural analyses of these compounds. All the structures are based on the antiparallel chain pleated sheet proposed by Pauling and Corey (1953) and have monoclinic unit cells with space group $P2_1$. In the alkaline earth salts, the planes of the carboxylate ions lie parallel to the chain axis; one-half of these ions are each in contact with one cation, and the other half each in contact with three carboxylate ions (see Figure 4-3). The cations are, in all cases, in contact with four carboxylate ions. On converting these salts in the solid state to β-poly(L-glutamic acid), the carboxyl groups maintain the orientation of the original carboxylate ions, but become intermeshed so that the intersheet spacing is reduced from 12.5–13 Å to about 8.8 Å. On the other hand, in β-poly(L-glutamic acid) prepared directly by precipitation from aqueous solutions of the sodium salt at low pH, carboxyl groups lie in planes normal to the chain directions; the side chains then interpenetrate more deeply, giving rise to an even smaller intersheet spacing (7.8 Å). One can speculate about several different ways in which the water molecules could work during the crystallization process to lead to the short intersheet spacing. Most likely, the water molecules partially shield the carboxyl ion groups on the side chains (both intra- and interchain-wise) from one another

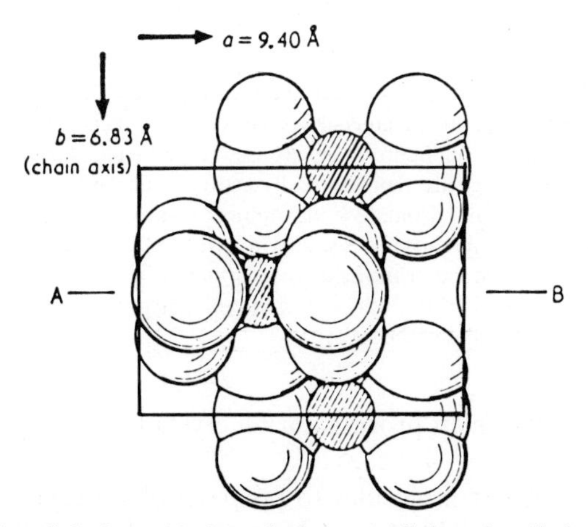

Fig. 4-3 View of medial plane *AB* through the monoclinic unit cell showing the packing of Ca^{2+} ions between carboxylate ions of poly(L-glutamic acid) chains. Keith et al. (1969*b*).

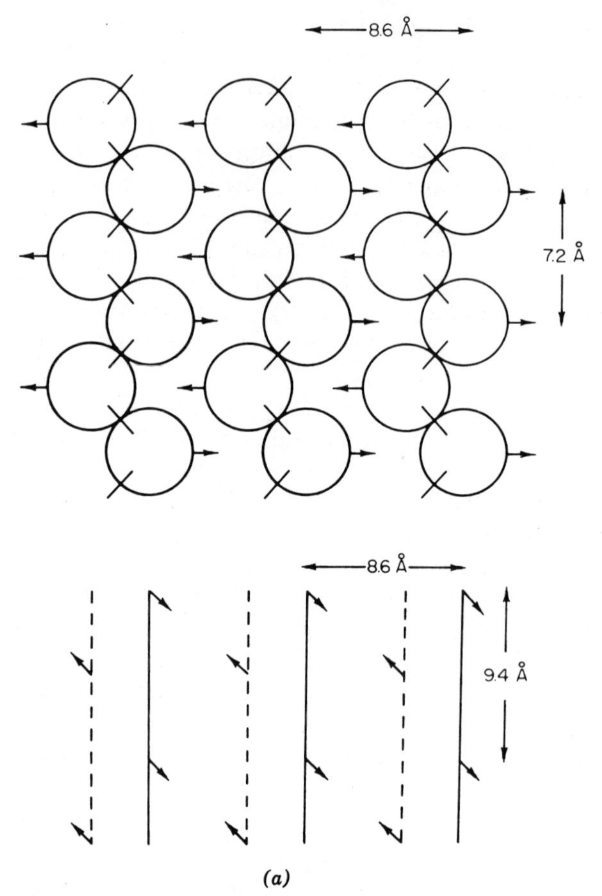

Fig. 4-4 Schematic illustration of two sheet-like forms of $(\text{Gly-Ala-Pro})_n$. In the top figures, the polyproline-II-like polypeptide chains are projected down their helix axes and are shown as circles, whereas, in the bottom figures, they are viewed perpendicularly and shown as full or dashed lines. Pyrrolidine rings are represented by arrows, and the lines in the top figures represent hydrogen bonds. Form I, from aqueous solution, is shown in (*a*) (from Segal and Traub, 1969), and form II, from organic solvents, is shown in (*b*) (from Doyle et al., 1971). Form III, from trifluorethanol, has a collagen-like triple helical conformation. $(\text{Gly-Gly-Pro})_n$ forms a structure very close to that illustrated in (*b*), but with additional hydrogen bonds between adjacent polypeptide chains on both sides of the sheets.

during the chain alignment process during crystallization, leading to a close packing.

The binding of water molecules to poly(L-prolyl-glycyl-L-proline) where no NH groups are available for bonding to solvent (the glycyl NH group is involved in an interchain hydrogen bond stabilizing the triple helix) suggests that the carbonyl oxygen is a relatively potent water hydrogen-bonding

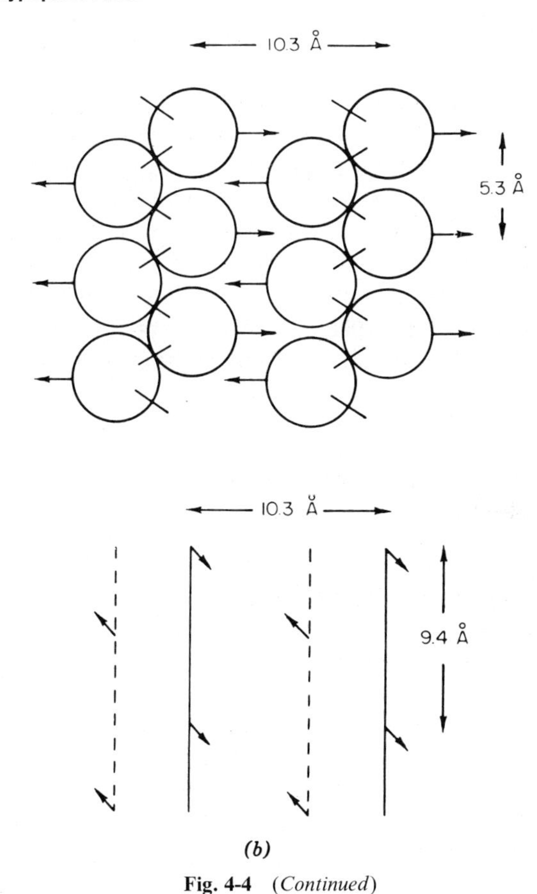

(b)

Fig. 4-4 (*Continued*)

site, as suggested from both dilute solution and theoretical studies (Swenson and Formek, 1967; Gibson and Scheraga, 1967; Hopfinger, 1971; Knof et al., 1972). Another polytripeptide, poly(L-alanyl-L-prolyglycine), has been shown to adopt three different forms in the solid state, depending on the solvent to which it is exposed (Doyle et al., 1971). The first is a hydrogen-bonded polyproline II-type sheet structure obtained from drying from aqueous solution. A more compact hydrogen-bonded polyproline II-type sheet structure results when the polymer is precipitated from organic solvents or aqueous solution, or when dried from ethylene glycol-hexafluoroisopropyl alcohol. These structures are illustrated in Figure 4-4. The third form is a triple-helical structure when dried from trifluoroethanol. More is said about the solution behavior of polytripeptides in Chapter 5. It should be noted that poly(L-alanyl-L-prolylglycine) in dilute solutions adopts an unordered conformation in water, some order in trifluoroethanol, and a high order in ethylene glycolhexafluoroethanol media.

Little detailed information concerning the precise nature of the interaction of polypeptides and solvent in moderately dilute solutions is directly available, although there have been many investigations of relatively low-concentration polypeptide solutions. Still, it is an interesting assumption to apply the polypeptide-solvent structural information found in the partially crystalline, condensed materials reported above to dilute solution solute-solvent organization. In the subsequent sections of this chapter, solution studies of a variety of polypeptides-solvent systems (from which some generalizations concerning solute-solvent organization can be made) are discussed.

III SOLVENT-DEPENDENT CONFORMATIONAL TRANSITIONS

Veis and co-workers (Veis et al., 1967) have considered systems in which a single type of potential polypeptide component-solvent interaction might be in direct competition with a similar limited type of solvent-solvent interaction. Three polyimino acids, poly(L-proline), poly(O-acetylhydroxy-L-proline) and poly(L-hydroxyproline), each of which undergo the well-characterized I-II conformational transition (Steinberg et al., 1960), were considered in these studies. The kinetic and thermodynamic behavior of I-II interconversion as a function of solvent composition was used to monitor polypeptide-solvent interactions. As an example of this approach, Table 4-3 contains the apparent I→II activation energy as a function of the content of formic acid in binary

Table 4-3 Activation Energies for the Form I→Form II Transition in Poly (O-acetyl-L-hydroxyproline) in Formic Acid-Diluent Mixtures

Solvent Mixture (v_1/v_2)	Apparent Activation Energy (50% conversion) (kcal/mole)
100% formic acid	9.3
80% formic acid—20% acetonitrile	16.5
90% formic acid—10% DMF	24.6
80% formic acid—20% DMF	22.4

Veis et al. (1967).

solvents for the I-II interconversion in poly(O-acetyl-L-hydroxyproline). The activation energies provide a direct measure of the extent and specificity of interaction between formic acid and the polyimino acid. From these studies, Veis et al. conclude (1) free energy gains through hydrogen bonding of formic acid molecules to peptide carbonyl oxygens drives I→II interconversion in poly(L-proline) and poly(O-acetyl-L-hydroxyproline). (2) The shift of poly(L-hydroxyproline) from form II to form I in anhydrous formic acid is due to intrachain hydrogen bond formation between peptide carbonyl oxygens and nearby side-chain hydroxy groups. This suggests that the single intrachain hydrogen-bond free energy gain coupled with any conformational free energy gains in adopting form I over form II (perhaps through hydrophobic interactions) is greater than free energy gains resulting from backbone CO and side-chain OH groups interacting with formic acid for the form II chain conformation. (3) A second specific interaction is evident from the reversion of poly(L-hydroxyproline) in formic acid back to form II on the addition of water. The water diluent may have several functions. One might be the formation of bridged hydrogen bonds between side-chain hydroxyls and backbone carbonyl oxygens, that is, —OH· · ·H—O—H· · · O═C—, in form II.

IV IR AND RAMAN STUDIES OF POLYPEPTIDE-SOLVENT INTERACTIONS

Some information concerning the hydrogen bonding of water in water-polypeptide systems can be gleaned from IR and Raman studies because OH stretching frequencies and intensities are quite sensitive to hydrogen bond formation (Pimentel and McClellan, 1971; Kollman and Allen, 1972). Additional information can be extracted from the temperature dependence of band frequencies and intensities.

There have not been many studies of aqueous solutions in which the feature of interest was the OH stretching region. The high intensity of water absorption in the IR makes such experiments difficult. Fisher and co-workers (McCable and Fisher, 1965, 1970; Subramanian and Fisher, 1972) developed a technique, based on difference spectroscopy, to obtain hydration spectra of polypeptides in dilute aqueous solutions. This technique has been employed by Fisher to study the interaction of water with poly(L-lysine) and poly(L-glutamic acid) during the pH-induced α-helix-coil transition common to these two biopolymers (Tinoco et al., 1962; Hermans, 1966). Fisher and co-workers conclude that the amount of water bound to random coil conformations is *not* very different from that of the α-helical structure. They also deduce that the random formation of intrachain >C═O. . . .H—N<

hydrogen bonds is energetically preferred to the formation of water-peptide group hydrogen bonds in random conformational states. These findings are not entirely consistent with current thoughts on protein hydration (see Chapter 5). Combelas et al. (1974) have carried out an IR study of the solvation of polypeptides dissolved in TFA: $CHCl_3$ mixtures that induce α-helix-coil transitions. The analysis of the fundamental amide and acid vibrations indicate complexing properties of the polypeptides not only for the backbone, but for polar side chains as well. The results suggest that the transition between helical chains and coiled chains may be induced by the formation of polymer-acid hydrogen-bonded complexes that compete with intramolecular association. There is no protonation of the peptide functions by acid, such as seems to take place in model amides (Combelas et al., 1969, 1974). This suggests that only peptides units belonging to the random coil and terminal helical regions have the potential to interact with the acid.

Most of the vibrational spectra data of water interacting with polypeptides have been determined from hydrated polypeptide films. Table 4-4 summarizes the OH frequencies of HDO in aqueous polypeptide films. Also reported in Table 4-4 are the OH stretching frequencies of pure water in

Table 4-4 IR Stretching Frequencies of Water

Sample	OH (Stretch in HDO, cm^{-1})	Bandwidth[a] (cm^{-1})
Water vapor[b]	3707	[c]
Liquid water, 22°C[d, e]	3400	255
Ice, 0°C[f]	3315	50
Polyglutamate, Na salt[g]	3385	280
Polylysine hydrobromide[g]	3420	200
Polyalanine[g]	3470	170
Polyproline[g, h]	3490	150

[a] Bandwidth for OH peak.
[b] Benedict et al. (1956).
[c] The inherent line width for low-pressure gas phase bands is very small, but vibrational-rotational couplings produce many closely spaced absorptions.
[d] Bayly et al. (1963).
[e] Falk and Ford (1966).
[f] The ice bands have been extrapolated to 0°C assuming a temperature coefficient of 0.23 cm^{-1}/deg for OH bands and 0.16 cm^{-1}/deg for OD bands (Ford and Falk, 1968). The temperature coefficients of the bands for liquid water are about three times as large.
[g] Kuntz et al. (1972), 35°C, all polypeptides in the L conformation. Band position and width depend on anion.
[h] Lozé and Josien (1969).

three different states. From this limited data, one would conclude that the interaction of water with polypeptide groups is very similar to that of liquid water. If anything, water-polypeptide interactions are, on the average, slightly weaker than liquid water-water interactions (all measurements made at approximately room temperature). Malcolm (1970, 1971) has found that the OH stretch of water in oriented hydrated films of polyalanine is shifted to higher frequencies than in pure water and is dichroic. Engel and co-workers (Strassmair et al., 1969; Ganser et al., 1970; Knof et al., 1972) and Swenson and Formek (1967) have focused on the interaction of the backbone CO groups in polypeptides with solvent, including water and water-alcohol mixtures. Zundel (1969) has written a monograph on the IR spectra of hydrated polystyrene sulfonic and phosphinic acids. Cation and anion effects are explored. The conclusions relevant to polypeptides with ionizable side chains are (1) alkali cations have small effects on water-stretching frequencies, whereas divalent and more highly charged cations can cause considerable shifts (50–200 cm^{-1}) to lower frequency; (2) the monovalent anions of the acid groups cause similar large shifts to lower frequency.

V NMR STUDIES OF POLYPEPTIDE-SOLVENT INTERACTIONS

Glasel (1970) has performed a set of NMR experiments in which he measured water mobility via the spin relaxation time for the deuterons of D_2O in solutions of biopolymers. From these studies, Glasel has postulated the following phenomenologic rules concerning D_2O-biopolymer group interactions

1. The following groups, in which M^+ and X^- are counterions

$$\rangle C{=}O, \qquad \rangle N{-}H, \qquad -\overset{\overset{\displaystyle O}{|}}{C}-O^- \ldots M^+, \qquad C{-}N^+H_3 \ldots X^-$$

do not form strong interactions with water.

2. $-\overset{\overset{\displaystyle O}{\|}}{C}-OH, \qquad C{-}N\langle^{H}_{H}$ do interact strongly with water.

3. Polymers that have side chains possessing formal charges will strongly interact with solvent only if the charges are intra- or intermolecularly wholly, or partly, neutralized through counterion effects.

4. When conformational fluctuations of the polymer are large and have characteristic times of the order of 10^{-3} sec., the D_2O-biopolymer interactions are destroyed.

Rules 1 and 3 are in contrast to a variety of other experimental results which suggest that water interacts quite strongly with charged macromolecules.

NMR spectroscopy has been used in numerous investigations (see Bradbury et al., 1968, 1973; Pauolillo et al., 1971, 1972; Bovey, 1972, and references therein) to monitor changes in the solution conformations of polypeptides as a function of solvent composition. Prevalent among these studies are those dealing with helix-coil transitions. Figure 4–5 demonstrates how NMR has been used to estimate the equilibrium distribution of helix and coil residues in a block copolypeptide as a function of TFA content in a TFA-$CDCl_3$ binary solvent mixture. Unfortunately, such experiments have not been designed to provide data directly relevant to deducing information concerning the molecular aspects of the peptide-group-solvent interaction. It is interesting to note, however, that quantitative application of some of the current theories developed to explain the α-CH double-peak spectrum associated with helix-coil transitions in polypeptides (Joubert et al., 1970; Bradbury and Fenn, 1969; Ferretti et al., 1969, 1970; Paolillo et al., 1974)

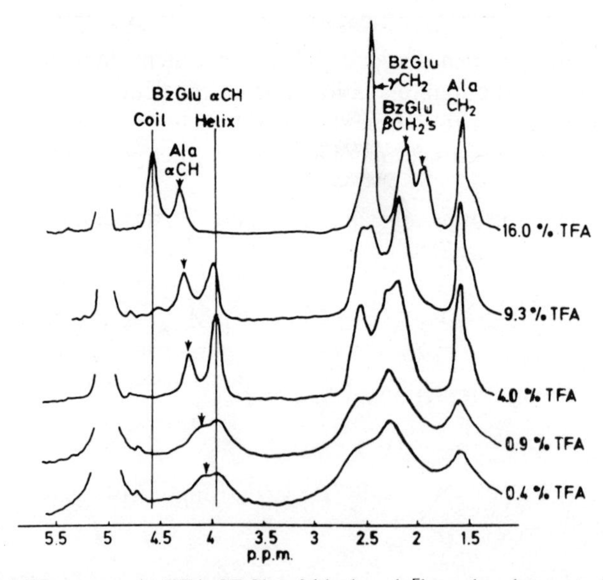

Fig. 4-5 220 MHz spectra in TFA-$CDCl_3$ of block poly[benzyl-L-glutamate (39)-L-alanine (46)-benzyl-L-glutamate (33).] Poly(L-alanine)α-CH is indicated by an arrow. [Reprinted with permission from Bradbury et al., *Rev. Pure Appl. Chem.*, 36, 53 (1973). Copyright by the International Union of Pure and Applied Chemistry.]

suggest that the solvation process, for all types of solvents, is a relatively slow process, $<3 \times 10^2$ sec^{-1}, as compared to the rate of helix coil interconversion, which is roughly 10^6 sec^{-1}.

VI MISCELLANEOUS STUDIES

There are several diverse and unrelated studies of polypeptides in solution that deserve mention because they provide some information about the specific nature of polypeptide solvation. Brumberger and Anderson (1972) have determined the radius of gyration and persistence length of poly(L-alanine) in dicloroacetic acid, DCA, and a 1:1 v/v mixture of trifluoroacetic acid and trifluoroethanol (TFA:TFE) using small angle X-ray scattering. They conclude that poly(L-alanine) exists in a relatively rigid, predominantly α-helical conformation in DCA and in an extended, more flexible form in TFA:TFE. Conformational changes are postulated with TFA and TFE molecules that do not occur in DCA. From a structural analysis of films cast from the solutions, Brumberger and Anderson further conclude that the solution conformations are retained to a considerable degree in the condensed phase. Nakajima and Murakami (1972) also studied conformational behavior of poly(L-alanine) in DCA using a variety of techniques, including optical rotatory dispersion. These workers conclude that poly(L-alanine) adopts an interrupted α-helical conformation with about 50% helicity in DCA. The Brumberger-Anderson findings and Nakajima-Murakami findings are only in marginal agreement. Parrish and Blout (1972) studied the conformational behavior of poly(L-alanine) in hexafluoroisopropanol by a variety of spectroscopic probes. They found that the polyamino acid adopts a regular, but distorted, α-helical conformation in this solvent. The workers postulate that the distortion in the α helix arises because the hydroxyl hydrogens of the solvent molecules strongly compete with the backbone amide hydrogens to form hydrogen bonds with the backbone carbonyl oxygens. Hence the carbonyl groups point slightly out from the helix axis and participate in bifracated hydrogen bonds with solvent and amide hydrogens.

Cosani et al. (1974), among others (Pedersen et al., 1971; Chow and Scheraga, 1971), have studied the coil to β transition of poly(L-lysine) in aqueous solution. They determined that the transition from an uncharged coil to an uncharged β form is endothermic and accompanied by a *positive* change of entropy, in spite of the fact that the polypeptide chain goes from an unordered state to an ordered one. The negative contribution to the entropy associated with the disorder-order transition of the peptide backbone is, presumably, overcome by a positive contribution due to a decrease in order in the water structure. Such a conclusion is consistent with the

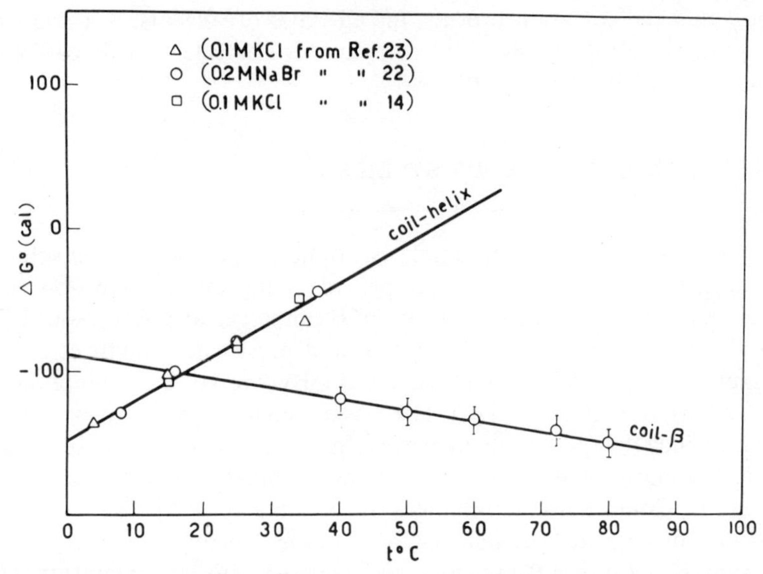

Fig. 4-6 Temperature dependence of $\triangle G_c^\circ$ values for the coil-β transition. Data related to the coil-α-helix transition that have been reproted in the literature are included for comparison. Cosani et al. (1974).

formation of "hydrophobic bonds" among side chains in the β structure that are entropy driven because of the decrease in order in the water structure (Nemethy and Scheraga, 1962). It is informative to compare the thermodynamic solution data on the coil to β transition of poly(L-lysine) with those reported on the α-helix-coil transition (Hermans, 1966; Ciferri et al., 1968; Baraskaya and Ptitsyn, 1971). The change in free energy ΔG_C°, as a function of solution temperature for both types of transitions is shown in Figure 4-6. The α-helix-coil transition is enthalpy driven, and the change in entropy is negative, with no significant variation in the order of the solvent structure during the transition. This indicates that side-chain "hydrophobic bonding" is more important in the β structure than in the α helix. From the data in Figure 4-6, it is plausible to suggest that backbone-water interactions in the β structure may contribute to stabilizing this structure in aqueous solution as compared to the coil and α-helical states.

REFERENCES

Baddiel, C. B., M. M. Breuer, and R. Stephens (1972). *J. Colloid Interface Sci.*, **40**, 429.

Barskaya, T. V. and O. B. Ptitsyn (1971). *Biopolymers*, **10**, 2181.

Bayly, J. G., V. B. Kartha, and W. H. Stevens (1963). *Infrared Phys.*, **3**, 211.

Benedict, W. S., N. Gailar, and E. K. Plyler (1956). *J. Chem. Phys.*, **24**, 1139.

Bovey, F. A. (1972). *High Resolution NMR of Macromolecules*. Academic Press, New York.

Bradbury, E. M., B. G. Carpenter, and H. Goldman (1968). *Biopolymers*, **6**, 837.

Bradbury, E. M., P. D. Cary, C. Crane-Robinson, and P. G. Hartman (1973). *Rev. Pure Appl. Chem.*, **36**, 53.

Bradbury, J. H. and M. D. Fenn (1969). *Austral. J. Chem.*, **22**, 357.

Breuer, M. M. and M. G. Kennerley (1971). *J. Colloid Interface Sci.*, **37**, 124.

Brumberger, H. and L. C. Anderson (1972). *Biopolymers*, **11**, 679.

Chow, P. Y. and H. A. Scheraga (1971). *Biopolymers*, **10**, 657.

Ciferri, A., D. Puett, L. Rajagh, and J. Hermans, Jr. (1968). *Biopolymers*, **6**, 1019.

Combelas, P., M. Avignon, C. Garrigow-LaGrange, and J. Lascombe (1974). In *Peptides, Polypeptides and Proteins* (E. R. Blout, F. A. Bovey, M. Goodman, and N. Lotan, Eds.). Wiley-Interscience, New York, p. 379.

Combelas, P., F. Cruege, J. Lascombe, C. Quivoron, and M. Rey-Lafon (1969). *J. Chem. Phys.*, **66**, 668.

Cosani, A. M. Terbojevich, L. Romanin-Jacur, and E. Peggion (1974). In *Peptides, Polypeptides and Proteins* (E. R. Blout, F. A. Bovey, M. Goodman, and N. Lotan, Eds.). Wiley-Interscience, New York, p. 166.

Corey, R. B. and L. Pauling (1953). *Proc. Roy. Soc.*, **B141**, 10.

Darke, A. and E. G. Finer (1975). *Biopolymers*, **14**, 441.

Doyle, B. B., W. Traub, G. P. Lorenzi, and E. R. Blout (1971). *Biochemistry*, **10**, 3052.

Elliott, A., R. D. B. Fraser, and T. P. MacRae (1965). *J. Mol. Biol.*, **11**, 821.

Falk, M. and T. A. Ford (1966). *Can. J. Chem.* **44**, 1699.

Ferretti, J. A. and B. Ninham (1969). *Macromolecules*, **2**, 30.

Ferretti, J. A., B. Ninham, and V. A. Parsegian (1970). *Macromolecules*, **3**, 34.

Ford, T. A. and M. Falk (1968). *Can. J. Chem.*, **46**, 3579.

Ganser, V., J. Engel, D. Winklmair, and G. Kruse (1970). *Biopolymers*, **9**, 329.

Gibson, K. D. and H. A. Scheraga (1967). *Proc. Natl. Acad. Sci. USA*, **58**, 420 (1967).

Glasel, J. A. (1970). *J. Am. Chem. Soc.*, **92**, 375.

Hermans, Jr., J. (1966). *J. Phys. Chem.*, **70**, 510.

Hopfinger, A. J. (1971). *Macromolecules*, **4**, 731 (1791).

Joubert, F. J., N. Lotan, and H. A. Scheraga (1970). *Biochemistry*, **9**, 2197.

Nakajima, A. and M. Murakami (1972). *Biopolymers*, **11**, 1295.

Keith, H. D., G. Giannoni, and F. J. Padden, Jr. (1969*a*). *Biopolymers*, **7**, 775.

Keith, H. D., F. J. Padden, Jr. and G. Giannoni (1969b). *J. Mol. Biol.* **43**, 423.

Knof, S., H. Strassmair, J. Engel, M. Rothe, and K. D. Steffen (1972). *Biopolymers*, **11**, 731.

Kollman, P. A. and L. C. Allen (1972). *Chem. Rev.* **72**, 283.

Kuntz, I. D., W. Rogers, and V. Ko (1972). See Kuntz & Kauzmann, (1974) in Chpt. 5.

Lozé, C. D. and M. L. Josien (1969). *Biopolymers*, **8**, 449.

Luzzati, V., M. Cesari, G. Spach, F. Masson, and J. M. Vincent (1961). *J. Mol. Biol.*, **3**, 566.

Malcolm, B. R. (1970). *Nature* (*Lond.*), **227**, 1358.

Malcolm, B. R. (1971). *J. Polym. Sci., Part C.*, **34**, 87.

McCabe, W. C. and H. F. Fisher (1965). *Nature* (*Lond.*), **207**, 1274.

McCabe, W. C. and H. F. Fisher (1970). *J. Phys. Chem.*, **74**, 2990.

Nemethy, G. and H. A. Scheraga (1962). *J. Phys. Chem.*, **66**, 1773.

Paolillo, L., P. A. Temussi, E. M. Bradbury, P. D. Car, C. Crane-Robinson, and P. G. Hartman (1974). In *Peptides, Polypeptides and Proteins* (E. R. Blout, F. A. Bovey, M. Goodman, and N. Lotan, Eds.). Wiley-Interscience, New York, p. 177.

Paolillo, L., P. A. Temussi, E. Trivellone, E. M. Bradbury, and C. Crane-Robinson (1971). *Biopolymers*, **10**, 2555.

Paolillo, L., P. A. Temussi, E. M. Bradbury, and C. Crane-Robinson (1972). *Biopolymers*, **11**, 2043.

Parrish, Jr., J. R. and E. R. Blout (1972). *Biopolymers*, **11**, 1001.

Parry, D. A. D. and A. Elliott (1965). *Nature* (*Lond.*), **206**, 616.

Parsons, D. F. and V. Martius (1965). *J. Mol. Biol.*, **10**, 530.

Pedersen, D., D. Gabriel, and J. Hermanns, Jr. (1971). *Biopolymers*, **10**, 2133.

Pimentel, G. and A. L. McClellan (1971). *Ann. Rev. Phys. Chem.*, **22**, 347.

Robinson, C., J. C. Ward, and R. B. Beevers (1958). *Disc. Faraday Soc.*, **25**, 29.

Saludjian, P., C. deLozé, and V. Luzzati (1963*a*). *C. R. Hebd. Séanc. Acad. Sci., Paris*, **256**, 4297; (1963*b*), 4514.

Sasisekharan, V. (1960). *J. Polymer Sci.*, **47**, 373.

Schmueli, U. and W. Traub (1965a). *J. Mol. Biol.*, **12**, 205.

Schmueli, U. and W. Traub (1965b). *Israel J. Chem.*, **3**, 42.

Schmueli, U. and W. Traub (1965b). *Israel J. Chem.*, **3**, 42.

Singer, S. J. (1962). *Adv. Protein Chem.*, **17**, 1.

Steinberg, I. Z., W. F. Harrington, A. Berger, M. Sela, and E. Katchalski (1960). *J. Am. Chem. Soc.*, **82**, 5263.

Strassmair, H., J. Engel, and G. Zundel (1969). *Biopolymers*, **8**, 237.

Subramanian, S. and H. F. Fisher (1972). *Biopolymers*, **11**, 1305.

Suwalsky, M. and W. Traub (1966). *Israel J. Chem.*, **3**, 247.

Swenson, C. A. and R. Formanek (1967). *J. Phys. Chem.*, **71**, 4073.

Tinoco, I., A. Halpern, and W. T. Simpson (1962). In *Polyamino Acids, Polypeptides and Proteins* (M. A. Stahmann, Ed.) University of Wisconsin Press, Madison, p. 147.

Traub, W. and U. Shmueli (1964). In *Proc. Sixth Int. Congr. Biochem.* New York, Vol. 2, p. 194.

Traub, W., U. Schmueli, M. Suwalsky, and A. Yonath, (1967). In *Conformation of Biopolymers* (G. N. Ramachandran, Ed.) Academic Press, New York, Vol. 2, p. 449.

Traub, W. and A. Yonath (1966). *J. Mol. Biol.*, **16**, 404.

Urnes, P. and P. Doty (1961). *Adv. Protein Chem.*, **16**, 401.

Veis, A., E. Kaufman, and C. C. W. Chao (1967). In *Conformation of Biopolymers* (G. N. Ramachandran, Ed.). Academic Press, New York, Vol. 2, p. 499.

Zundel, G. (1969). *Hydration and Intermolecular Interactions*. Academic Press, New York.

CHAPTER 5

Interaction of Water with Proteins

Since protein crystals are usually highly hydrated, protein structures determined in the solid state have generally been accepted as identical to those in solution. Many solution experiments suggest that a protein is surrounded by a "hydration shell" composed of several types of water. Thus the similarity of crystal and solution structures of a protein are, to a significant degree, dependent on the force fields generated by the crystalline and solution hydration shells surrounding the protein. Evidence suggests that these force fields are alike. This is consistent with spectroscopic findings that support similar solution and crystal protein structures. Still, small discrepancies are observed even at the present rough resolution limits. How important these differences can be in functional action is only now being considered.

I ASSOCIATION OF WATER MOLECULES WITH GLOBULAR PROTEINS

Solute-solvent association of water-soluble globular proteins has been the subject of many different investigations. The high interest in this particular class of intermolecular organizations is a consequence of the fact that all living systems exist in largely aqueous media. There are several recent reviews dealing with various aspects of the interaction of water with macromolecules in general, and globular proteins in particular (Eisenberg and Kauzmann, 1969; Steinhardt and Reynolds, 1969; Tait and Franks, 1971; Conway, 1972; Franks, 1972; Ling, 1972; Kuntz and Kauzmann, 1974; Lumry, 1974). In this section we attempt to summarize, unify, and compare information relevant to the structural organization of water molecules with a protein. A qualitative model describing the interaction of water molecules with a globular protein is presented. This model is meant to serve as a focal point in the discussions of the sometimes conflicting experimental findings concerning the nature of these interactions.

Figure 5-1 is a schematic cross-sectional view of the model chosen to describe the structural organization of water about a globular protein. There are three types of locally organized water molecules in this model in addition to the water molecules of bulk water. These types of water are discernible through thermodynamic and/or kinetic and/or spectroscopic probes.

There are a few water molecules (perhaps 10–20 per protein) that are held very strongly in specific interactive locations, such as in active sites, to bound metal ions, or in the interior of the protein. High-resolution X-ray and neutron studies of protein crystals have pinpointed the location of some of these water molecules. Sixteen tightly bound water molecules have been located in the interior of chymotrypsin (Birktoft and Blow, 1972); nine water molecules are found in the active site cavity of human carbonic anhydrase C (Liljas et al., 1972); eight water molecules have been located within the subtilisin molecule (see Kuntz and Kauzmann, 1974).

At this time, we can only speculate as to why these tightly bond water molecules are present. Those located within the protein might have been entrapped as the polypeptide chain(s) folded into the native state. The presence of these water molecules within the protein goes against the oil-drop model of globular proteins and introduces another degree of difficulty in any attempt to predict protein tertiary, or even secondary structure (Anfinsen and Scheraga, 1975). The water molecules located within the cavity of an active site of an enzyme protein may be necessary to the catalytic mechanism. This contention seems to be supported for lysozyme. Tightly bound water molecules have been located in the active site of lysozyme by Phillips and co-workers (1974), who fitted peaks in the isomorphous density

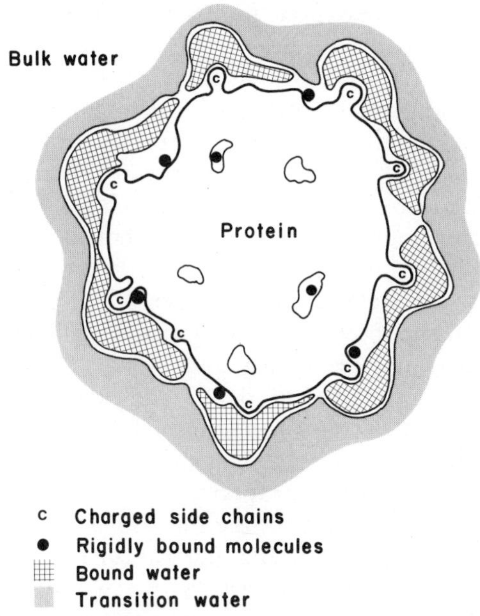

Fig. 5-1 Protein-Water Organization: A schematic cross-sectional representation.

map, and also by Moult and co-workers (1974), who fitted peaks in a suitably weighted difference map using calculated phases. In both studies, two or three water molecules, which are within 3.5 Å of carboxyl oxygen $O^{\delta 2}$ of Asp 52, must be displaced when the bacterial polysaccharide substrate is bound to the enzyme. The location of these water molecules, along with their displacement on substrate binding, is consistent with the mechanism of catalysis proposed by Phillips (1966). In turn, there is experimental evidence (Rupley and Gates, 1967; Dahlquist et al., 1968; Lin and Koshland, 1969; Parsons and Raftery, 1969) that supports Phillip's mechanism. Specifically, one proton of an initially tightly bound water molecule is donated to the carboxyl group of Glu 35, which is unionized because of its hydrophobic surrounding. The initial proton of Glu 35 is used to attack the glycosidic oxygen to initiate cleavage of one of the glycosidic bonds. The remaining OH^- of the water molecules combines with the carbonium ion of the C(l) carbon of the sugar ring whose linkage bond has been cleaved.

There has not been conclusive identification of the tightly bound water molecules by any technique other than X-ray diffraction. NMR experiments by Koenig and co-workers (Koenig and Schillinger, 1969*a*,*b*) suggest the existence of at least two relaxation processes in aqueous protein solutions (see Figure 5-2). One of these is due to rotation of the proteins. The second relaxation process, most discernible above 10 MHz for apotransferrin, is consistent with the presence of at least ten water molecules (the calculated number increases as increasing rotational freedom is assigned to the water molecules) firmly bound to the protein as it rotates.

The second class of water in aqueous-protein solutions can be termed "bound" water. It can be distinguished from the "tightly bound" water in that (1) there is much more bound than tightly bound water and (2) bound water is not as highly organized with respect to the protein as tightly bound water. As such, bound water does not undergo uniform motion with the protein. Bound water has been identified by virtually all the thermodynamic, kinetic, and spectroscopic techniques. Interestingly, X-ray diffraction has been one of the last techniques to identify the existence of bound water. Lipscomb (Lipscomb et al., 1968) has suggested, for carboxypeptidase, that some hundreds of water molecules are identifiable, with the electron density for the oxygen being about half the expected value. Over 100 water molecules have been detected at the surface of the rubredoxiin protein (Watenpaugh et al., 1972). J. J. Birktoft (see Kuntz and Kauzmann, 1974) reports that approximately 250 water molecules can be located at the surface of the subtilisin molecule. Schoenborn's (1972) neutron work on myoglobin also

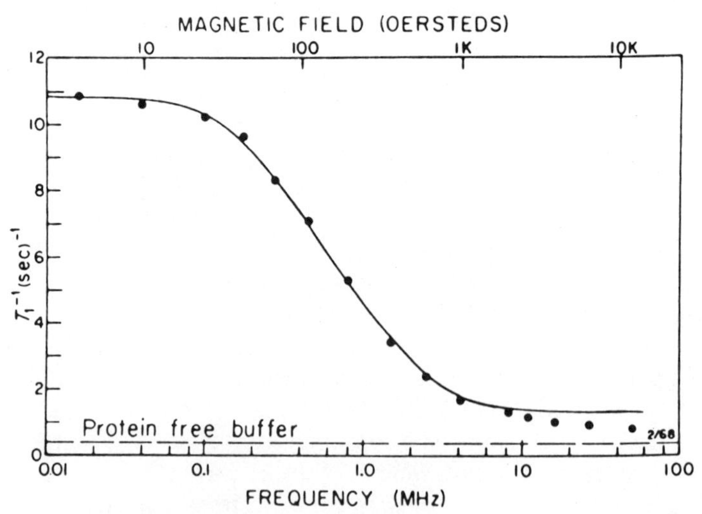

Fig. 5-2 Nuclear magnetic relaxation dispersion data for apotransferrin (25°C, OD_{280} = 171; 15 weight %; pH 7.7). The solid curve is calculated based on a single correlation time. It is clear from the data that there is a frequency dependence of T_1^{-1} above 10 MHz. [Reprinted from Koenig and Schillinger (1969*a*) by courtesy of Journal of Biological Chemistry.]

implicates the existence of a number of partially localized water molecules; more than 1000 can be found per myoglobin molecule.

Calorimetric experiments (Bull and Breese, 1968*a,b*; Mrevlishvili and Privalov, 1969; Berlin et al., 1970; Haly and Snaith, 1971) have characterized the bound water as unfreezable water. That is, there exists a moderate amount of water (see Table 5-1) in protein solutions that has a very low heat of fusion and a correspondingly low freezing point. Solutions containing this type of water possess a substantially increased configurational heat capacity that can be explained by water-protein interaction.

High-frequency dielectric dispersion experiments of protein solutions indicate two dispersion regions, 10^8 and 10^{10} Hz; the one at 10^8 Hz is of low amplitude. An example of these dispersions for packed lysozyme molecules

Table 5-1 Quantity of "Bound" Water in Aqueous-Protein Solutions

Protein	Quantity of Unfreezable Water in gH_2O/g protein Calorimetric	Bound water in gH_2O/g protein NMR
Casein	0.55 ± 0.05[a]	—
Hemoglobin		
Native	0.32[b]	0.42[e]
Denatured	0.34[b]	—
β-Lactoglobulin	0.55 ± 0.03[a]	—
Ovalbumin		
Native	0.32[b]	—
Denatured	0.33[b]	—
	<0.41[c]	
Serum albumin		
Native	0.32[b]	—
Denatured	0.33[b]	—
	0.49 ± 0.02[a]	
Lysozyme(native)	0.3[c]	0.34[e]
Myoglobin(native)	—	0.24[e]
Chymotrypsinogen(native)	—	0.34[e]
Ovalbumin(native)	—	0.33[e]
BSA(native)	0.34[e]	0.40[e]
	0.49[e]	

[a] Berlin et al. (1970).
[b] Mrevlishvili and Privalov (1969).
[c] Bull and Breese (1968*a,b*).
[d] Harvey and Hoekstra (1972).
[e] Kuntz (1971*a,b*).

is shown in Figure 5-3. There is general agreement that the higher-frequency dispersion is due to molecular rotation of bulk water, whereas the low-amplitude dispersion arises from the rotational contributions of bound water plus possible side-chain motion effects (Grant et al., 1968; Takashima, 1969; Harvey and Hoekstra, 1972; Grant and South, 1972).

NMR experiments suggest that the bound water molecules form, roughly, a monolayer about the protein and possess rotation times near 10^{-9} sec. at room temperature (Koenig and Schillinger, 1969*a,b*; Walter and Hope, 1971; Cooke and Kuntz, 1973).

Further, the NMR experiments indicate that the amount of bound water is virtually independent of the temperature when the experimental temperature is no higher than 5–10°C below the eutectic point of the bulk solution. The quantity of bound water can be determined from the integrated intensity of the NMR absorption signal. The results of such integrations are reported as part of Table 5-1. From the agreement of the limited data between the unfreezable water found by calorimetric measurements and bound water found by NMR, it is not unreasonable to conclude that both techniques measure the same water entity.

The infrared spectra of protein and polypeptide solutions and films have been most often used to determine secondary structure. However, a few studies have centered on the infrared spectrum of water in the presence of proteins. Buontempo et al. (1972) have reported the spectrum of water in protein films, noting that the spectrum is very similar to that of pure water. Small shifts to higher frequency can be observed for low water concentrations. Kuntz and co-workers (1972) have observed two OH stretching frequencies with bovine serum albumin films using HDO. A slight increase in bandwidth, as compared to pure water, was also noted. The frequency shift was less than 15 cm^{-1}, even at very low hydration levels. The similarity of IR spectra for the water of the protein films and that of pure water is surprising, since the surface of the protein contains many heterogenous hydrogen bonding sites, which should lead to a range of OH stretching frequencies. Model polypeptides provide a means of examining the various water binding sites, and a discussion is included in Chapter 4 that deals with polypeptide-solvent interactions.

When the available data is synthesized, some speculative inferences can be drawn. The enthalphy of the water-peptide hydrogen bond (water as the proton donor) is weaker than bulk water-water interaction. This follows from application of the Badger-Bauer rule, which assumes a linear relationship between IR shifts and hydrogen-bonding enthalphy (Pimentel and McClellan, 1971; Kollman and Allen, 1972). Johansson and Kollman (1972) conclude that the water-peptide interactions are approximately the same as water-water interactions in the gas phase. In protein films, most of the water

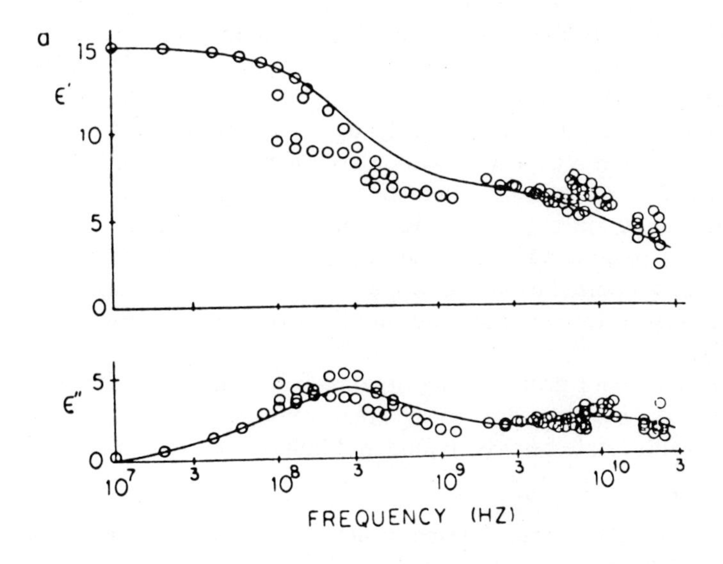

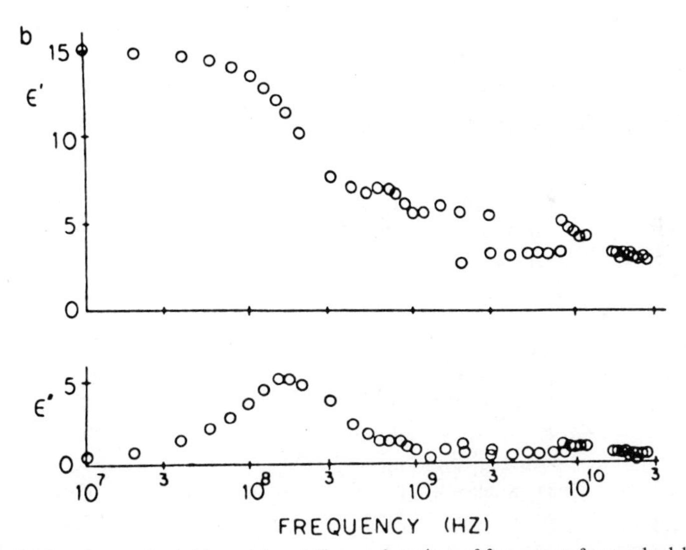

Fig. 5-3 Dielectric constant (ε') and loss (ε'') as a function of frequency for packed lysozyme samples; T = 25°C. (*a*) Samples contained 0.34 ± 0.03g of H_2O per gram of protein. (*b*) Samples contained 0.54 ± 0.04g of H_2O per gram of lysozyme. Solid curves are calculated from Debye equation for two dispersions. [From Harvey and Hoekstra, 1972; reprinted by courtesy of the Journal of Physical Chemistry.]

is associated with ionic sites rather than the amides of the side chains or backbones (Zundel, 1969; Kuntz et al., 1972; Breuer and Kennerley, 1971). Thus, in the model shown in Figure 5-1, a build-up in bound water is indicated about the charged side chains. Lastly, the typical protein hydrogen bonding environment that is exposed to water looks similar to that of bulk water. As a consequence, water-water interactions can remain important to low water concentrations, since a typical water molecule is thermodynamically ambivalent to interaction with other waters or with the protein. Table 5-2 summarizes reported estimates of the molecular thermodynamic quantities associated with the types of water in aqueous protein solutions, films, and crystals.

In the model for protein-water organization, a third type of water can be categorized. This is a hydration shell that overlays the bound water and acts as a transition state between bound and bulk water. Thus it can be termed

Table 5-2 Molecular Thermodynamics of the Type of Waters in Aqueous-Protein Solutions, Films, and Crystals

Type of Water	Binding Site or Interaction	ΔH (kcal/mole)	ΔS (eu)	ΔF (kcal/mole)
Rigidly Bound	Amide N—H			0.31[a]
	Amide C=O			1.88[a]
	N^+H_3 (lys)			5.15[a]
	COO^- (glu, asp)			4.20[a]
	OH (ser, tyr)			1.58[a]
	NH_2 (arg)			0.62[a]
	Activation to water motions			
Bound	Lysozyme powder			
	0.34 g H_2O/g 25°C	−2.9	−27	5.2[b]
	0.54 g H_2O/g	−0.4	−18	5.0[b]
	Bovine serum albumin, frozen solutions, −25°C	5.0	−4	6.2[c]
	Hemoglobin, solution	7.3	10	4.3[d]
Liquid bulk water		4.0	5.4	2.4[e]
Ice, 0°C		12.7	9.6	10.1[f]

[a] Hopfinger (1973).
[b] Harvey and Hoekstra (1972).
[c] Kuntz and Brassfield (1971).
[d] Calculated by Harvey and Hoekstra from data of Pennock and Schwan (1969).
[e] Collie et al. (1948).
[f] Eisenberg and Kauzmann (1969).

"transition water." This transition water can also be viewed as the last of the bound water that looses its self-integrity as the distance from the protein increases. Transition water is identical to the secondary layer, or layers, of loosely held water suggested by Kuntz and Kauzmann (1974) in relation to the interpretation of dielectric dispersion data. Transition water is estimated to have a rotation time only twice as slow as bulk water and therefore is difficult to identify.

II INTERACTION OF WATER MOLECULES WITH COLLAGEN AND ITS CONGENERIC MODELS

Hydrated collagen fibers have been thoroughly studied by high-resolution NMR in the hope of identifying both the quantity and organization of water about the individual tropocollagen molecules (Berendsen, 1962; Migchelsen et al., 1968; Dehl and Hoeve, 1969). The rationale for choosing collagens for NMR hydration studies is a consequence of the ordered, rod-like, solution conformation these molecules possess, in which it is not possible for all the amide (and imide) groups to participate in peptide hydrogen bonding. Some of the carbonyl oxygens and amide hydrogens are therefore potential water binding sites. Both the H_2O and D_2O NMR spectra from the fibers show two signals whose dependence on angular alignment in the magnetic field suggests that the water rotation in the fibers is somewhat anisotropic. Different workers have assigned different degrees of alignment of the water molecules relative to the tropocollagen molecule. The most recent NMR study of Dehl and Hoeve (1969) indicates that only a slight amount of alignment, with most molecules undergoing random reorientation, is sufficient to explain the relatively small splittings. Torchia et al (1974) have carried out ^{13}C-NMR studies of the α1-CB2 collagen fragment. They have shown that the translational diffusion constant of this molecule is consistent with a model in which the peptide is in a collagen-like conformation and surrounded by a single layer of bound water molecules.

One of the earlier reports of ordered water molecules about the collagen molecule came from a study of the dichroism of combination bands in the infrared spectrum (Fraser and MacRae, 1959). Recently, Suzuki and Fraser (1974) have reported additional measurements of infrared dichroism in partly hydrated sections of kangaroo-tail tendon. The IR-dichroic spectra of partially hydrated tendon is shown in Figure 5–4. The results of this investigation suggest that approximately 1.5 water molecules per tripeptide per chain are bound to the collagen molecule, and these water molecules are oriented, on the average, so that the angle between the fiber axis and the line joining the

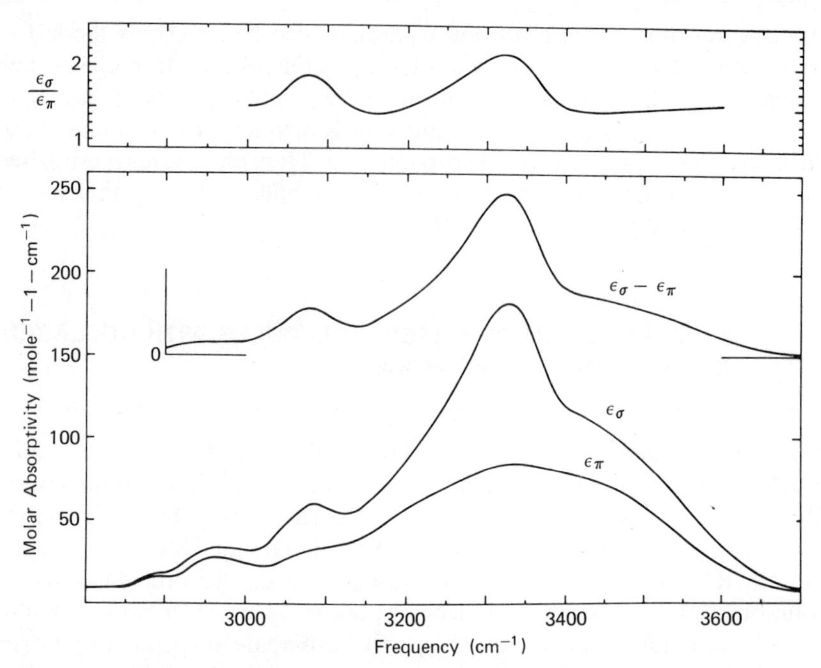

Fig. 5-4 Infrared spectrum of partly hydrated tendon section, which contains about one water molecule per residue, obtained with the electric vector vibrating parallel to the fiber axis (ε_π) and perpendicular to the fiber axis (ε_σ). Suzuki and Fraser (1974).

two hydrogens in the water molecule is $75 \pm 15°$. Such water orientation can be incorporated into models in which water molecules form pairs of hydrogen bonds between CO groups in adjacent chains in the collagen molecule. Burge et al. (1958) and Ramachandran and Chandrasekharan (1968) have proposed water-collagen models consistent with Suzuki and Fraser's findings in which the collagen conformation is topologically equivalent to the collagen-II structure proposed by Rich and Crick (1961).

Yonath and Traub (1969) have proposed a detailed model for the structure of the sequential polypeptide poly(propyl-Gly-Pro) from X-ray analysis. These polytripeptides organize into a collagen-like conformation. The presence of two ordered water molecules per asymmetric unit of three residues is part of this structural model. The predicted dichroism bands from this model suggest that the water molecules are oriented differently about the polypeptide than those in tendon. Surprisingly, other sequential tripeptides have not been used to study detailed water structuring about polypeptide chains, even though melt transition properties in various solvents have been extensively explored (see the next section).

Hiltner et al. (1974) have reported that three relaxation processes can be observed in the dynamic mechanical spectrum of native human dura mater collagen. The temperature at which these relaxation processes occur is quite sensitive to the water content of the sample material. This is demonstrated in Figure 5–5 for what Hiltner et al. term the β_2 relaxation process. These workers speculate that each of the transition processes characterize a given type of bound water. Unfortunately, there is no direct way to make a molecular assignment to the nature of these relaxation processes.

Water is part of the natural environment of collagen, whether in solution or in fibrils, and its substitution by other solvents has been found to substantially affect the denaturation temperature of the protein. The effects of various alcohols and other organic solvents on the thermal transition of collagen have been the subjects of several recent studies, which have, in fact, served to emphasize the influence of water on the stabilization and melting behavior of collagen, whether through intramolecular "hydrophobic bonds" (Schnell and Zahn, 1965; Schnell, 1968; Heidemann and Nill, 1969) or through implication in various other possible mechanisms (Russell and Cooper, 1969; Herbage et al., 1968, 1969; Bianchi et al., 1970). In concentrated aqueous solutions, the melting temperature of collagen varies greatly with water content, and an abrupt change at a concentration at which only bound water is present has been reported (Monaselidze and Bakradze, 1969). Measurements of enthalpy and entropy changes during denaturation in dilute solutions of collagens with different melting temperatures have also led to the conclusion that the stabilization of collagen is inadequately

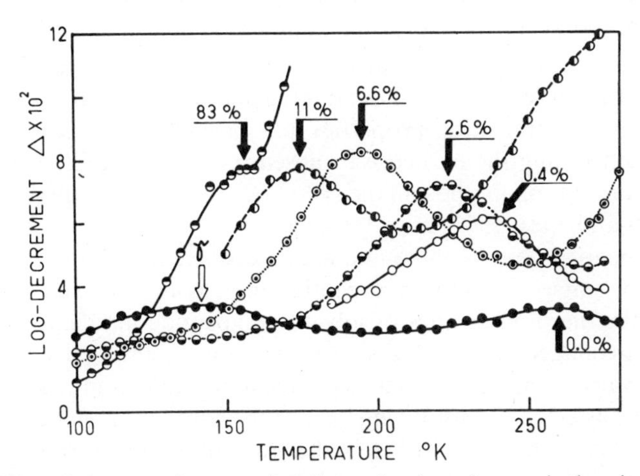

Fig. 5-5 Effect of water on a β_2 process in collagen showing a decrease in the relaxation temperature with increasing water content. Hilter et al. (1974).

accounted for by interchain hydrogen bonding and is probably dependent on a regular water structure adjacent to the molecule (Privalov, 1968; Privalov and Tiktopulo, 1970; Cooper, 1971). Stabilization of the collagen molecule by water is not consistent with the studies of synthetic collagen-like molecules presented in the next section. These triple-chain helical molecules "melt out" at lower temperatures in water than in the less polar 1,3-propanediol medium. It would seem that we are at an impasse in adequately explaining the structural and thermodynamic organization of water about collagen-like structures.

III SOLVENT-DEPENDENT MELT TRANSITIONS IN COLLAGEN AND COLLAGEN-LIKE MOLECULES

The high content of glycyl, prolyl, and hydroxyprolyl residues in collagen has encouraged the synthesis and study of polypeptides containing these residues as an aid to unraveling the structural and physicochemical complexities of collagen itself. About 15 years ago, interest turned to the synthesis of ordered sequential polytripeptide models for collagen, having, like the protein, glycine as every third residue as well as residues of one or both of the imino acids, proline and hydroxyproline (Kitaoka et al., 1958; Berger and Wolman, 1963; Debabov et al., 1963). Table 5-3 summarizes the solid-state conformation of the majority of potentially collagen-like synthetic polytripeptides and polyhexapeptides studied to date.

As early as 1961, Andreeva et al. (1961, 1963) were able to show that the polymer $(Gly\text{-}Pro\text{-}Hyp)_n$, near room temperature, possesses solution optical rotation and infrared spectra that resemble those of collagen. This finding, coupled with molecular weight-end group analysis, hydrogen-exchange studies, and melt-transition properties, has provided relatively convincing evidence concerning the existence or absence of collagen-like triple-helical structures in solution for most of the synthetic sequential polypeptide models (see Traub and Piez, 1971, for a review).

In this section, the role of the interaction of solvent with the triple-helical structure of some sequential polypeptide models is discussed in order to elucidate the thermodynamics of solvent-polypeptide interactions. When a solution of collagen is heated, a sharp transition occurs that can be monitored by various methods, such as optical rotation and viscosity. This transition is associated with the melting of the triple-chain helical structure. If the solution is then cooled below the transition temperature, properties characteristic of collagen are regained at a rate and to a degree dependent on the conditions. That a triple-chain helix is obtained can be shown by

Table 5-3 Conformations Reported for Polytripeptides Related to Collagen[a]

Amino acid sequence[b]	Conformation	Reference[c]
Gly-Pro-Hyp	Triple helix	Rogulenkova et al. (1964)
Gly-Pro-Pro	Triple helix	Yonath and Traub (1969)
Gly-Pro-Pro	Triple helix	Shibnev et al. (1965)
Gly(*O*-acetyl)Hyp-Pro	Triple helix	Traub and Yonath (1966)
Gly-Hyp-Pro	Triple helix	Andreeva et al. (1967)
Gly-Hyp-Hyp	Triple helix	Andreeva et al. (1970)
Gly-Pro-Leu	Not collagen-like	Kitaoka et al. (1958)
Gly-Pro-Ala	Triple helix	Traub and Yonath (1967)
Gly-Pro-Ala	Triple helix	Andreeva et al. (1967)
Gly-Pro(ε-tosyl)-Lys	Triple helix	Andreeva et al. (1967)
Gly-Pro-Lys-HCl	Triple helix	Traub et al. (1969)
Gly-Pro-Phe	Triple helix	Scatturin et al. (1967)
Gly-Pro-Ser	Triple helix	Andreeva et al. (1970)
Gly-Pro-Ser	Triple helix	Traub (1970)
Gly-Pro-Tyr	Triple helix	Andreeva et al. (1970)
Gly-Gly-Pro	Polyproline II	Traub (1969)
Gly-Ala-Pro (aqueous)[d]	Polyproline II	Segal and Traub (1969)
Gly-Ala-Pro (organic)[d]	Polyproline II	Schwartz et al. (1970)
Gly-Ala-Pro (organic)[d]	Polyproline II	Doyle et al. (1971)
Gly-Ala-Pro (TFE)[d]	Triple helix	Doyle et al. (1971)
Gly-Ala-Pro	Triple helix	Andreeva et al. (1970)
Gly-Ala-Hyp	Triple helix	Andreeva et al. (1970)
Gly-Ser-Hyp	Triple helix	Andreeva et al. (1970)
Gly-Ser-Pro	Polyproline II	Traub (1970)
Gly-Ala-Ala	β	Doyle et al. (1970)
Gly-Ala-Glu(OEt)	β	J. M. Anderson et al. (1970)
Ala-Pro-Pro	Not collagen-like	Andreeva et al. (1967)
Ala-Hyp-Hyp	Not collagen-like	Andreeva et al. (1967)
Random copolymers Gly-Imino	Not collagen-like	Andreeva et al. (1967)
Gly-Ala-Pro-Gly-Pro-Pro	Triple helix	Segal et al. (1969)
Gly-Pro-Ala-Gly-Pro-Pro	Triple helix	Segal et al. (1969)
Gly-Ala-Pro-Gly-Pro-Ala	Triple helix	Segal et al. (1969)
Gly-Ala-Ala-Gly-Pro-Pro	Triple helix	Segal et al. (1969)

[a] Only reports of solid state conformations based on X-ray studies are included.
[b] For ease of comparison, polytripeptide sequences quoted in this table, as well as in the text, show Gly in the first position. Readers can refer to the original works for details regarding the starting materials of the syntheses.
[c] The most detailed rather than the earliest reports of each group are quoted.
[d] The origin of the three different forms of Gly-Ala-Pro are described in the text.
Traub and Piez (1971).

X-ray diffraction. In general, the helix is regained only partially and in unbroken segments shorter than a collagen molecule. Figure 5-6 contains the temperature dependence of the ordered structure observed for the water-soluble guinea pig skin tropocollagen molecule as monitored by the two characteristic circular dichroism bands at 221.5 and 198 nm. (Brown et al., 1972). The melt transition is highly monophasic, occurring over a temperature interval of only 2°C. The synthetic polytripeptide and hexapeptides that adopt collagen-like structures also exhibit similar melt-transition curves. However, the transitions occur over wider temperature intervals and for a range of temperatures dependent on peptide sequence, solvent, and chain molecular weight. A typical example is shown in Figure 5-7.

Table 5-4 contains the triple-chain helix melt-transition temperatures of several collagen-like polypeptides in different solvents. Adjacent prolyl reside in the $(Gly\text{-}Pro\text{-}Pro)_n$ sequence stabilize the triple helix relative to the

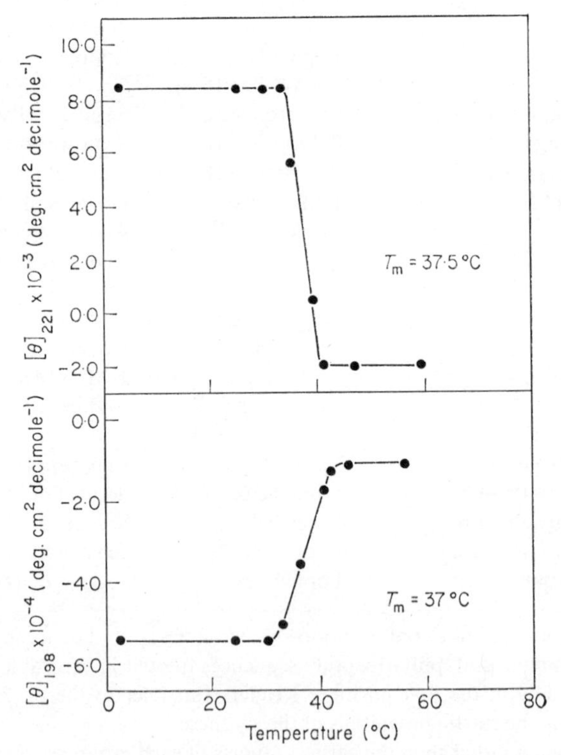

Fig. 5-6 Temperature dependence of the ordered collagen structure. Guinea pig skin collagen in water, concentration = 0.5 mg/ml. Brown et al. (1972).

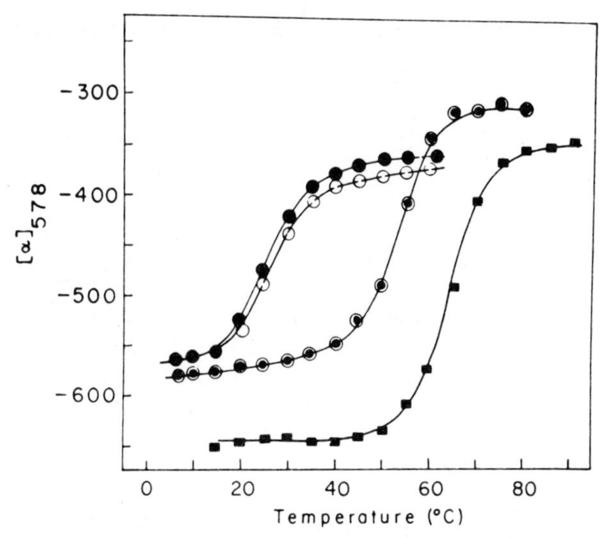

Fig. 5-7 Temperature dependence of specific rotation at 578 nm: (●) $(\text{Gly-Pro-Pro})_{10}$ in 0.1 *M* NaCl; (○) $(\text{Gly-Pro-Pro})_{10}$ in 10% acetic acid; (◉) $(\text{Gly-Pro-Pro})_{15}$ in 10% acetic acid; (■) $(\text{Gly-Pro-Pro})_{20}$ in 10% acetic acid. From Kobayashi et al. (1970) with permission.

$(\text{Gly-Pro-Ala})_n$ and $(\text{Gly-Pro-Ser})_n$ sequences. Brown et al. (1972) have attributed the increased stability to free energy gains through intrachain interactions. Segal (1969) has also discussed a possible hydrophobic interaction between adjacent prolyl residues on different chains, which in fact overlap appreciably in the $(\text{Gly-Pro-Pro})_n$ structure (Yonath and Traub, 1969). He suggests that such hydrophobic forces might account for the observation of a high denaturation temperature.

There is evidence that, in certain circumstances, water can destabilize a collagen fold conformation. It has been suggested (Segal, 1969) that water binding to free amino groups in residues following glycine may account for a relative instability of Gly-Amino-Pro sequences in a triple helix structure. Some additional support for this view appears to be provided by the solvent-dependent polymorphism of the polytripeptide $(\text{Gly-Ala-Pro})_n$ (Doyle et al., 1971). In water, this polymer was found to have a disordered conformation, and drying from aqueous solution led to an arrangement of polyproline-II-type chains in a sheet-like structure that is apparently stabilized by regularly interspersed water molecules, as shown in Figure 4-4 (Segal and Traub, 1969). Drying this form in vacuo results in a closer packed arrangement of polyproline-II-type chains, a structure that could also be derived by recovery from several organic solvents (Fig. 4-4) (Doyle et al., 1971). Yet

Table 5-4 Transition Temperatures (T_m) of Collagen-like Polypeptides

Polymer	n[a]	T_m(°C)	Solvent	Ref.
(Gly-Pro-Pro)$_n$[b]	—	~25	Water	Millionova (1964)
(Gly-Pro-Pro)$_n$	44	69	Water	Engel et al. (1966)
	27	55	Water	Brown et al. (1972)
	18	36	Water	
	66	67	Water	
	300	[c]	Water	
	66	89	1,3 propanediol	
	300	[c]	1,3 propanediol	
(Gly-Pro-Pro)$_{10}$	30	25	Water + 10% acetic acid	Kobayashi et al. (1970)
(Gly-Pro-Pro)$_{15}$	45	52	Water + 10% acetic acid	
(Gly-Pro-Pro)$_{20}$	60	65	Water + 10% acetic acid	
(Gly-Pro-Ala)$_n$[b]	~160	~45	Water	Heidemann and Bernhardt (1968)
	~80	~40	Water	
	~26	~25	Water	
(Gly-Pro-Ala)$_n$	96	58	1,3 propanediol	Brown et al. (1972)
	165	69	1,3 propanediol	
(Gly-Pro-Ser)$_n$	126	51	1,3 propanediol	Brown et al. (1972)
	225	69	1,3 propanediol	
(Gly-Ala-Pro-Gly-Pro-Pro)$_n$	16	20	Water	Segal (1969)
	30	26	Water	
	48	32	Water	
(Gly-Pro-Ala-Gly-Pro-Pro)$_n$	22	32	Water	Segal (1969)
	38	41	Water	
	48	49	Water	
(Gly-Ala-Pro-Gly-Pro-Ala)$_n$	19	49	Water + 2% acetic acid	Segal (1969)
	35	46	Water + 2% acetic acid	
(Gly-Ala-Ala-Gly-Pro-Pro)$_n$	18	19	Water	Segal (1969)
	28	35	Water	

[a] The average number of tripeptides per molecule is denoted by n.
[b] Values of T_m for (Gly-Pro-Hyp)$_n$ and (Gly-Pro-Ala)$_n$ were estimated from published thermal transition curves.
[c] The T_m was greater than 90°C and could not be estimated.

$(Gly\text{-}Ala\text{-}Pro)_n$ clearly is not precluded by steric hindrance from adopting a triple helix conformation any more than $(Gly\text{-}Pro\text{-}Pro)_n$ and other polypeptides that do adopt it. In fact, a third form of the polymer, recovered from triflourethanol solution, has turned out to be like triple-helical. While it is not clear by what mechanism the solvents direct the polymer into different conformations, it is noteworthy that only in the aqueous form are solvent molecules incorporated into the structure. The influence of solvent hydrogen bonding in determining conformation has been demonstrated in the comparable case of polyproline (Strassmair et al., 1969*a*,*b*; Engel, 1970). Brown et al. (1972) proposed a solvent-dependent melt-transition model for the triple-chain helix-disordered-state transition. The carbonyl oxygens, amide hydrogens, and side-chain hydroxyls, if present, were defined as the sites of solvent molecule binding. On this basis, a characteristic solvent-binding enthalpy, ΔH_{ps}, was determined for water and 1,3 propanediol by experimental calibration. Assuming the same values as for the T_m-data reported in Tables 5-4, (water + 10% acetic acid) and (water + 2% acetic acid) solvent mixtures we find

$$\begin{aligned}
\Delta H_{ps}\ (\text{water}) &= -2202\ \text{cal/mole/interaction} \\
\Delta H_{ps}\ (1{,}3\ \text{propanediol}) &= -2114\ \text{cal/mole/interaction} \\
\Delta H_{ps}\ (\text{water} + 10\%\ \text{acetic acid}) &= -2269\ \text{cal/mole/interaction} \\
\Delta H_{ps}\ (\text{water} + 2\%\ \text{acetic acid}) &= -2247\ \text{cal/mole/interaction}
\end{aligned}$$

These polypeptide-solvent interaction enthalpies fall into the category of moderate hydrogen bonds between individual solvent molecules and polypeptide solvent binding sites. It should be noted that the gain in ΔH_{ps} through the polypeptide binding site-solvent-molecule interactions suggests not only the existence of considerable bound solvent to the triple-chain helical structure, but also a competition of solvent molecules with interchain hydrogen bonds that ultimately might weaken the triple helix. This seems to be the case since the more negative ΔH_{ps} of water, as compared to 1,3 propanediol, leads to a lower melt temperature for the triple helix for all polymers studied in water and 1,3 propanediol (see Table 5-4). A striking observation is the closeness in magnitude of these characteristic solvation enthalpies. One should remember, nevertheless, that on a "per triple-chain helix" basis, these small energy variations can accumulate to substantial differences; for example, for $(Gly\text{-}Pro\text{-}Pro)_{10}$, the least solvent-interactive molecule reported in Table 5-4 according to the theory of Brown et al. (1972), the difference in solvation interaction enthalpy of water and 1,3 propanediol is 7.9 kcal/mole/triple helix.

A rather different case of water disrupting a collagen conformation was observed with (Gly-Pro-Ala)$_n$ of relatively low molecular weight. In aqueous solution at room temperature, it appeared to be disordered (Oriel and Blout, 1966), although at low temperature in an ethylene glycol-water mixture it showed evidence of a collagen-like structure (Brown et al., 1969). After being well dried from aqueous solution, the polymer was found to have a triple helix conformation, but with only the exposure to a moist atmosphere, the X-ray patterns of the specimens indicated an uncoiling of the molecules to form parallel polyproline II-type chains (Traub and Yonath, 1967).

IV MOLECULAR CHARACTERISTICS OF AQUEOUS HYDRATION OF PROTEINS AND POLYPEPTIDES

Probably the most straightforward approach to determining the amount of water absorbed by a solute molecule is to determine the absorption isotherms as a function of relative humidity. Bull carried out such experiments for collagen in 1944 (Bull, 1944; Bull and Breese, 1968*a*,*b*). Figure 5-8 demonstrates the water uptake by collagen as determined by Bull. Breuer and

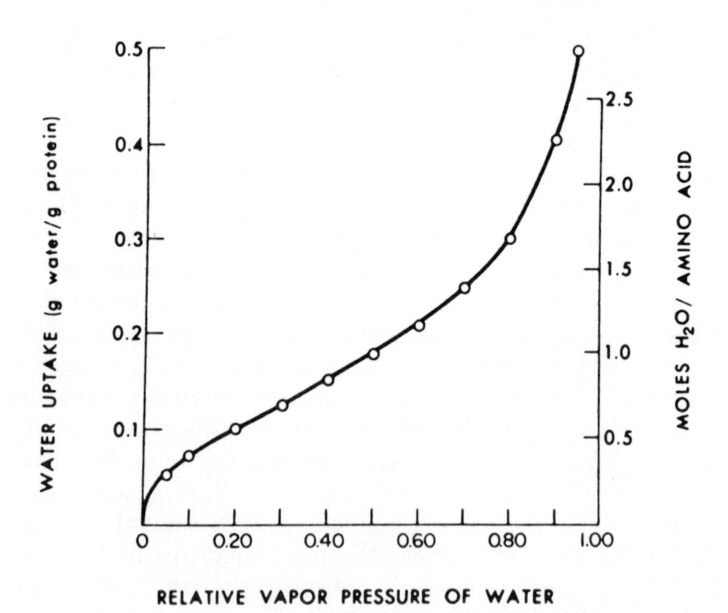

Fig. 5-8 Absorption isotherm for water uptake by collagen at 25°C. Water absorption is shown in grams of water per gram of protein and moles of water per mole of amino acid; see text. Data from Bull (1944).

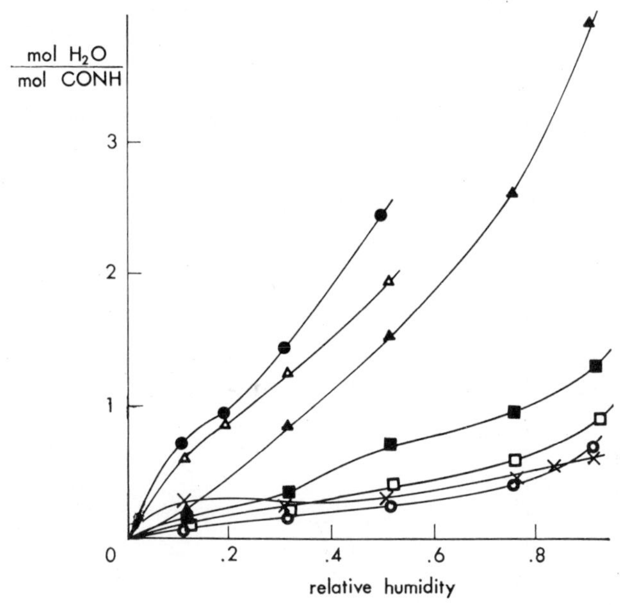

Fig. 5-9 Water-binding isotherms for some polypeptides. (○) Poly(L-alanine), (□) polyglycine I, (X) polyglycine II, (■) poly(L-glutamic acid), (▲) poly(L-lysine), (△) poly-(L-lysine·HBr), (●) sodium salt of poly(L-glutamic acid). Breuer and Kennerley (1971).

Kennerley (1971) determined the absorption isotherms of a series of homopolypeptide films, which are shown in Figure 5-9. Examples of the change in water binding as a function of side-chain group and chain conformation (polyglycine I [PGLY I] versus polyglycine II [PGLYII]) can be seen in this figure. A plot of the reciprocal quantities shown in Figure 5-9, when extrapolated to infinite relative humidity yields, roughly, the number of absorption sites, *n*, per amino acid residue. For ease of comparison, these quantities for the polypeptides are incorporated as part of Table 5-5. Several investigations (Blout and Lenormant, 1957; Lenormant et al., 1958; Shmueli and Traub, 1965; Kobyakov, 1969; Chirgadze and Ovsepyan, 1972) have focused on the influence of saturated water vapor pressure on the water uptake and structure of the hydrochlorides of poly(L-lysine), poly(L-glutamic acid), and its sodium salt. A general conclusion, consistent with Figures 5-8 and 5-9, is that water absorption is quite sensitive to the environmental relative humidity, and is not usually a linear function of relative humidity. Further, the chain conformation of the polypeptide can change as a function of absorbed water (see Table 4-2). Table 5-6 contains the values of water uptake for a number of proteins, peptides, and polypeptides, and a few

Table 5-5 Hydration Numbers for the Amino Acid Residues

Amino Acid Residues	Hydration Number (Number H_2O molecules per Residue)[a]				
	Pauling[b]	Bull[c]	Breuer and Kennerley[d]	Hopfinger[e]	Kuntz[f]
Ionic					
Asp-	1	6	—	6.5	6
Glu-	1	6	—	6.2	7
Tyr-	1	6	—	—	7
Arg^+	1	−7	—	2.3	3
His^+	1	6	—	2.8	4[i]
Lys^+	1	−7	4[g]	5.3	4[i]
Polar					
Asn	1	−7	—	2.2	2
Gln	1	−7	3	2.1	2[j]
Pro	1	6	—	0.9	3
Ser	1	6	—	1.7	2[j]
Thr	1	6	—	1.5	2[j]
Trp	1	6	—	1.9	2[j]
Asp	1	6	—	3.8	2
Glu	1	6	2[h]	3.6	2
Tyr	1	6	—	2.1	3
Arg	1	−7	—	2.3	3
Lys	1	−7	4	1.2	4
Nonpolar					
Ala	0	0	1	1.0	1
Gly	0	0	1	1.1	1
Phe	0	0	—	1.4	0
Val	0	0	—	0.9	1
Ile	0	0	—	0.8	1[j]
Leu	0	0	—	0.8	1[j]
Met	0	0	—	0.7	1[j]

[a] This is the average hydration capacity of a residue.
[b] Pauling (1945).
[c] Bull and Breese (1968*a*).
[d] Breuer and Kennerley (1971).
[e] Computed from the additive hydration shell parameters of Battershell and Hopfinger (1976) using alanine, assigned a hydration number of 1, as a standard.
[f] Kuntz (1971*a,b*).
[g] Poly (L-lysine)·HBr films were used.
[h] Poly (L-glutamic acid). No films were used.
[i] As Lys^+.
[j] Assumed values based on one water molecule per amide plus one water molecule per side-chain polar group.

Table 5-6 Water Uptake at 90% Relative Humidity

Compound	Temperature[a] (°C)	Moles H_2O[b] / Moles Monomer	g H_2O / g Protein	Reference[g]
A. Proteins[c]				
Bovine serum albumin	25	1.6	0.29[d]	(1)
Casein	24.2	1.8–1.9	0.32–0.35	(2)
Chymotrypsinogen	25	1.4	0.26[d]	(1)
Collagen	25	2.3	0.41	(3)
	—	2.0	0.36	(4)
Cytochrome *c*	25	1.9	0.35[d]	(1)
Gelatin	25	2.2	0.40	(3)
Hemoglobin	25	1.8	0.33[d]	(1)
	—	1.7	0.30	(5)
Insulin	25	1.2	0.21[d]	(1)
β-Lactoglobulin	25	1.6	0.29[d]	(1)
Lysozyme	17	1.6, 1.8	0.29, 0.33[e]	(6)
	25	1.2	0.22[d]	(1)
Myoglobin	25	2.1	0.38[d]	(1)
Ovalbumin	25	1.5	0.27	(3)
Ribonuclease	25	1.8	0.32[d]	(1)
Salmin	25	3.7	0.67	(3)
B. Peptides and polypeptides				
Tetraglycine	30	0.20[d]	0.07	(7)
Pentaglycine	30	0.19[d]	0.06	(7)
Hexaglycine	30	0.26[d]	0.085	(7)
Polyglycine I	30	0.55[d]	0.18	(7)
	31.5	0.88	0.28	(8)
Polyglycine II	30	0.55[d]	0.18	(7)
Polyalanine	31.5	0.65	0.13	(8)
Polyglycine-*dl*-alanine copolymer	—	1.10	0.31	(9)
Polyglutamic acid	31.5	1.32	0.16	(8)
Polyglutamate, Na^+	31.5	5–6	—[f]	(8)
Polylysine (NH_2)	31.5	3.90	0.48	(8)
Polylysine (NH_3^+)	31.5	4	—[f]	(8)
C. Others				
Polymethacrylic acid	—	0.78	0.16	(9)
Polyvinyl alcohol	—	0.93	0.38	(9)
Polyvinyl acetate	—	0.15	0.03	(9)
Polyvinylpyrrolidone	25	3.40		(9)
Nylon	25		0.08	(3)

[a] Temperature dependence of water uptake (see Reference 3) is approximately 1% per degree.
[b] Estimated for proteins using 100 for molecular weight of an "average" amino acid.
[c] Proteins are arranged alphabetically.
[d] Corrected from 92% relative humidity by subtracting 10%.
[e] First value, absorption; second value from desorption.
[f] Depends on counterion.
[g] Key to references: (1) Bull and Breese, 1968*a*; (2) Berlin et al., 1970; (3) Bull, 1944; (4) D'Arey and Watt, 1970; (5) McMeekin et al., 1950; (6) Hnojewyj and Reyerson, 1961; (7) Mellon et al., 1948; (8) Breuer and Kennerley, 1971; (9) McLaren and Rowen, 1951.

Kuntz and Kauzmann (1974).

synthetic polymers, at 90% relative humidity. From an inspection of Table 5-6, the qualitative ranking of the amount of uptake of water, and, indirectly, the strength of interaction with water would be:

$$(\text{charged-group } +) > (\text{counterion-charged-group})$$
$$> (\text{hydrogen bonding group}) > (\text{neutral group}) > (\text{aliphatic group}).$$

Using water absorption isotherm data, as well as other information, a few workers have proposed hydration number assignments for the amino acid residues (Pauling, 1945; Fisher, 1965; Bull and Breese, 1968*a*,*b*; Breuer and Kennerley, 1971; Kuntz, 1971*a*,*b*; Hopfinger, 1975). Table 5-5 summarizes the proposed data on amino acid residue hydration numbers. Pauling's data gives the right order of magnitude to fit the low humidity BET results, but only gives about one-fourth of the hydration levels found by NMR and calorimetric experiments. The hydration assignments of Bull are reported to predict, rather well, the water uptake of several proteins at 92% relative humidity (Bull and Breese, 1968*a*). Still, it is difficult to imagine a physical model that would predict negative hydration of the magnitude assigned in the Bull model. The Breuer and Kennerley data is incomplete and cannot be used to predict protein water uptake. However, the limited data agree well with those of Kuntz. Both sets of data were compiled from studies of water uptake by polypeptides. The hydration numbers suggested by Kuntz have been used to successfully predict the water absorption of several proteins, as shown in Table 5-7. The hydration numbers suggested by Hopfinger come from theoretical calculations carried out to describe the thermodynamic behavior of water-polypeptide group interactions (Hopfinger, 1971). The hydration numbers suggested by Hopfinger are similar to those of Kuntz, and predict values of the water hydration of proteins in good agreement with experiments, as shown in Table 5-7.

There is only a limited amount of data available on the molecular thermodynamics of solvent molecule–amino acid residue interactions. Bull (1944) carried out early absorption experiments from which the enthalpies and entropies of water absorption by three proteins have been estimated (Kuntz and Kauzmann, 1974). Berlin et al. (1970) determined the heats of desorption of four proteins using scanning calorimetry. These data are reported in Table 5-8. Tanford (1962) and Brandts (1965) obtained a measure of the "hydrophobic driving force" through determinations of the solubility of amino acids in various solvents. Goldsack and Chalifoux (1973) report the free energy change involved in transferring an amino acid side chain from water to methanol and assume that these free energy changes are measures of the "hydrophobic driving forces" of individual amino acid residues. These free energy differences have been labeled as hydrophobicity factors and are

Table 5-7 Observed and Predicted Protein Hydration

	Hydration (g H_2O/g protein)		
Protein (Native Conformation)	Observed[a]	Kuntz[b] Model	Hopfinger[c] Model
Lysozyme	0.34	0.36	0.32
Myoglobin	0.42	0.45	0.43
Chymotrypsinogen	0.34	0.39	0.37
Chymotrypsin	0.33	0.36	0.38
Ovalbumin	0.33	0.37	0.32
Bovine serum albumin (BSA)	0.40	0.45	0.43
Hemoglobin (denatured)	0.42	0.42	0.41
BSA + Urea	0.44	0.45	0.44
BSA, pH 3	0.30	0.32[d]	0.33[d]

[a] NMR freezing experiments.
[b] From data in Table 5-5 assuming all residues fully hydrated. "Buried" residue corrections were made for lysozyme that yielded a hydration value of 0.335.
[c] From data in Table 5-5 assuming all residues fully hydrated.
[d] Calculation assumes that all carboxyl groups are uncharged at pH 3.

Table 5-8 Heats of Desorption Determined by Scanning Calorimetry[a]

Protein	Composition (g H_2O/g protein)	ΔH_{vap} (kcal/mole H_2O)	Protein	Composition (g H_2O/g protein)	ΔH_{vap} (kcal/mole H_2O)
Casein	0.096	10.1	β-Lactoglobulin	0.090	10.3
	0.178	10.3		0.125	10.1
	0.187	12.2		0.157	9.5
	0.208	12.2		0.213	12.2
	0.238	11.3		0.220	12.2
	0.366	11.3			
	0.381	11.5			
			Bovine serum	0.055	12.1
Collagen	0.134	10.8	albumin	0.119	12.1
	0.240	12.2		0.178	11.3
	0.280	12.1		0.214	11.7
	0.348	13.0		0.453	12.1

[a] From Berlin et al. (1970).
Kuntz and Kauzmann (1974).

Table 5-9 Hydrophobicity Factors of Protein Side Chains

	Hydrophobicity Factor	
Amino Acid Side Chain	Goldsack Chalifoux[a]	Tanford[b,c] (kcal/mole)
Trp	3.00	3.40
Ile	2.95	2.60
Tyr	2.85	2.30
Phe	2.65	2.50
Pro	2.60	—
Leu	2.40	1.80
Val	1.70	1.50
Lys[d]	1.50	(1.00–1.50)
Met	1.30	1.30
Cys	1.00	—
Arg[d]	0.75	—
Ala	0.75	0.50
Thr	0.45	—
Gly	0	(0.00)
Ser	0	—
Asp	0	—
Glu[d]	0	—
His	0	—

[a] Goldsack and Chalifoux (1972).
[b] Tanford (1973).
[c] The figures given represent the additional free energy of transfer from water to an organic solvent that is generated when the side chain listed is substituted for the hydrogen atom of a glycyl residue. Most of the data are derived from experiments in which the organic solvent was ethanol or dioxane at 25°C (from Nozaki and Tanford, 1971).
[d] Charged form.

reported in Table 5-9. Tanford (1973) also reports a set of free energy changes that are measures of the "hydrophobic driving forces." These data have been included in Table 5-9. Although both the nonpolar solvents and physical meaning given to each of the two sets of free energy differences in Table 5-9 are marginally different, the consistency between corresponding free energies of like amino acid residues suggests that a common hydrophobic bonding potential has been measured in both cases.

V PROTEIN STRUCTURE IN SOLUTIONS AND CRYSTALS

The high aqueous solvent content, essential for the stability of protein crystals, was first noted in the pioneering diffraction photographs of pepsin (Bernal and Crawfoot, 1934). A recent survey (Matthews, 1968) shows that protein crystals have about 43% of solvent by volume, although the range is from 27 to 65%. This relatively high content of solvent has been the principal ab initio basis, supported by a good deal of chemical and spectroscopic data, for asserting that the conformation and activity of a protein in solution are very similar to that in crystals. Different accounts of this problem have been given by Edsall (1968), Rupley (1969), Drenth et al. (1971), Lipscomb (1972), and Matthews (1974).

Moreover, the solvent content greatly facilitates studies of enzyme–substrate activities in the solid state, and of three-dimensional structures of complexes of enzymes with ligands, such as poor substrates, products, inhibitors, and analogs. When the crystalline state modifies activities, the kinetic parameters are usually affected most by intermolecular contacts in crystals. The conformation of carboxypeptidase in solution and crystals is being extensively explored, since major structural and activity differences have been observed.

Chemical modification studies of the Aγ form of this enzyme with diazotized arsanilac acid (Johansen and Vallee, 1971; Johansen et al., 1972) first raised doubts about the similarity between the protein in solution and in crystals as well as of the catalytic mechanism proposed by Lipscomb and co-workers (Quiocho and Lipscomb, 1971). Specifically, the reagent modifies Tyr 248 (Johansen and Vallee, 1971). This residue moves by 12 Å to become directly involved in catalysis (Quiocho and Lipscomb, 1971). A difference in color for the modified enzyme in crystals (yellow) and in solution (red) was observed; this indicated conformational differences. In addition, it was found that the red color was due to an interaction between the modified tyrosyl residue and the zinc ion (Johansen and Vallee, 1971). A reasonable hypothesis to explain these findings is that, because of the intermolecular (crystal packing) interactions in the solid state, the movement of Tyr 248 is hindered, and it is not able to interact with the metal. Quiocho et al. (1972) and Johansen and Vallee (1973) have discussed the conformational differences between crystals and solutions for various forms of the enzyme. The enzyme used in the crystallographic studies, carboxypeptidase A_α (CPA_α) exhibits red color both in crystals and in solution when modified. The crystals are elongated along the *a* axis and have an activity that is one-third the enzymatic activity in solution (Quiocho et al., 1972; Johansen and Vallee, 1973). The A_α crystal unit cell parameters are $a = 51.41$ Å, $b = 59.89$ Å, $c = 47.19$ Å, and $\beta = 97°35'$. Examination of a three-dimensional space-filling molecular model of CPA_α reveals that the phenolic oxygen of Tyr 248 can approach within 2 Å of the Zn cofactor. This approach requires a movement by about 6 Å of the polypeptide chain in the general region of Tyr 248. Moreover, the position of Tyr 248 when bonded to Zn can just be seen in the electron density map of the crystal structure at a level that, when averaged over many unit cells, suggests that some 15–25% of the enzyme is in this form at pH 7.4 and 4°C (Lipscomb, 1973). It is probable that, when the Zn-Tyr 248 bond is present, the enzyme is catalytically inactive. However, crystals of all the generally available α, β, and γ forms of the arsanilazo enzyme are yellow and elongated along the b axis (Quiocho et al., 1972; Johansen and Vallee, 1973). The $A\gamma$ crystals have unit cell parameters of $a = 50.9$ Å, $b = 57.9$ Å, $c = 45.0$ Å and $\beta = 94°40'$. Furthermore, the ratio of the activity in crystals to that in solutions is 1/300 (Quiocho and Richards, 1966; Johansen and Vallee, 1973). Tyr 248, when modified, is obviously not able to come in close contact with the zinc ion in most crystal forms, the single exception being the one used in X-ray studies.

It is reasonable to suggest that the movement observed in the X-ray diffraction studies of the OH group of Tyr 248 by 12 Å when the very slowly hydrolyzed substrate Gly-Tyr is bound to A_α crystals, is hindered by intermolecular packing interactions in the crystalline forms entangled along the b axis. Yu and Jo (1973*a*) have used laser Raman scattering to determine that carboxypeptidase A undergoes a conformational change on crystallization from solution. The extent of the change and its relationship to modifications in activity as mentioned above are not yet clear. In contrast, these same workers (Yu and Jo, 1973*b*) conclude that lysozyme has the same main chain conformation in solution and in the solid state. An evolutionary related protein to lysozyme, α-lactalbumin, was found to undergo a major change in conformation on lyophilization, but, like lysozyme, no conformational change on crystallization (Yu, 1974). A laser Raman scattering investigation of insulin single crystals indicated conformational differences in the solid-state structure as compared to the solution conformations at acidic pH and pH 8 (Yu et al., 1974). Chirgadze and Ovsepyan (1973), using far infrared spectroscopy, report small conformational changes in a molecule of sperm-whale myoglobin in its native solid state for different pH values at room temperature. They were also able to monitor conformational changes in the protein at different stages of unfolding of the globule during its denaturation process. Although it is too early to make any generalizations about crystal and solution conformations of proteins, it would seem that (1) the greater the ratio of the number of residues per disulfide bond, the more likely conformational differences exist in solution and crystals, (2) the greater the number of subunits per globular protein, the more likely conformational differences, and (3) the higher the heat denaturation temperature, the less likely conformational differences.

REFERENCES

Anderson, J. M., W. B. Rippon, and A. G. Walton (1970). *Biochem. Biophys. Res. Commun.*, **39**, 802.

Anfinsen, C. B. and H. A. Scheraga (1975). In *Adv. in Protein Chem.* (C. B. Anfinsen, J. T. Edsall, and F. M. Richards, Eds.). Academic Press, New York, Vol. 29, p. 205.

Andreeva, N. S., V. A. Debabov, M. I. Millionova, V. A. Shibnev, and Yu. N. Chirgadze (1961). *Biofizaka*, **6**, 244.

Andreeva, N. S., N. G. Esipova, M. I. Millionova, V. N. Rogulenkova, and V. A. Shibnev (1967). In *Conformation of Biopolymers* (G. N. Ramachandran, Ed.). Academic Press, New York, p. 469.

Andreeva, N. S., N. G. Esipova, M. I. Millionova, V. N. Rogulenkova, V. G. Tumanyan, and V. A. Shibnev (1970). *Biofizika*, **15**, 198.

Andreeva, N. S., M. I. Millionova, and Yu. N. Chirgadze (1963). In *Aspects of Protein Structure* (G. N. Ramachandran, Ed.). Academic Press, New York, p. 137.

Berger, A. and Y. Wolman (1963). *Proc. Int. Congr. Biochem., 5th*, 1961, Vol. 9, p. 82.

Berendsen, H. J. (1962). *J. Chem. Phys.*, **36**, 3297.

Berlin, E., P. G. Kliman, and M. J. Pallansch (1970). *J. Colloid Interface Sci.*, **34**, 488.

Bernal, J. D. and D. Crawfoot (1934). *Nature*, **133**, 794.

Bianchi, E., R. Rampone, A. Tealdi, and A. Ciferri (1970). *J. Biol. Chem.*, **245**, 3341.

Birktoft, J. J. and D. Blow (1972). *J. Mol. Biol.*, **68**, 187.

Blout, E. R. and H. Lenormant (1957). *Nature (Lond.)*, **179**, 960.

Brandts, J. (1965). *Biological Macromolecules*. Marcel Dekker Inc., New York.

Breuer, M. M. and M. G. Kennerley (1971). *J. Colloid Interface Sci.*, **37**, 124.

Brown, F. R., J. P. Carver, and E. R. Blout (1969). *J. Mol. Biol.*, **39**, 307.

Brown, F. R., A. J. Hopfinger, and E. R. Blout (1972). *J. Mol. Biol.*, **63**, 101.

Burge, R. E., P. M. Cowan, and S. McGavin (1958). In *Recent Advances in Gelatin and Glue Research* (G. Stainsby, Ed.). Pergamon Press, New York, p. 25.

Bull, H. B. (1944). *J. Am. Chem. Soc.*, **66**, 1499.

Bull, H. B. and K. Breese (1968*a*). *Arch. Biochem. Biophys.*, **128**, 488.

Bull, H. B. and K. Breese (1968*b*). *Arch. Biochem. Biophys.*, **128**, 497.

Buontempo, U., G. Careri, and P. Fasella (1972). *Biopolymers*, **11**, 519.

Chirgadze, Y. N. and A. M. Ovsepyan (1972). *Biopolymers*, **11**, 2179.

Chirgadze, Y. N. and A. M. Ovsepyan (1973). *Biopolymers*, **12**, 637.

Collie, C. H., J. B. Hasted, and D. M. Ritson (1948). *Proc. Phys. Soc.*, **60**, 145.

Conway, B. E. (1972). *Rev. Macromol. Chem.*, **7**, 113.

Cooke, R. and I. D. Kuntz (1973). *Annu. Rev. Biophys. Bioeng.* (in press).

Cooper, A. (1971). *J. Mol. Biol.*, **55**, 123.

Dahlquist, F. W., T. Rand-Meir, and M. A. Raftery (1968). *Proc. Natl. Acad. Sci. USA*, **61**, 1194.

D'Arcy, R. L. and I. C. Watt (1970). *Trans. Faraday Soc.*, **66**, 1236.

Debabov, V. G., T. D. Kozarenko, and V. A. Shibnev (1963). *Proc. Int. Congr. Biochem., 5th*, 1961, Vol. 9, p. 63.

Dehl, R. E. and C. A. J. Hoeve (1969). *J. Chem. Phys.*, **50**, 3245.

Doyle, B. B., W. Traub, G. P. Lorenzi, and E. R. Blout (1971). *Biochemistry*, **10**, 3052.

Doyle, B. B., W. Traub, G. P. Lorenzi, F. R. Brown, and E. R. Blout (1970). *J. Mol. Biol.*, **51**, 47.

Drenth, J., W. G. J. Hol, J. N. Jansonius, and R. Koekoek (1971). In *Cold Spring Harbor Symp. Quant. Biol.*, **36**, 387.

Edsall, J. T. (1968). In *Structural Chemistry and Molecular Biology* (A. Rich and N. Davidson, Eds.). Freeman, San Francisco, p. 88.

Eisenberg, D. and W. Kauzmann (1969). *The Structure and Properties of Water*. Oxford Univ. Press (Clarendon), London and New York.

Engel, J. (1970). In *Chemistry and Molecular Biology of the Intercellular Matrix* (E. A. Balazs, Ed.). Academic Press, New York, Vol. 1, p. 127.

Engel, J., J. Kurtz, E. Katchalski, and A. Berger, (1966). *J. Mol. Biol.*, **17**, 255.

Fisher, H. F. (1965). *Biochem. Biophys. Acta*, **109**, 544.

Fraser, R. D. B. and T. P. MacRae (1959). *Nature (Lond.)*, **183**, 170.

Franks, F. (1972). *Water, A Comprehensive Treatise*. Plenum, New York.

Goldsack, D. E. and R. C. Chalifoux (1973). *J. Theor. Biol.*, **39**, 645.

Grant, E. H., S. E. Keefe, and S. Takashima (1968). *J. Phys. Chem.*, **72**, 4373.

Grant, E. H. and G. P. South (1972). *Adv. Mol. Relaxation Processes*, **3**, 355.

Haly, A. R. and J. W. Snaith (1971). *Biopolymers*, **10**, 1681.

Harvey, S. and P. Hoekstra (1972). *J. Phys. Chem.*, **76**, 2987.

Heidemann, E. and H. W. Bernhardt (1968). *Nature (Lond.)*, **220**, 1326.

Heidemann, E. and H. W. Nill (1969). *Kolloid-Z. Z. Polym.*, **232**, 674.

Herbage, D., A. Hanus, and G. Vallet (1968). *Biochim. Biophys. Acta*, **168**, 544.

Herbage, D., A. Huc, and G. Vallet (1969). *Biochim. Biophys. Acta*, **194**, 325.

Hiltner, A., S. Nomura, and E. Baer (1974). In *Peptides, Polypeptides and Proteins* (E. R. Blout, F. A. Bovey, M. Goodman, and N. Lotan, Eds.). Wiley-Interscience, New York, p. 485.

Hnojewj, W. S. and L. H. Reyerson (1961). *J. Phys. Chem.*, **65**, 1694.

Hopfinger, A. J. (1971). *Macromolecules*, **4**, 731.

Hopfinger, A. J. (1973). *Conformational Properties of Macromolecules*, Academic Press, New York.

Hopfinger, A. J. (1975). Data prepared for this manuscript.

Hopfinger, A. J. and R. D. Battershell (1976). *J. Med. Chem.*, **19**, 569.

Johansson, A. and P. A. Kollman (1972). *J. Am. Chem. Soc.*, **94**, 6196.

Johansen, J. T., D. M. Livingston, and B. L. Vallee (1972). *Biochemistry*, **11**, 2585.

Johansen, J. T. and B. L. Vallee (1971). *Proc. Natl. Acad. Sci. USA*, **68**, 2534.

Johansen, J. T. and B. L. Vallee (1973). *Proc. Natl. Acad. Sci. USA*, **70**, 2006.

Kitaoka, H., S. Sakakibara, and H. Tani (1958). *Bull. Chem. Soc. Jap.*, **31**, 802.

Kobayashi, Y., R. Sakai, K. Kakuichi, and T. Isemura (1970). *Biopolymers*, **9**, 415.

Kobyakov, V. V. (1969). In *Properties and Functions of Macromolecules and Macromolecular Systems* (in Russian). Nauka, Moscow, p. 42.

Koenig, S. H. and W. E. Schillinger (1969*a*). *J. Biol. Chem.*, **244**, 3283.

Koenig, S. H. and W. E. Schillinger (1969*b*). *J. Biol. Chem.*, **244**, 6520.

Kollman, P. A. and L. C. Allen (1972). *Chem. Rev.*, **72**, 283.

Kuntz, I. D. (1971*a*). *J. Am. Chem. Soc.*, **93**, 514.

Kuntz, I. D. (1971*b*). *J. Am. Chem. Soc.*, **93**, 516.

Kuntz, I. D. and T. S. Brassfield (1971). *Arch. Biochem. Biophys.*, **142**, 660.

Kuntz, I. D. and W. Kauzmann, (1974). In *Advances in Protein Chemistry* (C. B. Anfinsen, J. T. Edsall, and F. M. Richards, Eds.). Academic Press, New York, Vol. 28, p. 239.

Kuntz, I. D., W. Rogers, and V. Ko (1972). (in preparation).

Lenormant, H., A. Baudras, and E. R. Blout (1958). *J. Am. Chem. Soc.*, **80**, 6191.

Liljas, A., K. K. Kannan, P. C. Bergsten, I. Waara, K. Fridborg, B. Standberg, U. Carlbom, L. Jarup, S. Lougren, and M. Petef (1972). *Nature* (*New Biol.*), **235**, 131.

Lin, T.-Y. and D. E. Koshland (1969). *J. Biol. Chem.*, **244**, 505.

Ling, G. N. (1972). In *Water and Aqueous Solutions* (R. A. Horne, Ed.). Wiley-Interscience, New York, p. 663.

Lipscomb, W. N. (1972). *Chem. Soc. Rev.*, **1**, 319.

Lipscomb, W. N. (1973). *Proc. Natl. Acad. Sci. USA*, **70**, 3797.

Lipscomb, W. N., J. A. Hartsuck, G. N. Reeke, F. A. Quiocho, R. N. Bethge, M. L. Ludwig, T. A. Steitz, H. Muirhead, and J. C. Coppola (1968). *Brookhaven Symp. Biol.*, **21**, 24.

Lumry, R. (1974). *J. Food Sci.*, **38**, 744.

Matthews, B. W. (1968). *J. Mol. Biol.*, **33**, 491.

Matthews, B. W. (1974). *The Proteins*, 3rd ed. (H. Neurath and R. L. Hill, Eds.). Academic Press, New York.

McLaren, A. D. and J. W. Rowen (1951). *J. Polym. Sci.* **7**, 289.

McMeekin, T. L., M. L. Groves, and N. J. Hipp (1950). *J. Am. Chem. Soc.*, **72**, 3662.

Mellon, E. F., A. H. Korn, and S. R. Hoover (1948). *J. Am. Chem. Soc.*, **72**, 3040.

Migchelsen, C., H. J. Berendsen, and A. Rupprecht (1968). *J. Mol. Biol.*, **37**, 235.

Millionova, M. I. (1964). *Biofizika*, **9**, 145.

Monaselidze, D. R. and N. G. Bakradze (1969). *Dokl. Akad. Nauk SSSR*, **189**, 899.

Moult et al., Reported by M. Levitt (1974). In *Peptides, Polypeptides and Proteins* (E. R. Blout, F. A. Bovey, M. Goodman, and N. Lotan, Eds.). Wiley-Interscience, New York, p. 99.

Mrevlishvili, G. M. and P. L. Privalov (1969). In *Water in Biological Systems* (L. P. Kayuskin, Ed.). Plenum, New York, p. 63.

Nozaki, Y. and C. Tanford (1971). *J. Biol. Chem.*, **246**, 2211.

Oriel, P. J. and E. R. Blout (1966). *J. Am. Chem. Soc.*, **88**, 2041.

Pauling, L. (1945). *J. Am. Chem. Soc.*, **67**, 555.

Parson, S. M. and M. A. Raftery (1969). *Biochemistry*, **8**, 701.

Pennock, B. E. and H. P. Schwan (1969). *J. Phys. Chem.*, **73**, 2600.

Phillips, D. C. (1966). *Scientific American*, **215**, 78.

Phillips, D. C., Reported by M. Levitt (1974). In *Peptides, Polypeptides and Proteins* (E. R. Blout, F. A. Bovey, M. Goodman, and N. Lotan, Eds.). Wiley-Interscience, New York, p. 99.

Pimentel, G. and A. L. McClellan (1971). *Annu. Rev. Phys. Chem.*, **22**, 347.

Privalov, P. L. (1968). *Biofizika*, **13**, 955.

Privalov, P. L. and E. I. Tiktopulo (1970). *Biopolymers*, **9**, 127.

Quiocho, F. A. and W. N. Lipscomb (1971). *Adv. Protein Chem.*, **25**, 1.

Quiocho, F. A., C. H. McMurray, and W. N. Lipscomb (1972). *Proc. Natl. Acad. Sci. USA*, **69**, 2850.

Quiocho, F. A. and F. M. Richards (1966). *Biochemistry*, **5**, 4062.

Ramachandran, G. N. and R. Chandrasekharan (1968). *Biopolymers*, **6**, 1649.

Rich, A. and F. H. C. Crick (1961). *J. Mol. Biol.*, **3**, 483.

Rogulenkova, V. N., M. I. Millionova, and N. S. Andreeva, (1964). *J. Mol. Biol.*, **9**, 253.

Rupley, J. A. (1969). In *Structure and Stability of Biological Macromolecules* (S. N. Timasheff and G. D. Fasman, Eds.). Dekker, New York, p. 291.

Rupley, R. U. and V. Gates (1967). *Proc. Natl. Acad. Sci. USA*, **57**, 496.

Russell, A. E. and D. R. Cooper (1969). *Biochem. J.*, **8**, 3980.

Scatturin, A., A. Del Pra, A. M. Tamburro, and E. Scoffone (1967). *Chim. Ind. (Milan)*, **49**, 970.

Schmueli, U. and W. Traub (1965). *J. Mol. Biol.*, **12**, 205.

Schnell, J. (1968). *Arch. Biochem. Biophys.*, **127**, 496.

Schnell, J. and H. Zahn (1965). *Makromol. Chem.*, **84**, 192.

Schoenborn, B. P. (1972). *Cold Spring Harbor Symp. Quant. Biol.*, **36**, 569.

Schwartz, A., J. C. Andries, and A. G. Walton (1970). *Nature (Lond.)*, **226**, 161.

Segal, D. M. (1969). *J. Mol. Biol.*, **43**, 497.

Segal, D. M. and W. Traub (1969). *J. Mol. Biol.*, **43**, 487.

Segal, D. M., W. Traub, and A. Yonath (1969). *J. Mol. Biol.*, **43**, 519.

Shibnev, V. A., V. N. Rogulenkova, and N. S. Andreeva (1965). *Biofizika*, **10**, 164.

Steinhardt, J. and J. A. Reynolds (1969). *Multiple Equilibria in Proteins*. Academic Press, New York.

Strassmair, H., J. Engel, and G. Zundel (1969*b*). *Biopolymers*, **8**, 237.

Strassmair, H., S. Know, and J. Engel (1969*a*). *Hoppeseyler's Z. Physiol. Chem.*, **350**, 1153.

Suzuki, E. and R. D. B. Fraser (1974). In *Peptides, Polypeptides and Proteins* (E. R. Blout, F. A. Bovey, M. Goodman, and N. Lotan, Eds.). Wiley-Interscience, New York, p. 449.

Tait, M. J. and F. Franks (1971). *Nature (Lond.)*, **230**, 91.

Takashima, S. (1969). In *Physical Principles and Techniques of Proteins Chemistry* (S. J. Leach, Ed.). Academic Press, New York, Part A, p. 291.

Tanford, C. J. (1962). *J. Am. Chem. Soc.*, **84**, 4240.

Tanford, C. J. (1973). *The Hydrophobic Effect*. Wiley-Interscience, New York.

Torchia, D. A., J. R. Lyerla, Jr., and A. J. Quattrone (1974). In *Peptides, Polypeptides and Proteins* (E. R. Blout, F. A. Bovey, M. Goodman, and N. Lotan, Eds.). Wiley-Interscience, New York, p. 436.

Traub, W. (1969). *J. Mol. Biol.*, **43**, 479.

Traub, W. (1970). Unpublished data.

Traub, W. and K. A. Piez (1971). In *Adv. in Prot. Chem.* (C. B. Anfinsen, Jr., J. T. Edsall, and F. M. Richards, Eds.). Academic Press, New York, Vol. 25, p. 243.

Traub, W. and A. Yonath (1966). *J. Mol. Biol.*, **16**, 404.

Traub, W. and A. Yonath (1967). *J. Mol. Biol.*, **25**, 351.

Traub, W., A. Yonath, and D. M. Segal (1969). *Nature (Lond.)*, **221**, 914.

Walter, J. A. and A. B. Hope (1971). *Aust. J. Biol. Scil*, **24**, 497.

Watenpaugh, K. D., L. C. Sieker, J. R. Herriott, and L. H. Jensen (1972). *Cold Spring Harbor Symp. Quant. Biol.*, **36**, 359.

Yonath, A. and W. Traub (1969). *J. Mol. Biol.*, **43**, 461.

Yu, N. T. (1974). *J. Am. Chem. Soc.*, **96**, 4664.

Yu, N. T., B. H. Jo, C. C. Robert, C. C. Chang, and J. D. Huber (1974). *Arch. Biochem. Biophys.*, **160**, 614.

Yu, N. T. and D. H. Jo (1973*a*). *J. Am. Chem. Soc.*, **95**, 5033.

Yu, N. T. and B. H. Jo (1973*b*). *Arch. Biochem. Biophys.*, **156**, 469.

Zundel, G. (1969). *Hydration and Intermolecular Interactions*. Academic Press, New York.

CHAPTER 6

Drug Molecule–Protein Interactions

We now know that the receptor sites for drugs nearly always reside in soluble enzymes, multiprotein assemblies, and in membranes (probably membrane proteins). An example of each type of these three drug-receptor systems is presented here. In all three cases, at least part of the drug activity involves conformational changes in the protein to which it is bound. Unlike the results presented in most other chapters, detailed molecular structure descriptions of the interactions are not available. Structural alteration is measured through spectroscopic and activity probes. Most of the drug–protein systems mentioned in this chapter are taken from the excellent review of Levitzki (1973).

I INTRODUCTION AND MECHANISMS OF DRUG ACTION

The discovery of protein flexibility and the finding that binding of small ligands to the protein can induce specific conformational changes have played a central role in the understanding of the mechanism of enzyme action (Koshland and Neet, 1968). Direct evidence for conformational changes in proteins induced by ligands has been obtained from X-ray crystallography. Thus, for example, carboxypeptidase-A (Recke et al., 1967) and lactic dehydrogenase (Smiley et al., 1971) change their conformation in a very specific way by binding a substrate. The catalytic groups at the active site are then aligned properly for binding of the ligand, a typical "induced fit" process (Koshland, 1958). The recent studies by Levitt (1974*a,b*) of the detailed interaction of the binding of hexa-N-acetylglucosamine substrate to lysozyme using energy calculations indicates that even extremely subtle changes in molecular geometry can significantly alter binding behavior. Active sites are not the only receptor sites on the molecule. Regulatory sites (allosteric sites), distinct from the active site, have also been identified in many multi-subunit enzymes. These sites are found to play a key role in regulating the enzyme activity by binding the regulatory ligands.

The mechanism by which the regulatory ligand modulates the activity of the active sites involves conformational changes induced in the protein by the bound ligand. Thus effector molecules can either "switch on" or "switch off" a reaction catalyzed by an enzyme.

It is generally accepted that drugs can exert their function by a variety of mechanisms. One of the difficulties in drug research is the identification of the target organ and the specific receptors involved in the drug studied. Only in a limited number of cases has the drug receptor been defined; not until then could the molecular basis of its action be studied. One can distinguish among three classes of drugs according to the nature of the target affected.

***a.* Competitive inhibitors:** Drugs may compete either for active sites or allosteric sites, thus modulating enzyme activity or any other receptor activity. This class of drugs also includes drugs that act as affinity labels for either active sites or regulatory sites.

***b.* Allosteric effectors:** Drug receptors may exist that do not bind any ligand other than the drug molecule. When such receptors become bound to a drug, the effect of the drug manifests itself in multiple biochemical pathways analogous to the action of a hormone.

c. **General structural modifiers:** The effect of such drugs would be nonspecific and may cause some structural changes in a variety of organs by nonspecific interactions. It seems that ethanol would fall into this category.

It is interesting that, whenever a drug receptor has been identified, it has turned out to be a protein. However, it is possible that, since knowledge of protein chemistry is much more developed than that of lipid or nucleic acid chemistry, other nonprotein receptors still await discovery. It is tempting, however, to assume that all receptor sites for drugs, as well as for other ligands, are proteins since their three-dimensional architecture provides a wide range of properties necessary to construct an immense variety of specific biologic activities. However, the interaction of proteins with other components such as lipids and polysaccharides may indeed affect the architecture of the receptor, thus increasing the versatility of proteins in performing different tasks and in construction of a variety of receptors. At this stage it seems quite clear that the understanding of drug action requires the detailed study of the drug–receptor interaction. Since the most well-understood receptors are soluble proteins, mainly enzymes, most of the detailed mechanistic studies on drug action have been performed on enzymes. Although the most important receptors are membrane bound or involve more complicated structures such as ribosomes, their interaction with a ligand is a more difficult one to study. Nevertheless, it seems that the detailed understanding of the interaction of a soluble protein with a drug or any other ligand is a prerequisite for the understanding of drug action on a receptor integrated in a membrane or in a large organelle. In the remainder of this chapter, we present examples of drug–protein systems that have been studied in sufficient detail to expose some detailed interactive–structural features. Still, much of the discussion is more biochemical than structural as compared to most other sections of this book. This reflects the state of the art in regard to understanding drug–protein interactions.

II DRUG-INDUCED CONFORMATIONAL CHANGES IN CTP SYNTHETASE

The interaction of the enzyme CTP synthetase with the drug 6-diazo-5-oxo-norleucine (DON) reveals some interesting aspects of drug-induced conformational changes. DON and DONV (5-diazo-4-oxo-norvaline) inhibit glutamine- and asparagine- requiring enzymes both in bacterial (Gibson et al., 1967) and in mammalian (Handschumacher et al., 1968) systems. The two compounds very strongly inhibit the growth of neoplastic tissues

(Handschumacher et al., 1968) that often lack glutamine synthetase and asparagine synthetase. The use of these drugs could be of therapeutic value and may complement asparaginase treatment of some types of cancer.

DON reacts with the reactive SH group in the glutamine site, blocking the glutamylation of the enzyme and thus eliminating the glutamine activity. In the absence of added ligands, the protein exists as a dimer (Long et al., 1970; Levitzki and Koshland, 1972); since only one of the two glutamine sites reacts, a single rate constant is observed. Binding of GTP to the enzyme dimer increases the rate of DON inactivation eightfold. When ATP and UTP are added at saturating concentration, the enzyme dimer (108,000 molecular weight) aggregates to a tetramer of 216,000 molecular weight (Long et al., 1970; Levitzki and Koshland, 1972). DON reacts with only two of the four sites on the tetramer. The rate constants of the two sites are different. The first of these sites shows a rate constant similar to that of the dimer in the presence of GTP. The reaction of the first DON molecule induces a conformational change with a negatively cooperative effect manifested by the slower reaction of the DON with the second drug site. When the allosteric activator GTP is added to the assembled tetramer, the two drug sites become equivalent—the negative cooperative effect is abolished. A summary of the rate constants characterizing the interaction of the drug sites with DON is given in Table 6-1. Distinct differences between the

Table 6-1 The Effect of Ligands on the Rate of DON Inactivation

Ligand	Molecular Species of CTP Synthetase[a]	k_{DON}, min^{-1}
Mg^{2+}	108, 000 (dimer)	0.52×10^3
Mg^{2+} + GTP	108,000 (dimer)	4.0×10^3
Mg^{2+} + ATP + UTP	216,000 (tetramer)	$k_1 = 3.5 \times 10^3$
		$k_2 = 0.35 \times 10^3$
Mg^{2+} + ATP + UTP + GTP	216,000 (tetramer)	$k_1 = k_2 = 3.5 + 10^3$

[a] CTP synthetase (7×10^{-6} *M* in sites) was inactivated with DON (8.9×10^{-5} *M*) in the presence of 0.02 *M* imidazole acetate pH 7.2 at 0°C. The different ligands present were at the following concentrations: $[Mg^{2+}] = 0.01$ *M*; $[GTP] = 2.1 \times 10^{-3}$ *M*; $[ATP] = 1.6 \times 10^{-3}$ *M*; $[UTP] = 1.6 \times 10^{-3}$ *M*. Aliquots were removed at different times to check both glutamine and ammonia activity.

Abbreviations: CTP, cytidine-5′-triphosphate; DON, 6-diazo-5-oxo-norleucine; DONV, 5-diazo-4-oxo-norleucine; DTNB, 5,5′-dithiobis-(2-nitro-benzoic acid); CTPS, CTP synthetase; GDH, glutamic dehydrogenase.
Levitzki (1973).

DON–enzyme and the enzyme–glutamine complex can be demonstrated by following the kinetics of the reaction with DTNB. As seen in Figure 6-1, the DON enzyme exhibits a different spectrum of SH reactivities than the glutamylated enzyme. The DON effect on the conformation of the enzyme is much more dramatic than the effect of glutamine when compared with the SH reactivities of the native deglutaminated enzyme (Figure 6-1). The origin of this tremendous difference between the DON- and the glutamine-induced changes could be due to the minor but significant differences in the structure of the two molecules. All the groups in the glutamyl moiety that can potentially make bonds with residues on the protein are displaced one carbon atom further from the surface of the protein in the case of DON. Therefore, those interactions typical of the glutamyl moiety can be modified because of their dislocation. It has indeed become apparent that small changes in a protein molecule can have dramatic effects on the structure of the rest of the molecule. For example, the binding of oxygen to hemoglobin, which induces a 0.8 Å movement of the iron atom, triggers profound conformational changes in the rest of the molecule (Perutz, 1970).

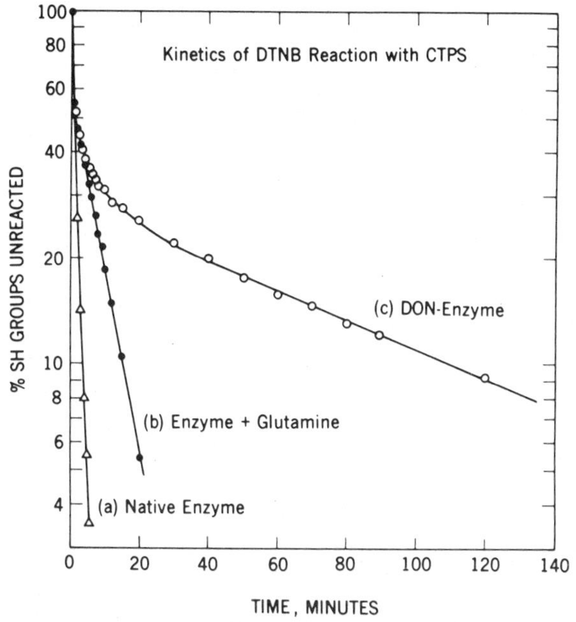

Fig. 6-1 Semilog plot of reaction of DTNB with CTP synthetase, DON-CTP synthetase, and glutamyl-CTP synthetase. CTP synthetase freed of β-mercaptoethanol and all ligands on a G-25 column were treated as indicated in each case and then treated with a 100-fold excess of DTNB. OD was followed at 412 m at pH 8.0 (ε_M = 13, 600). Levitzki (1973).

Another possibility is that, if the interactions of the drug with the receptor site are stronger than with the natural ligand, the conformational changes induced are more profound. Since all the ligands except GTP are involved in the CTP synthetase reaction, all the ligand binding sites should be aligned as represented schematically in Figure 6-2. Therefore, binding of DON at the glutamylation site could affect the binding of the other ligands. Surprisingly, the DON-modified enzyme possesses normal ammonia activity, and the kinetic parameters for ATP, UTP and NH_3 are identical in the modified and the unmodified enzyme. Furthermore, direct binding measurements of the ligands ATP and UTP reveal the identity of the native and the DON enzyme. Thus, even though the glutamylation site is in close proximity to the ammonia site and to the sites of ATP and UTP, DON binding to the glutamine site does not affect either the binding or the catalytic properties of these sites. However, the binding of the drug in one subunit affects the binding of the second mole to the *neighboring* subunit, as described earlier. The binding of 1 mole of DON induces a conformational change across the *pp* binding domain in "switching off" the second DON site. Furthermore, when the isologous dimers are assembled in tetramers, a DON-induced conformational change across the *qq* binding domains affects the reactivity of the second DON site across the plane of dissociation (Figure 6-3). Indeed, the second DON molecule reacts more slowly in the assembled tetramer (Table 6-1). The allosteric DON within the dimer also abolishes the negative cooperation between the DON sites. The latter effect is probably because of interactions across the *qq* binding domains (Figure 6-3).

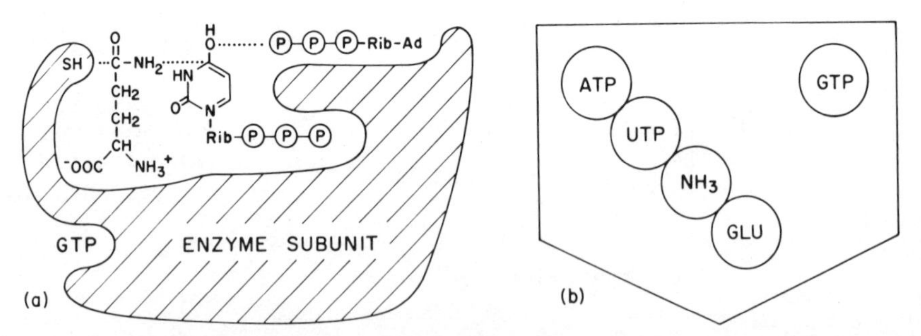

Fig. 6-2 Schematic illustration of active and allosteric sites of CTP synthetase. The sites are arranged so that the sequence of proposed reactions can be executed with minimal rearrangement of substrates. The allosteric site for GTP is remote from the series of active sites, but no evidence to assign its distance from the glutamine site is available. The amide nitrogen of glutamine occupies the NH_3 site. DON occupies the glutamyl subsites but not the ammonia site, and therefore DON-CTPS has full ammonia activity. Levitzki (1973).

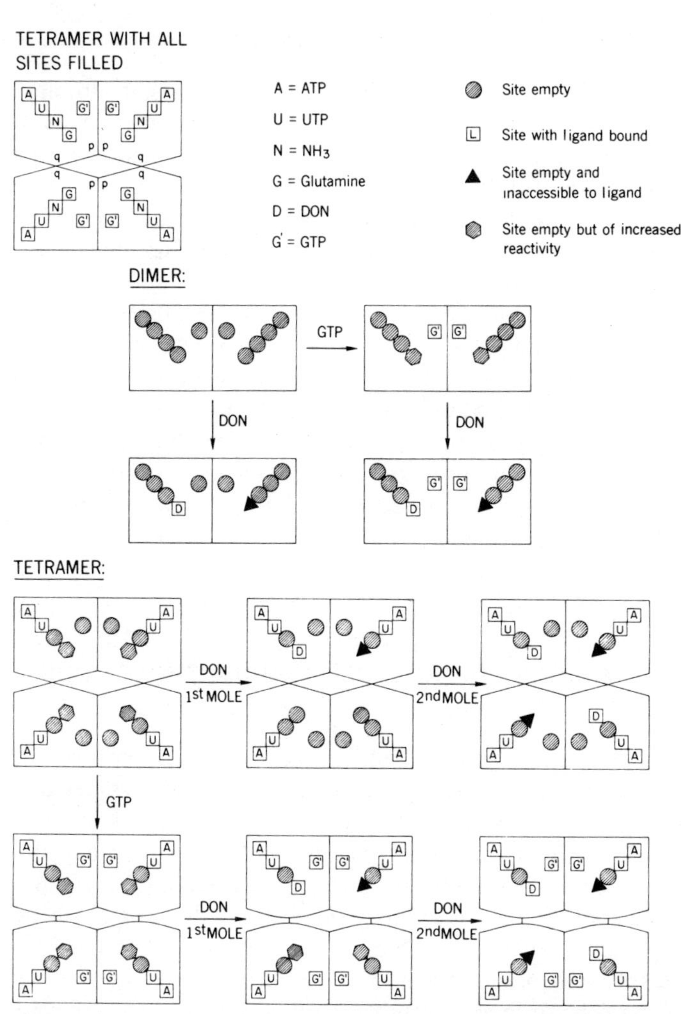

Fig. 6-3 The conformational changes induced by DON and GTP. (*a*) *Dimer*: The interaction of DON with the enzyme dimer labels only one of the two glutamine sites both in the absence and the presence of GTP. The presence of GTP increases the *rate* of inactivation, as is schematically indicated by the conformational change in the shape of the glutamine subsite. However, the presence of GTP does not affect the "switching off" of one of the DON sites induced by the binding of DON to the other subunit. (*b*) *Tetramer*: The assembly of the enzyme dimers to tetramers induces a conformational change in the glutamine subsite as is indicated by the change in shape of the site. The first mole of DON reacts at a similar rate to the site on the dimer in the presence of GTP. The interaction of the first DON "switches off" the second glutamine site on the *same* dimer and induces a more moderate conformational change in the glutamine sites across the dissociation plane. Thus the second mole of DON reacts more slowly.

Again, the reaction of one of the subunits with DON "switches off" the second glutamine site on the second dimer. In the presence of the allosteric activator GTP, the two dimers constructing the enzyme tetramer are "uncoupled," and the two glutamine sites react with the DON independently with the high rate constant. Again however, only half of the glutamine sites are labeled, since the GTP does not abolish the intradimer subunit interactions. Levitzki (1973).

The striking feature of the DON-induced changes is the fact that, although profound conformational changes occur, as is indicated from the reactivity toward DTNB, they are expressed functionally in only one section of the active site. The chemistry of the CTP–synthetase reaction indicates a close geometric alignment of ATP, UTP, and glutamine. Even so, the interaction of DON with the glutamine site leaves the other ligand sites unaffected. The conformational effects are *channeled* specifically to vacant glutamine sites in other subunits. Also, the allosteric activator GTP affects only the reactivity of the glutamine site toward DON and remains without any effect on the adjacent ammonia, ATP, or UTP sites. A summary of the interaction of CTP synthetase with DON and GTP is given in Figure 6-3.

This system is an example of one in which the drug interacts with a known receptor for a known natural ligand. The second example to be discussed is a case in which the drug interacts with an enzyme, at a site, the natural ligand for which has not been defined, and is not part of the active site as is in the case of CTP synthetase.

III GLUTAMIC DEHYDROGENASE–LIGAND INTERACTIONS

γ-Aminobutyric acid, or its precursor glutamic acid, play an important role in the central nervous system (Kravitz et al., 1965), and thus enzymes that control their levels may be the target for some drugs that are known to affect behavior. Indeed, it was found that chlorpromazine, desipramine, and amitriptyline are inhibitors of glutamic dehydrogenase (GDH) (Fahiem and Shemisa, 1970). The conformational changes induced by chlorpromazine are detectable by a variety of techniques (Shemisa and Fehiem, 1971). Changes in the absorption spectrum of the enzyme and the fluorescence spectrum are induced by the drug. The drug also induces the dissociation of the enzyme from its polymerized state to the free hexamers. The presence of a "drug site" on the enzyme molecule is an extremely interesting finding. Furthermore, a structure–activity study of the effects of a series of isosteres of promazine on GDH shows that the effect on GDH correlates closely with their antipsychotic activity in vivo (Shemisa and Fahien, 1971). Of great interest therefore is whether a natural chlorpromazine-like allosteric effector exists in the mammalian brain.

The dissociation constant of the chlorpromazine–GDH complex is ten times lower than the level of the drug in the brain after administration of a pharmacologic dose (Shemisa and Fahien, 1971; Salzman and Brodie, 1956). Therefore, it seems possible that chlorpromazine can affect behavior via the inhibition of glutamic dehydrogenase in the brain. Glutamic dehydrogenase is also present at high levels in other tissues such as liver and kidney, and thus it may be that side effects of chlorpromazine treatment, such as jaundice, can also be traced to the strong interaction with glutamic dehydrogenase and the induced structural changes occurring in the enzyme.

IV CONFORMATIONAL CHANGES IN MULTIPROTEIN ASSEMBLY INDUCED BY DRUGS: THE EFFECT OF STREPTOMYCIN ON THE BACTERIAL RIBOSOME

Ribosomes are organelles of an intermediate level of complexity. They can be obtained as a homogeneous population, their composition is known, and their physicochemical and biologic properties can be studied in solution. They are much more complex than multisubunit enzymes and multienzyme complexes, but their level of architectural complexity is much lower than that of membranes.

Streptomycin and dihydrostreptomycin bind exclusively to the 30S subunit of the ribosome from *E. coli* (Kaji and Tamaka, 1968). Up to two molecules of streptomycin per ribosome are bound. No binding to the 50S subunit has been detected. The streptomycin receptor site can be identified as a specific protein of the "core" particle of the ribosome (Staelhein and Meselson, 1966; Traub et al., 1966). It is now known that streptomycin sensitivity, resistance, and dependence are all structural variations of the streptomycin-binding protein (Traub and Nomura, 1968). The ribosome is composed of at least 25 distinct proteins (Traub and Nomura, 1968), only one of which (P_{10}) is responsible for streptomycin sensitivity. 30S particles assembled in the absence of P_{10} are devoid of streptomycin sensitivity. It should be stressed, however, that it has never been demonstrated that it is the P_{10} protein that binds the drug, although it is a logical working hypothesis. The function of the protein P_{10} is not yet uniquely defined. The protein is not necessary for the assembly of the 16S RNA with the ribosomal proteins to form a 30S particle. It seems that the role played by P_{10} in the ribosome assembly is to influence the error frequency occurring during translation. The translational ambiguity is an inherent property of bacterial ribosomes, and the misreading of the code is enhanced on the binding of streptomycin to the P_{10}-containing assembly. One has to bear in mind that the 30S particle is composed of at least 25 distinct proteins assembled in a unique design. Only one of the proteins, designated P_{10} (Traub and Nomura, 1968), interacts with the drug, leading to a malfunctioning ribosome. It is very likely that the drug is topographically remote from the essential sites involved in the 30S activity. It follows that the effect of the drug must be transmitted via an induced conformational change from the streptomycin binding site in the 30S "core" to the relevant sites involved in the biologic activity. Since the 30S particle is a giant assembly of subunits compared to known multisubunit enzymes, this conformational effect must be extremely directional and specific. It is attractive to speculate that, if nature has chosen a specific streptomycin-binding protein as part of the *E. coli* ribosome, a mechanism of conformational transmission from the drug site was also devised.

Streptomycin induces the following effects in a cell-free system of a streptomycin-sensitive bacteria.

1. Inhibition of translation (Erdos and Ullman, 1959; Schwartz, 1965).
2. Stimulation of the incorporation of amino acids not coded for in the added messenger, namely, misreading of the genetic code. For example, in the presence of streptomycin, poly U-directed synthesis of phenylalanine is inhibited, while at the same time the incorporation of isoleucine or serine is stimulated to produce a net stimulation of peptide bond synthesis (Davies and Davis, 1968).
3. Stimulation of polypeptide synthesis directed by unnatural messenger-like templates such as denatured DNA, r-RNA, t-RNA, etc (McCarthy et al., 1966; Morgan et al., 1967).

Using a hydrogen–tritium exchange rate as a conformational probe, Sherman and Simpson (1969) showed that low streptomycin to ribosome ratios stimulate the tritium exchange rate. This result is interpreted as indicating the "loosening" of ribosomal structure. On increasing the streptomycin to ribosome ratios, a maximal exchange rate was reached. At higher ratios of streptomycin to ribosome, the ribosomal structure begins to "retighten," as indicated by a decrease in the exchange rate. In this range of streptomycin to ribosome ratios, inhibition of translation and misreadings of the code begin. Thus it seems that the gross conformational changes observed are intimately connected with the inhibition of polypeptide biosynthesis and the misreadings of the genetic code. It is also suggested that the two conformations, the "loose" and the "tight," are induced by the binding of the drug to the two different sites. This suggestion is not consistent with the claim that only one site on the ribosome is involved (Weisblum and Davis, 1968) in streptomycin inhibition of protein synthesis. However, the finding is consistent with the binding studies of Kaji and Tanaka (1968), who found two binding sites per ribosome. The finding of two binding sites per ribosome may be taken as an indication that the "core" protein involved in streptomycin binding possesses at least two sites. There is no method to determine precisely the fraction of active ribosomes; therefore, the stoichiometry of the ribosome–streptomycin complex is not uniquely defined.

V DRUG ACTION ON RECEPTORS INTEGRATED IN MEMBRANES

Active transport of Na^+ is known to be coupled in some cases to the activity of Na^+- and K^+-dependent Mg-ATP phosphohydrolase (ATPase) (Albers, 1967). The erythrocyte membrane and a variety of other preparations from different tissues possess this type of sodium–potassium pump. The activity of ATP hydrolysis is integrated within membranes from a variety of sources (Hokin, 1970; Mizuno et al., 1969; Kepner and Macy, 1968; Kahlenberg et al., 1969). The phosphorylated enzyme, embedded in the membrane, mediates the active transport by a conformational change. The conformational change that is believed to occur in the enzyme on phosphorylation (E to E*) is necessary for the inhibition of its activity by cardioactive glycosides such as ouabain. Whether the phosphorylation or just the binding of ATP, Mg^{2+}, and Na^+ (Hoffman, 1969) is a prerequisite to ouabain binding has yet to be clarified. The enzyme is probably oriented across the membrane, since the maximal activity (erythrocyte ghosts) is observed when ATP is inside, and when Na^+ is high inside and K^+ is high outside (Hoffman, 1969; Glynn, 1962). The receptor for the drugs from the cardiac steroid family resides in the (Na^+ + K^+)-ATPase moiety. The binding of ouabain to the enzyme occurs in the presence of ATP, Na^+, and Mg^{2+} and abolishes the K^+-dependent dephosphorylation of the enzyme. Furthermore, when the enzyme preparation is treated with the drug first, the enzyme can be phosphorylated by orthophosphate (Kahlenberg et al., 1969), whereas the native enzyme is devoid of such a property (Post et al., 1965). The phosphorylation sites in the ouabain-treated preparation seem to be identical to those phosphorylated by ATP when the native enzyme is incubated with the nucleotide (Kahlenberg et al., 1969).

Although the drug abolishes the K^+-dependent step, this does not fully explain the inhibition of either the enzymatic activity or the active transport. However, the drug-induced phosphorylation by free orthophosphate means that the *free energy* difference between the phosphorylated and the nonphophorylated forms of the enzyme has been abolished. Therefore, the enzyme is "locked" in a particular inactive conformation induced by the drug, and both the ATPase and the transport activities are inhibited.

The phosphorylation of the sodium–potassium ATPase by orthophosphate may be taken as the strongest evidence for the drug-induced conformational change. It has been suggested that the enzyme is composed of two types of subunits, only one of which is capable of being phosphorylated (Nakao et al., 1969). It is visualized that the enzyme is embedded across the membrane so that the subunit containing the phosphorylation sites points inside, whereas the ouabain site resides in the subunit pointing outside. Once the drug binds to the drug site outside, a conformational change is transmitted from the outer subunit to the inner one, causing inhibition.

VI INTERACTION OF POLYCYCLIC AROMATIC HYDROCARBONS WITH HISTONE IV

Carcinogenic ligands probably fall into the general structural modifier class of drugs like ethanol. Perhaps all of man's socioenvironmental compounds fit into this category. In this section we concentrate on one particular structural class of environmental contaminants, some polycyclic aromatic hydrocarbons, and histone IV. Chapter 9 deals with histones in detail and it suffices here to state that the histones are chromatin proteins that associate with DNA. The polycyclic aromatic hydrocarbons generally possess large extinction coefficients and high quantum yields of fluorescence. Thus they are good structural probes to use in studying intermolecular associations in which they are participants. Small et al. (1973) have made use of these probe properties for 3,4-benzpyrene, pyrene, anthracine, and 9-methylanthracene to study their interactions with histone IV. These workers find that all the polycyclic aromatic hydrocarbons bind to histone IV, but only after it has been allowed to undergo a slow characteristic change in the presence of divalent phosphate anions. This is thought to be an order-inducing conformational transition. The hydrocarbons do not bind to the disordered form of histone IV found at low pH, or to a characteristic form obtained as a result of temperature suppression of the particular slow conformational change and allowing the fast conformational change to proceed. Other workers (Lesko et al., 1968; Kodama et al., 1968; O'Brien et al., 1969; Sluyser, 1968) have observed the binding of aromatic hydrocarbons to histones, but without information regarding the structural chemistry of the association.

One can speculate about how these carcinogenic agents carry out their life-destroying function. It would seem that a conformational ordering process in the histones as a result of complexing with DNA might provide the necessary ligand binding sites for the aromatic hydrocarbons. Once bound, these carcinogenic ligands may appear as defects to the ribosome, resulting in transcription errors.

REFERENCES

Albers, R. W. (1967). *Ann. Rev. Biochem.*, **30**, 727.

Albers, R. W., G. J. Koval, and G. J. Siegel (1968). *Mol. Pharmacol.*, **4**, 324.

Davies, J. and B. D. Davis (1968). *J. Biol. Chem.*, **243**, 3312.

Erdos, T. and A. Ulmann (1959). *Nature*, **183**, 618.

Fahien, L. A. and O. A. Shemisa (1970). *Mol. Pharmacol.*, **6**, 156.

Gibson, R., J. Pittard, and E. Reich (1967). *Biochim. Biophys. Acta*, **136**, 573.

Glynn, I. M., (1962). *J. Physiol.*, **160**, 18p.

Handschumacher, R. E., C. J. Bates, P. K. Chang, A. T. Andrews, and G. A. Fischer (1968). *Science*, **161**, 62.

Hoffman, J. F. (1969). *J. Gen. Physiol.*, **54**, 3435.

Hokin, L. F., J. F. Danielli, J. F. Moran, and D. J. Triggle Eds. (1970). In *Fundamental Concepts in Drug-Receptor Interaction* Academic Press, New York, p. 205.

Kahlenberg, A., N. C. Dulac, J. F. Dixon, P. R. Glassworthy, and L. E. Hokin (1969). *Arch. Biochem. Biophys.*, **131**, 253.

Kaji, H. and Y. Tanaka (1968). *J. Mol. Biol.*, **32**, 221.

Kepner, G. R. and R. I. Macey (1968). *Biochim. Biophys. Acta*, **163**, 188.

Kodama, M., Y. Tagashira, and C. Nagata (1968). *J. Biochem.*, **64**, 81.

Koshland, D. E. Jr. (1958). *Proc. Natl. Acad. Sci., USA*, **44**, 98.

Koshland, D. E. Jr. and D. E. Neet (1968). *Annu. Rev. Biochem.*, **37**, 359.

Kravitz, E. A., P. B. Molinoff, and Z. W. Hall (1965). *Proc. Natl. Acad. Sci. USA*, **54**, 778.

Lesko, S. A. Jr., A. Smith, P. O. P. Tso, and R. S. Umans (1968). *Biochemistry* **7**, 434.

Levitt, M. (1974*a*). *J. Mol. Biol.*, **82**, 393.

Levitt, M. (1974*b*). In *Peptides, Polypeptides, and Proteins* (E. R. Blout, F. A. Bovey, M. Goodman, and N. Lotan, Eds.). Wiley-Interscience, New York, p. 99.

Levitzki, A. (1973). *Mod. Pharmacol*, **1**, pt. 1, 305.

Levitzki, A., and D. E. Koshland, Jr. (1972). *Biochemistry*, **11**, 247.

Long, C. W., A. Levitzki, and D. E. Koshland, Jr. (1970). *J. Biol. Chem.*, **245**, 80.

McCarthy, B. J., J. J. Holland, and C. A. Buck (1966). *Cold Spring Harbor Symp. Quant. Biol.*, **31**, 683.

Mizuno, N., K. Nagano, T. Nakao, Y. Toshimo, M. Fujita, and M. Nakao (1969). *Biochim. Biophys. Acta*. **108**, 311.

Morgan, A. R., R. D. Wells, and H. G. Khorana (1967). *J. Mol. Biol.*, **26**, 477.

Nakao, M., K. Nagano, M. Matsui, N. Mizuno, T. Nakao, and Y. Tashima (1969). In *The Molecular Basis of Membrane Action* (D. C. Tosteson, Ed.). Prentice-Hall, Englewood, N. J., p. 539.

O'Brien, R. L., R. Stanton, and R. L. Craig (1969). *Biochim. Biophys. Acta*, **186**, 414.

Perutz, M. F. (1970). *Nature*, **228**, 726.

Post, R. I., A. K. Sen and A. S. Rosenthal (1965). *J. Biol. Chem.*, **240**, 1437.

Recke, G. N., J. A. Hartsuch, M. L. Ludwig, F. A. Quiocho, T. A. Steitz, and W. N. Lipscomb (1967). *Proc. Natl. Acad. Sci. USA*, **58**, 2220.

Salzman, N. P. and B. B. Brodie (1956). *J. Pharmacol. Expt. Ther.*, **118**, 46.

Schwartz, J. H. (1965). *Proc. Natl. Acad. Sci. USA*, **53**, 1133.

Shemisa, O. A., and L. A. Fahien (1971). *Mol. Pharmacol.*, **7**, 8.

Sherman, M. I., and M. V. Simpson (1969). *Proc. Natl. Acad. Sci. USA*, **64**, 1388.

Sluyser, M. (1968). *Biochim. Biophys. Acta*, **154**, 606.

Small, E. W., A. M. Craig, and I. Isenberg (1973). *Biopolymers*, **12**, 1149.

Smiley, I. E., R. Koekoek, M. J. Adams, and M. G. Rossman (1971). *J. Mol. Biol.*, **55**, 467.
Staehlein, T. and M. Meselson (1966). *J. Mol. Biol.*, **19**, 207.
Traub, P., E. Hosokawa, and M. Nomura (1966). *J. Mol. Biol.*, **19**, 211.
Traub, P. and M. Nomura (1968). *Science*, **160**, 198.
Weisbium, B. and J. Davis (1968). *Bacteriol. Rev.*, **32**, 493.

CHAPTER 7

Hydration Properties of DNA

The hydration of DNA is an important factor in the stability of its secondary structure. Since DNA is a polyelectrolyte as well as a water-binding macromolecule, there is a sometimes competitive, sometimes cooperative, set of interactions among water molecules, ions, and DNA that simultaneously dictate molecular organization. In this chapter, we emphasize the hydration interactions. Different degrees of hydration are shown to produce the three different DNA configurations seen by X-ray diffraction studies and denoted A (Fuller et al., 1965), B (Langridge et al., 1960), and C (Marvin et al., 1961). The most prominent hydration form is actually denoted by B′. This form is similar to the normal B form (Langridge et al., 1960), but with a smaller rotational angle between successive base pairs in the helix (Bram, 1971; Ivanov et al., 1973). To sustain this B′ form requires a high degree of hydration.

I INTRODUCTION

An obvious property of DNA that would be expected to have some influence on secondary structure and that changes markedly as the concentration of salt is varied is the preferential solvation of this macromolecule. Hearst and co-workers (Hearst and Vinograd, 1961*a,b*; Hearst, 1965; Tunis and Hearst 1968*a*) as well as Cohen and Eisenberg (1968) have demonstrated that, as the water activity, A_w, decreases with increasing concentrations of salts of Na^+, Li^+, and Cs^+ in aqueous solutions of DNA, the preferential solvation or net hydration of the macromolecule, Γ (in moles of H_2O/mole of nucleotide), decreases. This effect is largely independent of the nature of the cation and dependent mainly on the value of A_w for a given DNA. There is a small dependence on base composition, with AT pairs being hydrated to the extent of 2 additional moles of water compared to GC pairs (Tunis and Hearst, 1968*b*). The observed transformations in the secondary structure of DNA can be correlated with its state of hydration, as first reported by Wolf et al. (1972). A rough correlation has also been pointed out by Zimmer and Luck (1973), who observed that the lowering of the positive CD band in the spectrum of DNA in aqueous electrolyte solutions followed the same course as the hydration of DNA in these solutions reported by Chattoraj and Bull (1971).

II THE PRIMARY HYDRATION SHELL

On the basis of the data and discussions of Wang (1955), Hearst (1965), Tunis and Hearst (1968*a,b*), Cohen and Eisenberg (1968), Falk et al. (1962, 1963*a,b*, 1970), and Wolf and Hanlon (1975), the hydration of DNA can be characterized in terms of two discrete water layers, a primary and a secondary shell. The primary layer consists of those water molecules immediately adjacent to the DNA duplex. This layer is in a chemically different state than liquid water. Its infrared spectrum indicates that the OH frequencies are not identical with those of liquid water, and it is impermeable to electrolyte in the sense that ions cannot randomly diffuse about in this layer with the same activation energy characteristic of liquid water.

Within this primary shell, the water molecules fit into two classes. One of these consists of water molecules that are directly bound to sites in the DNA duplex, according to the scheme of Falk et al. (1962, 1963*a,b*). If the quantitative data of Falk et al. are applicable to the solution form of DNA, the number of molecules in this class should be about 11–12. Five of these are located at base sites in the grooves of the helix and have a lower affinity than the remaining six. The four water molecules bound at furanose oxygens and phosphodiester linkages are probably more tightly bound than those associated with the base sites. The remaining two waters associated at the ionic phosphate site as part of its hydration shell have the highest affinity. Removal of these last two molecules generally results in complete structural disruption of DNA (Falk et al. 1963*b*).

The other class of water molecules found in the primary hydration layer is distinguishable only insofar as the hydration layers of DNA do not exhibit the infrared spectral characteristics of liquid water until about 18 water molecules have been adsorbed per nucleotide residue. Since only 11–12 of these 20 can be directly site bound to DNA, this leaves about six to seven water molecules forming an interface region that is hydrogen bonded to the site-bound waters and is spectrally perturbed as a consequence. Presumably, all members of this second set of water molecules in the primary hydration shell would be expected to have lower affinity for DNA than those molecules bound directly to the DNA sites.

In Figure 7-1, a typical difference profile in the OH stretching region of DNA between 93 and 3% relative humidity is compared with the corresponding absorbance profiles of HDO in liquid water recorded by the method of Falk and Ford (1966). The difference spectrum for DNA between 93 and 3% relative humidity corresponds to a cross section of the hydration shell, including water molecules closely associated with the biopolymer and others removed by several molecular diameters from the biopolymer surface. The general similarity between the spectral profiles of water of hydration and of liquid water suggests the absence of regions of ordered water in the hydration shell of DNA (Falk et al., 1963*a*). The distribution of OH stretching frequencies of HDO is related to the distribution of hydrogen bond strengths (Falk and Ford, 1966). Evidently these distributions differ very little for the bulk of water of hydration and for liquid water at the corresponding temperature.

In Figure 7-2 are plotted the sequence of four difference profiles corresponding to the spectrum of HDO desorbed in the intervals between 93, 82, 62, 33, and 3% relative humidity. The peak frequencies of these difference profiles, averaged from measurements on four DNA films, show that consecutive layers of water surrounding the macromolecule have similar distributions of hydrogen bond strengths. It should be noted that even at 3% relative humidity DNA holds one molecule of water per nucleotide and possibly more (Killion and Reyerson, 1966). Even though hydrogen bond energies of water–biopolymer and water–water interactions are comparable, the mobility of water molecules near the biopolymer surface may be expected to be lower than in bulk water, since the biopolymer surface will act as a momentum sink for thermal fluctuations.

The effect of slow cooling/heating on the spectrum in the OD stretching region for a typical film of partially deuterated DNA of low water content

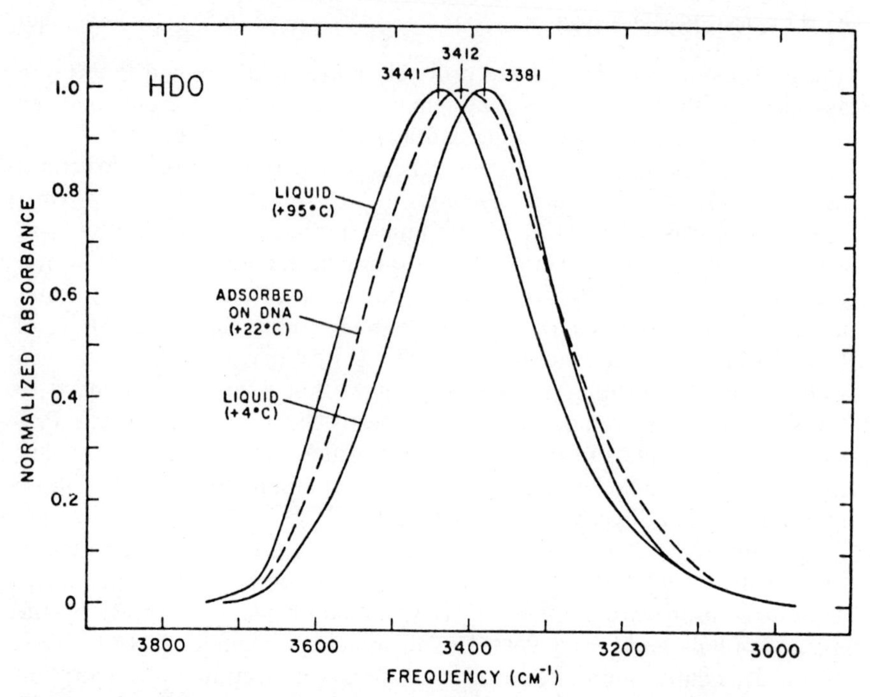

Fig. 7-1 Comparison of the OH stretching profiles of 1 mole % of HDO in D_2O: (*a*) liquid water at 4 and 95%, (*b*) absorbed water on DNA at room temperature (difference profile corresponding to HDO desorbed between 93 and 3% relative humidity). Falk et al (1970).

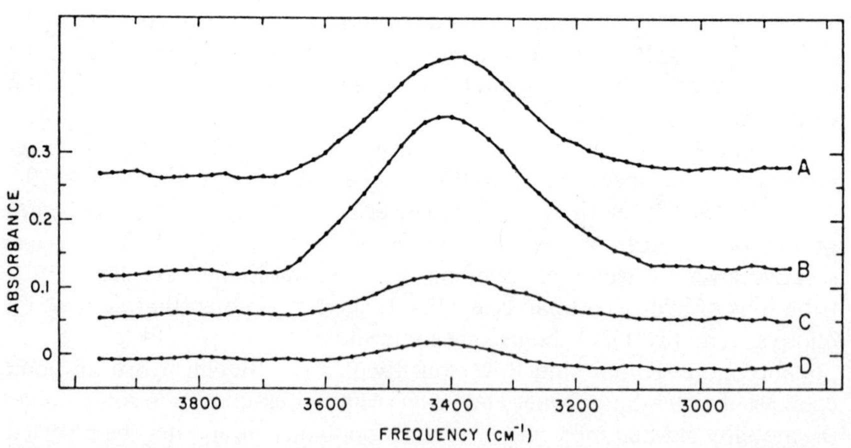

Fig. 7-2 Difference profiles in the OH stretching region, corresponding to HDO desorbed from DNA in the relative humidity ranges of 93 to 82% (A), 82 to 62% (B), 62 to 33% (C), and 33 to 3% (D). A, B, and C have been displaced upward by 0.05, 0.1, and 0.25 absorbance units, respectively. Falk et al. (1970).

(about nine water molecules per nucleotide) is completely reversible with temperature, indicating a total absence of crystallization. The peak intensity suggests a wide distribution of hydrogen bond strengths, characteristic of the liquid and amorphous states of water.

The effect of cooling on the spectrum of a DNA film of a somewhat higher water content (about 14 water molecules per nucleotide) indicates that the band shape undergoes a sudden change, usually between -10 and $-20°$, corresponding to crystallization of a part of the water. The crystalline regions correspond to local pockets of ice I. The ice band appears upon cooling for all DNA films with a water content greater than about 13 molecules of water per nucleotide. These results suggest that a portion of the water molecules in the primary hydration layer of about ten water molecules per nucleotide are incapable of crystallization, even when the surrounding water crystallizes into ice I. The directly DNA-bound water molecules are the most likely noncrystallizable solvent species. An additional intermediate layer of about three water molecules per nucleotide crystallizes with difficulty and tends to supercool if the cooling rate is high. These are probably some of the other water molecules of the primary hydration shell that interface with another layer of liquid structure.

The present results indicate that, at least for DNA, there is no "ice-like" ordering of water molecules about the biopolymer. The innermost, least mobile part of the hydration shell is in fact entirely incapable of crystallizing into the ice structure. Evidently, the preferred configuration of water next to the biopolymer is incompatible with the structure of ice. There is no spectral evidence of "quasi-crystallization" of any part of the hydration shell at room temperature, and at low temperatures it is the water far from the biopolymer surface that freezes into ice. The hydration shell of DNA is not "ice-like" in the structural sense.

III THE SECONDARY HYDRATION SHELL AND SOLVATION MECHANISMS

Surrounding the primary shell is a secondary shell whose water structure is essentially that of liquid water found in the bulk solvent. This layer is freely permeable to the ions of the electrolyte. Most of this secondary layer may be more apparent than real, since it would include the domain of the solvent in which the electrolyte content is lower than in the bulk solvent because of the Donnan effect.

It is probable that the apparent dehydration of the proposed hydration-layer model, as reflected in the decreased values of the preferential solvation as the electrolyte content of the solution increases, proceeds by several mechanisms. The simplest of these is an effect on the Donnan contribution to the preferential solvation (Hearst, 1965; Cohen and Eisenberg, 1968) due to increased electrolyte concentration. This should be a nonspecific effect (as contrasted to the site-binding mechanism discussed below) and would influence only the secondary hydration layer.

Another mechanism could involve the direct interaction of the cation of the electrolyte with the negatively charged phosphates, partially neutralizing one another's effective charge. This process might release the more loosely bound water from the primary shell in the vicinity of the phosphate groups because of the usual reversal of electrostrictive effects found for low-molecular- weight solutes. Cohen and Eisenberg claim that this does not account for even a part of the increase in the partial specific volume of DNA with increase in electrolyte concentration (Cohen and Eisenberg, 1968). It should be noted that the increase in the charge neutralization of the polyelectrolyte will also decrease the Donnan contribution to the preferential hydration (Hearst, 1965; Cohen and Eisenberg, 1968).

A third mechanism that may be operative in DNA solutions involves the removal of water from the primary shell because of the reduction of the water activity in these concentrated electrolyte solutions. As has been extensively discussed by Hearst and co-workers (Hearst and Vinograd, 1961*a,b*; Hearst, 1965; Tunis and Hearst, 1968*a,b*) and Cohen and Eisenberg (1968), the degree of hydration of all species in solution is modulated by A_w, and the hydration of the polymer component decreases with decreasing A_w, as has been experimentally observed by these workers.

Given the assumption that the partial specific volume of DNA does not change on denaturation, one can estimate the change of hydration on denaturation. In addition, the change in hydration with temperature and the heat of hydration can be calculated for native and denatured DNA, according to the method of Vinograd et al. (1965). Table 7-1 gives these

Table 7-1 Results of Temperature-Dependent Buoyant Density Studies on Native and Denatured DNA in Potassium Trifluoroacetate.

DNA	$(d\rho_0/dT)$, g/ml-°C	ΔH, cal/mole H_2O	$(\partial\Gamma/\partial F)a_w$, mole H_2O/mole nucleotide, °C	Γ, mole H_2O/mole nucleotide 25°C	54.5°C
Native	11.8×10^{-4}	397	−0.024	6.9	6.2
Denatured	10.0×10^{-1}	341	−0.015	5.3	4.9
3 DNA's in CsCl[a]		409	−0.028	6.6 − 7.4	

[a] Data of Vinegrad et al. (1965).

values for native and denatured DNA in potassium trifluoroacetate (Tunis and Hearst, 1968*b*). A value of 0.562 cc/g for the partial specific volume of KDNA was used in calculating hydrations (Hearst, 1962). "Denatured DNA" in these calculations means the fully denatured form, densities being extrapolated to 25°C from above the T_m. The hydration of the intermediate forms appears to be approximately proportional to the degree of structure as measured by OD_{260} (Tunis and Hearst, 1968*b*).

IV THE DNA SECONDARY STRUCTURE–HYDRATION SHELL INTERRELATIONSHIP

Using the proposed hydration model and mechanisms of dehydration, together with the specific ion effects, it is possible to explain changes in DNA secondary structure by proposing that the maintenance of the "B" conformation in aqueous solution is dependent not only on the absence of perturbing cationic influences but also on the preservation of an intact primary hydration shell consisting of tightly and loosely bound water molecules. As this primary shell is stripped away, conformational adjustments of both an inter- and intranucleotide nature ensue, leading mainly to the C structure, with some A structure formed in the presence of some salts.

At high water activities, in the absence of perturbing ionic effects of added electrolyte, both the secondary and the primary hydration shells are intact. As the salt concentration increases, with increasing site binding of cations, the dehydration of DNA probably proceeds initially by all of the mechanisms described in the preceding paragraphs. The secondary shell is primarily affected, although there is probably some removal of the more weakly bound water molecules of the primary layer because of ion pair formation and the lowering of the water activity of the solvent. This process continues until a Γ of 13 $\pm$ 1 mole of H_2O/mole of nucleotide is reached which corresponds to all of the secondary shell, and most of the loosely bound primary layer water molecules being removed. This leaves only the tightly bound water molecules of the primary shell. As the latter are removed by the further lowering of the water activity of the solvent, the fractional B content falls precipitously, which directly reflects the sensitivity of the transformations to the state of integrity of the primary hydration shell. The critical number of water molecules that, when lost from the hydration layers, leads to the complete loss of the B form can be estimated from the normalized percent B′ data of Wolf and Hanlon (1975) and is on the order of approximately 18 water molecules per nucleotide residue. This corresponds reasonably well with the estimate of Falk et al. (1962, 1963*a*,*b*, 1970) of the number of water molecules in the hydration shell of DNA whose infrared spectra differ from that of liquid water.

The C and A DNA structures are realized with loss of water in the primary hydration layer, which retains only the water molecules associated with the phosphate group. The value of 6–8 moles of water/mole of nucleotide for the 0% B intercept using the ammonium salt is significantly higher (Wolf and Hanlon, 1975) than that of other cationic sources. This, in addition to greater effectiveness of the ion in inducing C structure, suggests that some rather unique ways of binding to the DNA structure may be available to this cation (Wolf and Hanlon, 1975). Herskovits and co-workers (1961*a,b*; 1962) have studied the influence of alcohol–water solvents on the hydrodynamical and optical properties of DNA. The low-molecular-weight alcohols were employed because of their ability to mix well with water while possessing low polarities. The denaturing effect of the alcohols on DNA was seen to be reversible. Similar conclusions were drawn regarding changes in the DNA secondary structure by monitoring the circular dichroism of DNA in water–ethanol mixtures (Ivanov et al., 1973; Brams and Mommaerts, 1964; Girod et al., 1973). Tunis-Schneider and Maestie (1970), from a comparison of the circular dichroism spectra of DNA in films at different humidities with those in water–alcohol solvents, deduced that, at ~80% volume fraction of ethanol and 5×10^{-4} *M* NaCl, the DNA molecule adopts an A form. However, this finding is premised on the assignment that the CD spectra in 80% ethanol are distorted by the scattering of light because of the aggregation of DNA molecules. Frisman et al. (1974) have studied the influence of alcohol–water solvents on the conformation of native DNA using the methods of flow birefringence and viscometry. Conformational transitions were observed at low alcohol concentrations corresponding to the destruction of at least part of the primary hydration layer of water. This is evident from the rapid drops in viscosity $[\eta]$ for the $[\eta]$ versus alcohol (vol %) plots in Figure 7-3. A gradual change in the secondary structure of the DNA molecule was also inferred from the steady decrease in $[\eta]$ at high ethanol concentrations. This is consistent with the findings of Tunis-Schneider and Maestre (1970).

On the basis of their wide-angle X-ray scattering experiments with DNA gels in 6 *M* LiCl, Maniatis et al. (1974) have concluded that calf thymus DNA at this electrolyte concentration is in the B conformation. In contrast, Wolf and Hanlon (1975) estimate that, at this concentration of LiCl, DNA in dilute solution has only 13% B′ character, with the remainder of the nucleotide residues in the C (72%) and the A (15%) forms. The hydration of DNA in gels may thus be significantly higher than it is in a dilute solution of DNA in 6 *M* LiCl. The fractional B content, since it depends on DNA hydration, might be expected to correspondingly vary with "phase" state.

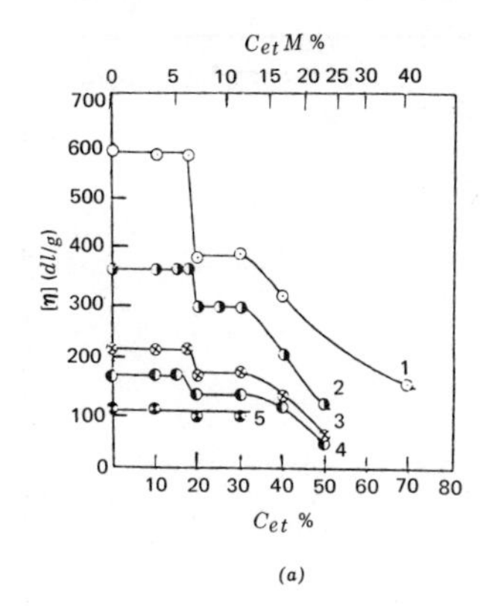

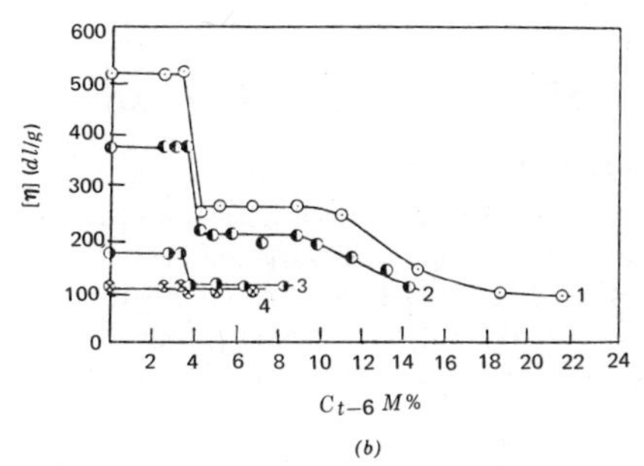

Fig. 7-3 (*a*) Dependence of $[\eta]$ for DNA on the percent of ethanol (vol %) in the solvent at different ionic strengths: 1–4 × 10^{-4}; 2–1 × 10^{-3}; 3–4 × 10^{-3}; 4–10^{-2}; 5–1 × 10^{-1} *M* NaCl. The upper scale corresponds to mol % of ethanol. (*b*) Dependence of $[\eta]$ for DNA on the percent of tert-butanol (mol %) solvent at different ionic strengths: 1–5 × 10^{-4}; 2–1 × 10^{-3}; 3–1 × 10^{-2}; 4–10^{-1} *M* NaCl. Frisman et al. (1974).

The degree of DNA-water hydration, and even the rate of water uptake, as a function of relative humidity is nearly independent of DNA complexing with basic polypeptides and, based on limited data, histones (Sowalsky & Traub, 1968; 1972). The effects of the relative humidity on the X-ray spacings of the first equatorial reflections of DNA and three of its complexes are presented in Figure 7-4. The observed sharp maxima in the complexes, in contrast to DNA that goes gradually into solution when immersed in water (Feughelman et al., 1955), may indicate crosslinks between the molecules as has been suggested for several DNA complexes (Suwalsky, et al., 1969; Luzzati and Nicolaieff, 1963; Raukas, 1965). The first equatorial reflection of spermheads increases from 19.1 Å at 35% relative humidity, to 26.3 Å at 100%, whereas, for polylysine-DNA, the spacing of the corresponding reflection at 0% and 100% relative humidity was 18.2 Å and 27.4 Å, respectively, and for polyarginine-DNA, 17.7 Å and 25.3 Å, respectively.

The transformations in DNA secondary structure appear to be initimately related both to ionic interactions and the state of hydration of the nucleic acid. This relationship can be quantitatively expressed in terms of the minimal hydration of approximately 18 moles of water/mole of nucleotide required

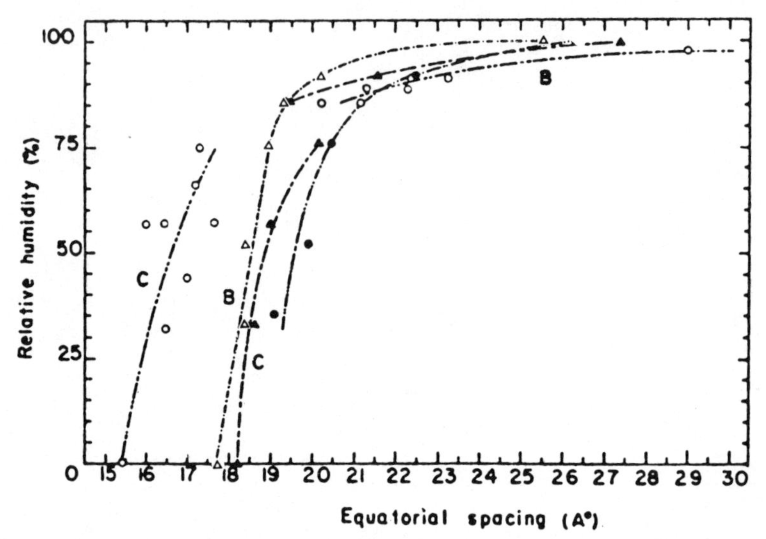

Fig. 7-4 Effect of relative humidity on the spacing of the first equatorial reflection of DNA, salmon spermheads, and of the complexes of poly-L-lysine and poly-L-arginine with DNA. DNA conformation of each material is indicated by B or C. Data for free DNA are from Feughelman et al. (1955), and Marvin et al. (1961). ——————0 DNA, ————▲ poly-L-lysine-DNA, –·–·△ poly-L-arginine-DNA, –··–··–● salman spermheads. Suwalsky and Traub (1972).

to maintain the ionically unperturbed B conformation. Reduction of this hydration to about 4 moles of H_2O leads, for most salts, to the production of the C and A forms. This change should result in DNA molecules that are more easily packaged (Marvin et al., 1961) in biological structures such as chromosomes and phage heads. This suggests that the physiologic mechanisms controlling the hydration of nuclear constituents may be involved in regulating cell division processes.

REFERENCES

Bram, S. (1971). *J. Mol. Biol.*, **58**, 277.

Brams, J. and W. H. N. M. Mommaerts (1964). *J. Mol. Biol.*, **10**, 73.

Chattoraj, D. K. and H. B. Bull (1971). *Arch Biochem. Biophys.*, **142**, 363.

Cohen, G. and H. Eisenberg (1968). *Biopolymers*, **6**, 1077.

Falk, M. and T. A. Ford (1966). *Can. J. Chem.*, **44**, 1699.

Falk, M., K. A. Hartman, Jr., and R. C. Lord (1962). *J. Am. Chem. Soc.*, **84**, 3843.

Falk, M., K. A. Hartman, Jr., and R. C. Lord (1963*a*). *J. Am. Chem. Soc.*, **85**, 387.

Falk, M., K. A. Hartman, Jr., and R. C. Lord (1963*b*). *J. Am. Chem. Soc.*, **85**, 391.

Falk, M., A. G. Poole, and C. G. Goymour (1970). *Can. J. Chem.*, **48**, 1536.

Feughelman, M., R. Langridge, W. E. Seeds, A. R. Stokes, H. R. Wilson, C. W. Hooper, M. H. F. Wilkins, R. K. Barely, and L. D. Hamilton (1955). *Nature*, **175**, 834.

Frisman, V., S. V. Veselkov, S. V. Slonitsky, L. S. Karavaev and V. I. Voroblev (1974). *Biopolymers*, **13**, 2169.

Fuller, W., M. H. F. Wilkins, H. R. Wilson, and L. D. Hamilton (1965). *J. Mol. Biol.*, **12**, 60.

Girod, J. C. Johnson, S. K. Huntington, and M. F. Maestre (1973). *Biochemistry*, **12**, 5092.

Hearst, J. E. (1962). *J. Mol. Biol.*, **4**, 415.

Hearst, J. E. (1965). *Biopolymers*, **3**, 57.

Hearst, J. E. and J. Vinograd (1961*a*). *Proc. Natl. Acad. Sci. USA*, **47**, 825.

Hearst, J. E. and J. Vinograd (1961*b*). *Proc. Natl. Acad. Sci. USA*, **47**, 1005.

Herskovits, T. (1962). *Arch. Biochem. Biophys.*, **97**, 474.

Herskovits, T. and E. Geiduschek (1961). *Arch. Biochem. Biophys.*, **95**, 114.

Herskovits, T., S. Singer, and E. Geiduschek (1961). *Arch. Biochem. Biophys.*, **94**.

Ivanov, V. I., L. E. Minchenkova, A. K. Schyolkina, and A. I. Poletayev (1973). *Biopolymers*, **12**, 89.

Killion, P. J. and L. H. Reyerson (1966). *J. Colloid Interface Sci.*, **22**, 582.

Langridge, R., D. A. Marvin, W. E. Seeds, W. E. Wilson, H. R. Hooper, M. H. F. Wilkins, and L. D. Hamilton (1960). *J. Mol. Biol.*, **2**, 38.

Luzzati, V. and A. Nicolaieff (1963). *J. Mol. Biol.*, **7**, 142.

Maniatis, T., J. Venable, Jr., and L. S. Lerman (1974). *J. Mol. Biol.*, **84**, 37.

Marvin, D. A., M. Spencer, M. H. F. Wilkins, and L. D. Hamilton (1961). *J. Mol. Biol.*, **3**, 547.

Raukas, E. (1965). *Biofizaka* (Eng. transl.), **10**, 455.

Suwalsky, M. and W. Traub (1968). *Israel J. Chem.*, **6**, Vp.

Suwalsky, M. and W. Traub (1972). *Biopolymers*, **11**, 2223.

Suwalsky, M., W. Traub, V. Shmueli, and J. A. Subirana (1969). *J. Mol. Biol.*, **42**, 363.

Tunis, M. J. and J. E. Hearst (1968*a*). *Biopolymers*, **6**, 1218.

Tunis, M. J. and J. E. Hearst (1968*b*). *Biopolymers*, **6**, 1325.

Tunis-Schneider, M. and M. Maestre (1970). *J. Mol. Biol.*, **52**, 521.

Vinograd, J., R. Greenwald, and J. E. Hearst (1965). *Biopolymers*, **3**, 109.

Wang, J. H. (1955). *J. Am. Chem. Soc.*, **77**, 258.

Wolf. B., A. Chan, and S. Hanlon (1972). *Fed. Proc. Fed. Am. Soc. Exp. Biol.*, **31**, 923.

Wolf, B. and S. Hanlon (1975). *Biochemistry*, **14**, 1661.

Zimmer, C., and G. Luck (1973). *Biochim. Biophys. Acta*, **312**, 215.

CHAPTER 8

DNA Intercalation Processes*

The intercalation of portions of certain molecules between base pairs of DNA and oligomers of nucleic acids is, relative to other classes of intermolecular interactions, well understood. This stems in part from the very obvious role molecular structure plays in dictating these interactions. Large planar groups are required for intercalation. However, it is now clear that many compounds that intercalate, and some that do not, also bind outside the double helix. The most important aspect of these binding/intercalating processes as they relate to carcinogenic versus anticancerous properties still remains a mystery. Clearly understanding the molecular thermodynamics and resulting structural organization is essential to explaining biologic function and using it to our advantage in these complexes.

* Also termed DNA *Interpolation* Processes

I ACTINOMYCIN-D–NUCLEIC ACID INTERACTIONS

A Proposed Interaction Models

Actinomycin D (Act-D), shown in Figure 8-1, is an important antibiotic that binds to double-stranded DNA and selectively inhibits RNA synthesis (Kirk, 1960; Kersten et al., 1960; Reich et al., 1961). The interaction between Act-D and DNA has been widely studied by a variety of techniques [e.g., see the reviews by Reich and Goldberg (1964); Goldberg and Friedman (1971); and Sobell (1973)]. There is a general requirement for the presence of a guanine base when Act-D binds to DNA. This specificity was illustrated in the structure of the cocrystalline complex of Act-D obtained by Sobell and co-workers (Sobell et al., 1971) where it was found that Act-D cocrystallized with two deoxyguanosine molecules. On the basis of this crystalline complex (Jain and Sobell, 1972), Sobell and Jain (1972) proposed a sterochemical model for the binding of the drug to DNA. This model involves an intercalation of the phenoxazone ring between adjacent base pairs of the double helix with the cyclic pentapeptide groups in the minor groove of the double helix consistent with the intercalation model proposed by Müller and Crothers (1968) on the basis of kinetic and hydrodynamic studies. In the crystalline complex,

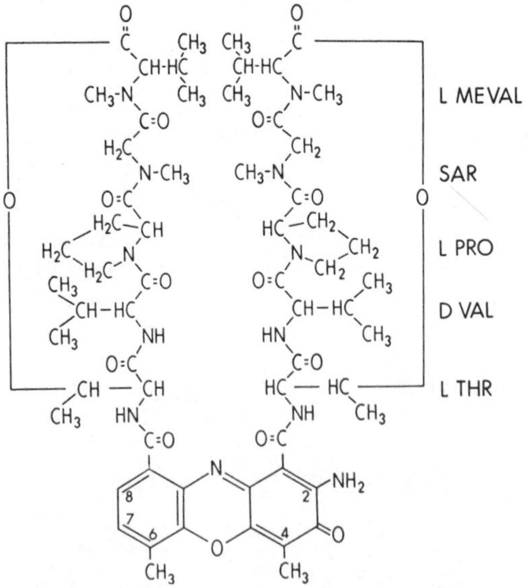

Fig. 8-1 Structural formula of actinomycin D. Abbreviations: Thr, threonine; Val, valine; Pro, proline; Sar, sarcosine; MeVal, methylvaline. Krugh (1974).

each of the guanine rings forms two hydrogen bonds with the L-threonine residue of one of the pentapeptide rings. From this observation, Sobell and Jain (1972) proposed that the general requirement for guanine in the binding of Act-D to DNA is a result of the hydrogen bond formation between the guanine base and the pentapeptide ring. This suggestion is partially supported by the early data of Wells and Larson (1970) in which they found that Act-D did bind the strongest to poly (dG-dC)·poly(dG-dC), but these workers also observed that Act-D could bind almost as tightly to other DNAs that contained only one guanine base at the intercalation site. It is also significant that Wells and Larson (1970) found that poly d(A-T-C)·poly d(G-A-T) did not bind Act-D, even though it contains guanine bases. Wells and Larson have concluded that the binding of Act-D is dependent on the conformation of the DNA molecule, which in turn is a function of the base sequence at the binding site. Krugh and Neely (Krugh, 1972; Krugh and Neely, 1973*a*,*b*) have evidence to suggest that the phenoxazone ring is able to discriminate between a guanine base and an adenine base, which suggests that the requirement for a presence of a guanine base is a result of the structure of the phenoxazone ring.

Sobell and Jain have proposed models for the 1:2 Act-D:d-pGpC and 1:2 Act-D:d-ApTpGpCpApT complexes (Sobell and Jain, 1972). In their model, the phenoxazone ring intercalates between GC and CG Watson-Crick base pairs, with the pentapeptide lactone rings extending over two base pairs on either side of the intercalation site. The symmetry axis relating the Act-D pentapeptide lactone ring coincides with the symmetry axis relating the sugar–phosphate backbone and base sequence of the double helix. The Sobell-Jain proposal has led to a systematic spectral investigation of complex formation between Act-D and d-pG (Arison and Hoogsteen, 1970; Krugh and Neely, 1973*b*; Patel, 1974*a*), d-pGpC (Krugh, 1972; Schara and Müller, 1972; Krugh and Neely (1973*b*); Patel, 1974*b*), and d-ApTpGpCpApT (Patel, 1974*c*) in aqueous solution. In particular, the NMR studies of Patel have been very useful in monitoring the intercalation and hydrogen bond interactions proposed to account for the specificity and stability of polypeptide–nucleic acid complexes (Sobell, 1973).

B 1 : 2 Act-D : d-pG and 1 : 2 Act-D : d-pGpC Complexes in Solution

Patel (1974*d*) has obtained proton and phosphorus NMR spectra as Act-D is gradually added to d-pGpC (or d-pG) aqueous solution. Beyond a 1:2 ratio of Act-D:d-pGpC (or Act-D:d-pG), the antibiotic precipitates. The data indicate that the stoichiometry of the water-soluble complex is probably 1 equivalent of antibiotic to 2 equivalents of dinucleotide (or mononucleotide). There is a shift in the proton resonances of the antibiotic between free and complexed forms. The lifetime of this exchange process is estimated to be ≪ 2 msec.

The purine and phenoxazone rings stack in the 1:2 Act-D:d-pG complex (Patel, 1974*a*). The stacking geometry in the complex has been evaluated on the basis of chemical shifts. Compared to a unique orientation of stacked phenoxazone and purine rings in the structure of the complex in the crystal (Jain and Sobell, 1972), a range of stacking geometries is suggested in aqueous solutions (Patel, 1974*a*). The experimental proton complexation shifts are consistent with the Sobell-Jain model (1972) of the stacking geometries in the 1:2 Act-D:d-pGpC complex. The geometry of a GC base pair–phenoxazone ring complex is illustrated in Figure 8-2. Schematic representations of the structures of 1:2 Act-D:d-pCpG and 1:2 Act-D:d-pGpC as derived form the Sobell-Jain proposal (1972) and supported by spectral measurements (Krugh, 1972; Schara and Müller, 1972; Krugh and Neely, 1973*b*) are presented in Figure 8-3. For the complex with d-pCpG, only the purine rings stack with the phenoxazone ring. In the complex with d-pGpC, however, the phenoxazone ring is intercalated between the GC and CG Watson-Crick base pairs. The intercalation results in a large ^{31}P downfield chemical shift of the phosphate resonances at the intercalating site (Patel, 1974*b*). It is proposed (Patel, 1974*d*) that the rates of rotation of the amino groups of A-NH_2 and G-NH_2 about the C—N bonds are decreased to an intermediate NMR exchange condition in the 1:2 Act-D:d-pG complex, compared to fast rotation on the NMR time scale for the uncomplexed species in solution. Such restricted rotations of amino groups could arise if

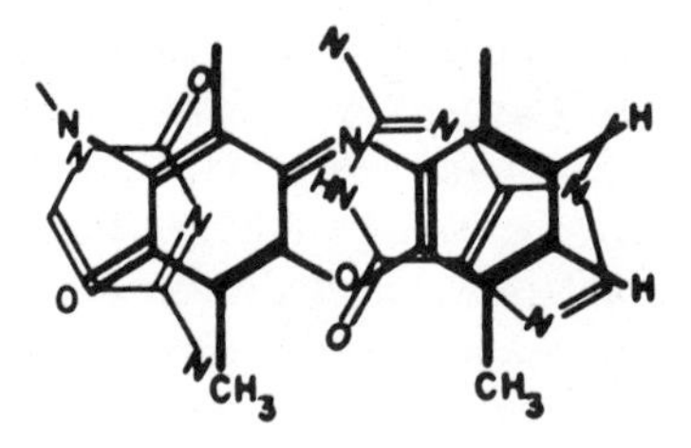

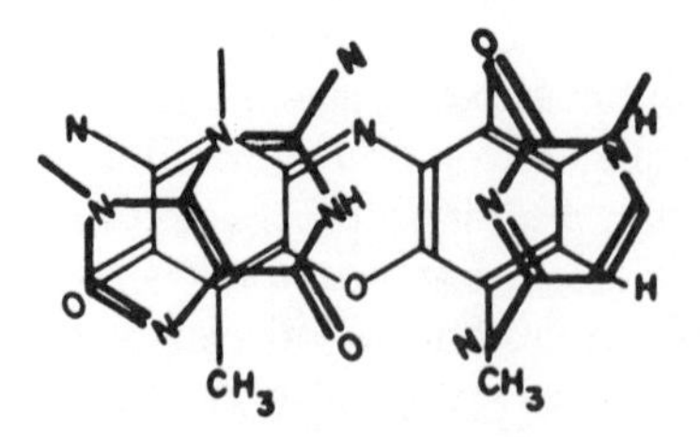

Fig. 8-2 The stacking orientations of the Watson-Crick GC base pairs above and below the phenoxazone ring of the antibiotic. Patel (1974*d*).

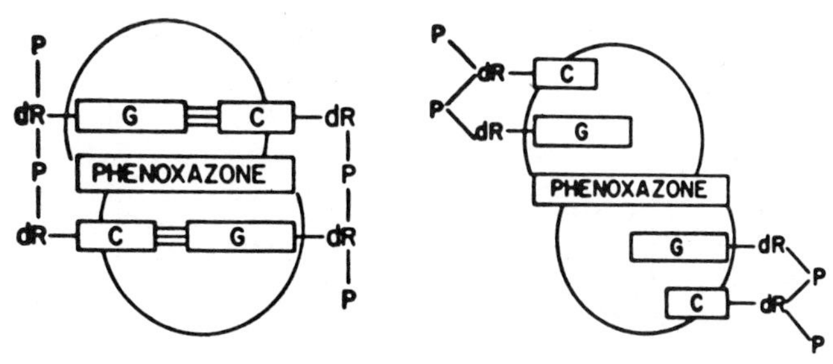

Fig. 8-3 (Left) Intercalation of the phenoxazone ring between GpC sequences in 1:2 Act-D:d-pGpC complex. (Right) The stacking of the guanosine rings with the phenoxazone ring in the 1:2 Act-D:d-dCpG complex. Patel (1974*d*).

their exchangeable protons participate in hydrogen bonds or are buried (Patel, 1974*a*). The temperature coefficient of the G-NH_2 protons at 6.4 ppm in the 1:2 Act-D:d-pG complex has a value of 2.4×10^{-3} ppm/°C, which is characteristic of hydrogen-bonded or buried protons (Patel, 1974*a*). The G-NH_2 proton of d-pGpC undergoes a downfield shift upon complexation with Act-D in aqueous solution. This shift, which persists over a temperature range of 10°–50°C, is attributed to the participation of the G-NH_2 protons in intermolecular hydrogen bonding with acceptor groups (Patel, 1974*b*).

C 1 : 2 Act-D : d-ApTpGpCpApT Complex in Solution

Complex formation between Act-D and the hexanucleotide d-ApTpGpC · pApT has been investigated by ^{1}H and ^{31}P NMR spectroscopy (Patel, 1974*c*). The proton spectra monitor the Watson-Crick base pairs (G-N_1H and T-N_3H resonances), and the phosphorus spectra monitor the sugar–phosphate backbone.

The proton NMR spectral changes resulting from the gradual addition of Act-D to d-ApTpGpCpApT in water (pH 7) at 0°C are shown in Figure 8-4. With the addition of 0.5 equivalent of the antibiotic, which amounts to one Act-D molecule per double-stranded hexanucleotide, chemically shifted resonances are observed that have the twofold symmetry of the oligonucleotide double helix removed in the spectrum of the complex. Addition of Act-D greater than 0.5 equivalent results in no additional spectral changes. The addition of 0.3 equivalent of Act-D (i.e., 0.6 Act-D molecules per double-stranded hexanucleotide) to d-ApTpGpCpApT yields a spectrum that is a superposition of the spectra of the hexanucleotide and the complex (Figure 8-4). This effect suggests that the exchange of the antibiotic between double-helical hexanucleotides is slow on the NMR time scale at

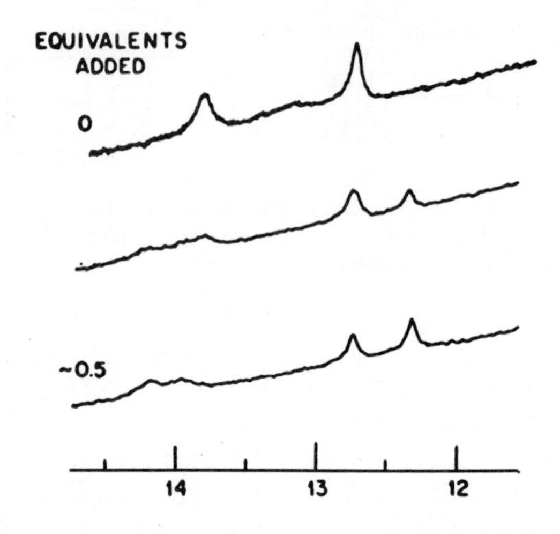

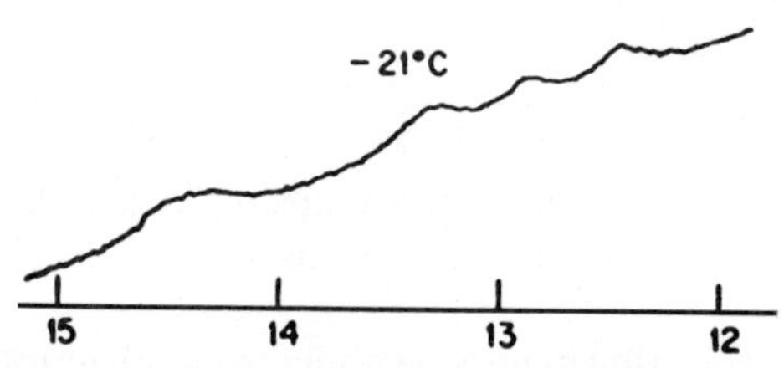

Fig. 8-4 (Top) The high-resolution 300 MHz proton NMR spectra of d-ApTpGpCpApT in H_2O (pH 7) at 0° C as a function of Act-D concentration. (Bottom) 1:2 Act-D:d-ApTpGpCpApT in H_2O:MeOH = 3:2 at −21° C. Patel (1974*d*).

0°C. Complex stability, therefore, increases on the order (1:2 Act-D:d-pG) < (1:2 Act-D:d-pGpC) < (1:2 Act-D:d-ApTpGpCpApT). The slow exchange of Act-D between free and complexed d-ApGpCpApApT, as indicated in the ^{31}P NMR spectra, suggests a tightly bound complex. One of the GpC phosphates faces the monopolar benzenoid ring of the phenoxazone, and the GpC phosphate on the other chain faces the polar quinonoid ring of the phenoxazone. A hydrogen bond through a water molecule has been pro-

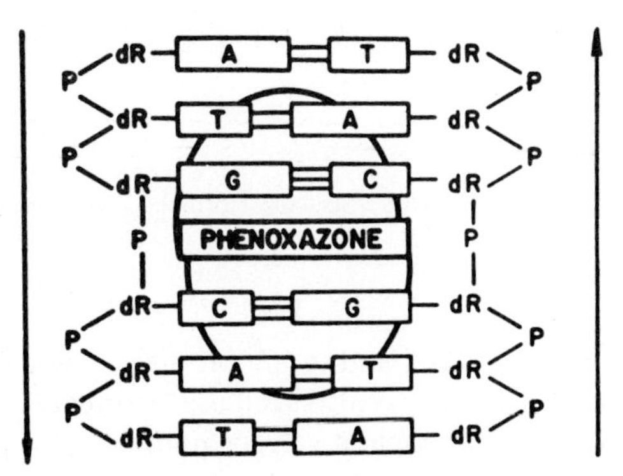

Fig. 8-5 A schematic of the 1:2 Act-D:d-ApTpGpCpApT complex with intercalation of the antibiotic between the $(GC)_{central}$ base pairs. Patel (1974*d*).

posed for the 2-amino group of the phenoxazone ring and its neighboring phosphate group (Sobell and Jain, 1972). Upon complex formation, a significant nonequivalence, 0.4 ppm, is observed for the G-N_1H protons of the $(GC)_{central}$ base pairs; a 0.2 ppm nonequivalence is observed for the T-N_3H protons of the (AT) internal base pairs, whereas the smallest nonequivalence is observed for the T-N_3H protons of the (AT) terminal base pairs (Patel, 1974*c*). The data strongly suggest that the asymmetrically substituted phenoxazone ring of Act-D binds at the $(GC)_{central}$ Watson-Crick base pairs in double-helical d-ApTpGpCp- ApT, as illustrated in Figure 8-5. Sobell and Jain (1972) postulated that the site of intercalation would also be between the GpC sequences in d-ApTpGpCpApT. The chemical shift changes that occur upon complexation also reflect the perturbed ring current contributions of neighboring base pairs when the helix unwinds to incorporate the intercalating residue. The unwinding process is discussed in greater detail in a later section. There is evidence that the 1:2 Act-D:d-ApTpGpCpApT complex melts at a higher temperature than the double-stranded d-ApT · pGpCpApT. The GpN_1H resonances of the (GC) central base pairs observed in the spectrum of the complex at 25°C are broadened at this temperature in the spectrum of the hexanucleotide.

II INTERCALATION OF ETHIDIUM BROMIDE

Ethidium bromide (see Figure 8-6) is in a class of phenanthridinium drugs that has been found to be useful in the treatment of certain trypanosome infections (Newton, 1964). Its medicinal action stems from the ability of the

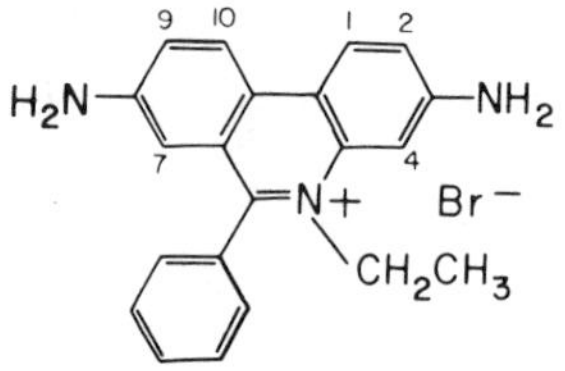

Fig. 8-6 Structural formula of ethidium bromide. Tsai et al. (1975).

drug to bind to DNA and RNA and to inhibit nucleic acid function (Elliot, 1963; Waring, 1964). This drug has also been used to create covalently closed circular DNA molecules that contain various numbers of supercoils in their structure (Wang, 1969). These molecules have served as useful probes in understanding the nature of protein–DNA interactions and control mechanisms that regulate RNA transcription (Maniatis and Ptashne, 1973; Saucier and Wang, 1972).

Ethidium bromide–nucleic acid complexes have been extensively studied by a variety of techniques (Elliot, 1963; Le Pecq et al., 1964; Waring, 1965, 1966; Ward et al., 1965; Le Pecq and Paoletti, 1967; Dalgleish, 1971). It is generally accepted that the strong mode of binding of ethidium bromide to double-stranded nucleic acids occurs by intercalation of the planar phenanthridinium ring between adjacent base pairs. Viscometric and hydrodynamic measurements with linear DNA molecules indicate a lengthening and stiffening of the helix in the presence of increasing concentrations of the drug (Waring, 1964), an interpretation directly supported by electron-microscopic measurements (Freifelder, 1971). The ability of this drug to unwind circular supercoiled DNA is also considered to be another sensitive criteria for intercalative binding (Crawford and Waring, 1967; Baur and Vinograd, 1968). This is thought to reflect unwinding of the DNA helix by the drug at the immediate site of intercalation and is discussed in detail later in this chapter. Fiber X-ray diffraction studies have been carried out on the complex, and a tentative molecular model for ethidium–DNA interaction has been proposed (Fuller and Waring, 1964). However, the intercalation model is not universally accepted, and other models have been proposed (Pritchard et al., 1966), including one in which the aromatic heterocyclic rings are located on the outside of the helix (Gurskii, 1966).

Tsai and co-workers (1974) have reported the successful cocrystallization of ethidium with several self-complementary dinucleoside monophosphates and have solved the three-dimensional structure of one of these, [ethidium: 5-iodouridyly] (3′–5′) adenonsinel, to atomic resolution by X-ray crystallography (Tsai et al., 1975). Figures 8-7 and 8-8 show a portion of the ethidium: 5-iodouridylyl-(3′–5′) adenosine crystal structure (hereafter denoted ethidium: ioU-A). The structure consists of two ioU-A molecules (dark solid bonds) held together by adenine·uracil Watson-Crick base pairing.

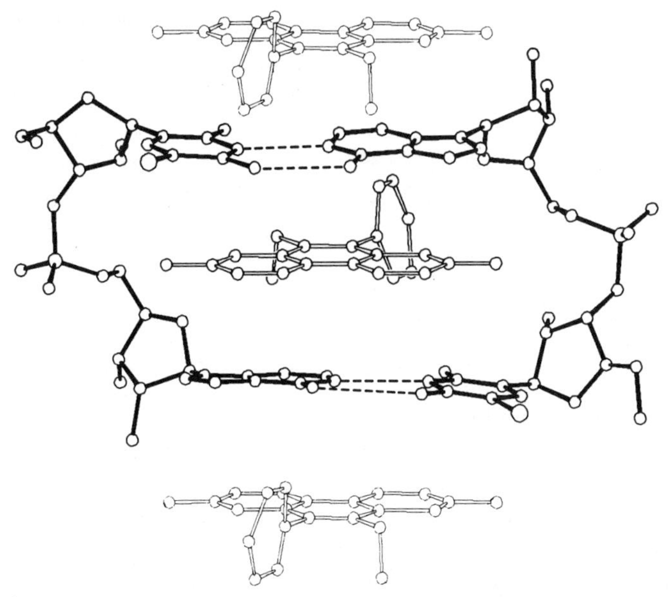

Fig. 8-7 A computer-drawn illustration of a portion of the ethidium:ioU-A crystal structure viewed approximately parrallel to the planes of the adenine-uracil base pairs and ethidium molecules. ioU-A molecules are drawn with dark solid bonds; the intercalative ethidium molecule is shown with dark open bonds; stacked ethidium molecule(s) are drawn with light open bonds. Hydrogen bonding between adenine·uracil base pairs has been indicated by broken lines. Tsai et al. (1975).

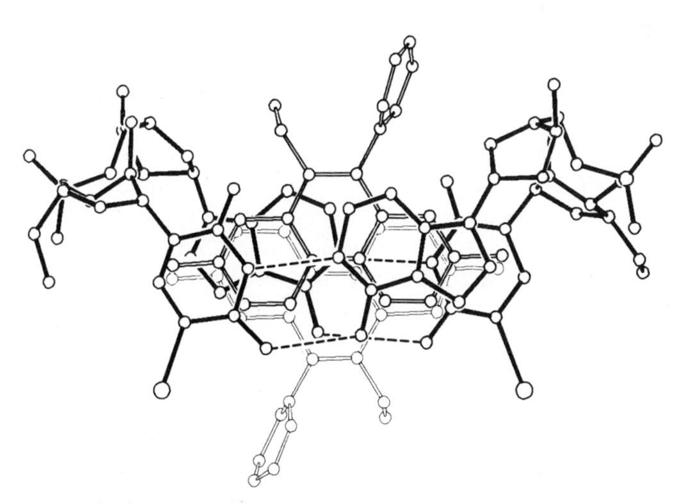

Fig. 8-8 A computer-drawn illustration of the ethidium:ioU-A complex viewed perpendicular to the planes of the adenine·uracil base pairs and ethidium molecules. This figure illustrates the noncrystallographic twofold symmetry that exists in this model drug–nucleic acid interaction. Tsai et. al. (1975).

Adjacent base pairs within this paired ioU-A structure and between neighboring ioU-A molecules in adjoining unit cells are separated by 6.8 Å. This separation results from intercalative binding by one ethidium molecule (dark open bonds) and stacking by other ethidium molecules (light open bonds) above (and below) the base pairs. The intercalative ethidium is oriented such that its phenyl and ethyl substituents lie in the narrow groove of the miniature ioU-A double helix. Both amino groups on this ethidium molecule are juxtaposed to the adenosine 0-5′ phosphodiester oxygen, and their contacts (3.3 Å; 3.5 Å) suggest possible weak electrostatic and hydrogen bonding interactions. The stacked ethidium lies in the opposite direction; its phenyl and ethyl groups neighbor iodine atoms on uracil residues. The amino groups on this ethidium molecule are not in immediate apposition to the charged phosphate groups, and instead form hydrogen bonds to neighboring water molecules. The distance that separates glycosidic carbon atoms between base pairs is of particular interest; this distance corresponds to the interchain separation in DNA and RNA. This value is 10.4 Å for both base pairs and may be significantly shorter than that found in DNA and RNA and in single crystal studies of self-complementary dinucleoside monophosphates (Fuller et al., 1965; Rosenberg et al., 1973; Day et al., 1973). The relative angular orientation of base pairs within the paired dinucleotide, as estimated from the relative twist between vectors connecting glycosidic carbon atoms within each base pair, is about 7°, which is significantly smaller than the corresponding twist angle in DNA (36°), double helical RNA (32.7°), and in the single crystal dinucleoside monophosphate studies (about 32°). The small angular twist between base pairs reflects the presence of the intercalative ethidium molecule. This corresponds to an unwinding of double-helical nucleic acid polymers at the immediate site of drug intercalation. Conformational changes in the sugar–phosphate chains accompany this unwinding. These conformational changes partly reflect the differences in ribose sugar ring puckering that are observed (both iodouridine residues have 3C′ endo sugar conformations, whereas both adenosine residues have C2′ endo sugar conformations), as well as small but systematic changes in the torsional angles that describe the phosphodiester linkage and the C4′–C5′ bond.

A considerable amount of water structure surrounds the columns of ethidium: ioU-A complexes stacked in the *y* direction of the crystal lattice. Twenty water molecules have been located in the asymmetric unit; many form hydrogen-bonded water–water tetrahedral—like structures and water–hydroxyl linkages to the sugar–phosphate chains. The complex is heavily solvated in the crystal lattice, which suggests that the association of these compounds in the solid state is not significantly different from their solution associations (Krugh et al., 1975). In addition, the binding of ethidium to this dinucleoside monophosphate demonstrates what may be a more general principle in the binding of small symmetric (or pseudosymmetric) ligands to DNA—the use of symmetry in drug–DNA interactions.

Although the binding of ethidium to DNA and RNA shows no detectable base specificity, spectral studies with ethidium and a series of ribodinucleoside monophosphates and deoxyribodinucleotides carried out recently have indicated a marked preference for ethidium binding. Krugh et al. (1975) have examined solution complexes of ethidium bromide with nine different deoxydinucleotides and the four self-complementary ribodinucleoside monophosphates, as well as mixtures of complementary and noncomplementary deoxydinucleotides, as models for the binding of the drug to DNA and RNA. They conclude that ethidium bromide is able to discriminate and preferentially bind to deoxydinucleotide and ribodinucleoside monophosphate sequence isomers. The CpG binding curve is unique to both the deoxy and ribose dinucleotide titrations at 25°C; this nucleotide binds the strongest of all the nucleotides tested. The ribodinucleoside monophosphates generally appear to bind slightly stronger than their deoxy analogs. However, it isn't quite fair to directly compare the deoxy and ribodinucleotide data, because the terminal phosphate of the dinucleotides increases the solubility of the compounds and decreases the strength of the interaction. A sigmoidal shape of the titration curves indicates that ethidium bromide forms a complex with more than one CpG molecule in a cooperative manner. The similarity of the circular dichroism and visible spectra of the complexes of ethidium bromide with certain dinucleotides to the spectra reported for DNA and RNA provides a reasonable indication that the dinucleotides (both deoxy and ribo) serve as good model compounds for the nucleic acids. Corresponding to the solid state, it is reasonable that, in solution, complementary nucleotides form a minihelix around an ethidium bromide molecule.

Krugh et al. (1975) conclude that ethidium bromide may bind to the various sequences available as intercalation sites on DNA and RNA with significantly different binding constants. This is consistent with the formation of crystalline complexes that have been isolated. Complexes have readily been obtained between ethidium: C-G and ethidium: U-A, as well as with ioC-G and ioU-A. However, repeated attempts to obtain complexes with the reverse sequences (i.e., G-C and A-U) have been unsuccessful. This may indicate a sequence specificity in these dinucleotide studies of the pyrimidine-purine type, and it is possible that a similar sequence specificity (although, clearly, not an absolute one) exists in ethidium: DNA binding.

Prior to crystallographic studies, Fuller and Waring (1964) made two interesting predictions concerning the sterochemistry of ethidium intercalation. The first concerns the magnitude of the unwinding angle that accompanies ethidium intercalation; this was estimated in their model to be about $-12°$. The second concerns the relative orientation of the ethidium molecule with respect to the base pairs; the phenyl and ethyl groups of ethidium arc postulated to lie in the wide groove of the DNA helix. As discussed previously, the relative angular orientation between adenine·iodouracil base pairs in the intercalated ethidium: ioU-A structure is about 7°. If the crystallographic studies involving oligomeric nucleotide by Tsai et al. (1975) can be carried over to the RNA and DNA polymer studies, the unwinding angle is −26° for RNA intercalation and −29° for DNA intercalation, values considerably larger than the Fuller-Waring figure. DNA molecules containing various degrees of superhelicity have suggested the unwinding angle to be about −26° for ethidium intercalation (Wang, 1974). An important variable in understanding the polynucleotide conformation in intercalation is the ribose sugar ring puckering that has been assumed to be C2′ endo for all nucleotide residues in the Fuller-Waring model. The ethidium: ioU-A structure, however, demonstrates a mixture of C2′ and C3′ endo sugar conformations, and this explains in part the angular unwinding difference that is observed. Lastly, the intercalative ethidium is oriented in a manner exactly opposite to the Fuller-Waring model; the phenyl and ethyl group of ethidium are found to lie in the narrow groove of the miniature double helix.

III AMINOACRIDINES AND GENERAL CHARACTERISTICS OF INTERCALATION

A Biologic Significance

A reliable picture is emerging of the nature of the complex between nucleic acids and the aminoacridines, an important group of bacteriostatic and mutagenic agents. It has been shown, for example, that the inhibition of protein synthesis by proflavine may be connected with its ability to interact with transfer RNA in particular (Finkelstein and Weinstein, 1967). However, other causes are sometimes possible for the biologic effects of the aminoacridines. It appears that the inhibition of *E. coli* RNA-polymerase by proflavine can occur not only by virtue of its interaction with the DNA primer, which is necessary for the reaction, but also by direct interaction between the proflavine and the binding sites on the RNA polymerase that interact with the nucleotide bases of the primer DNA and with the substrate nucleotide triphosphates (Nicholson and Peacocke, 1966). The antibacterial action of the aminoacridines has long been the subject of intensive study (Albert, 1966; Hinshelwood, 1946; Dean and Hinshelwood, 1966). Their ability to induce mutations, apparently by causing deletion or insertion of a single nucleotide in DNA, has also been of high interest (Brenner et al., 1961; Crick et al., 1961). Aminoacridines, with their cationic charge and three flat aromatic rings, have structural features similar to those of other compounds whose interaction with DNA is of much interest, such as carcinogens (both polycyclic hydrocarbons and benzacridines); certain antibiotics; nucleic acid derivatives (purines, nucleosides); histologic dyes, for example, those with three flat, fused rings such as pyronin, toluidine blue, and triphenylmethane dyes such as methyl green; phenanthridine trypanocides, for example, ethidium bromide; and, of course, other acridine derivatives, many of which are noted for their antimalarial activity, for example, atebrin. See Figure 8-9 for the molecular structures of some of these compounds. As

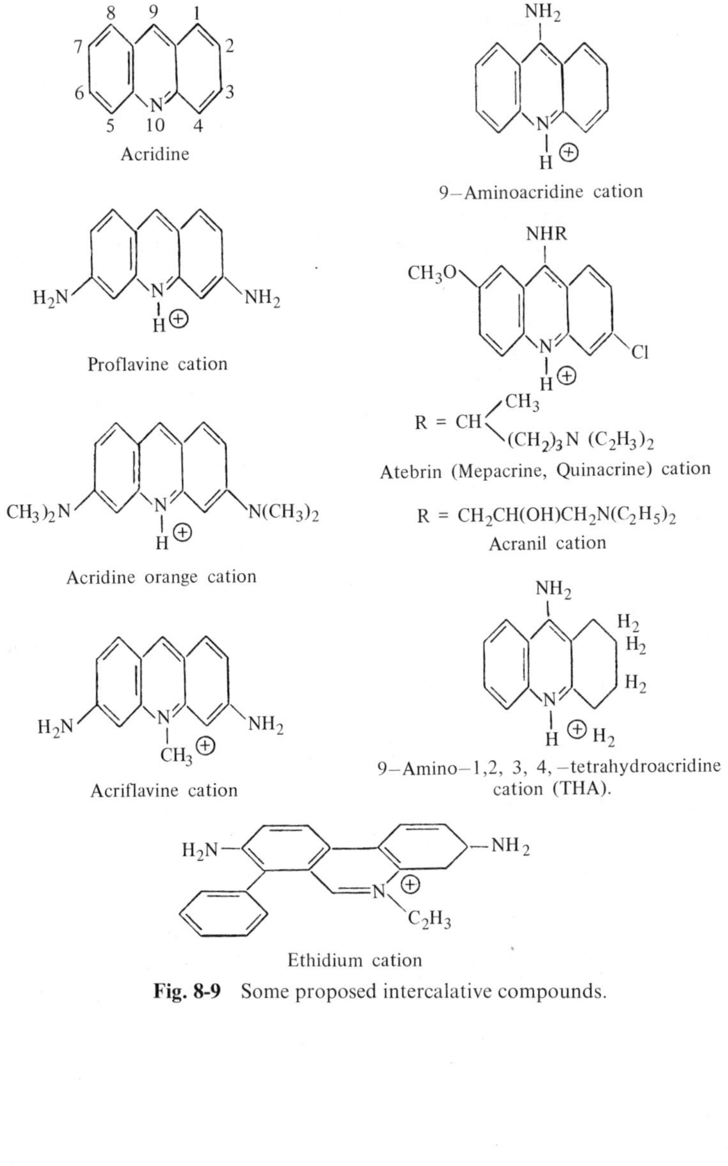
8 9 1
7 2
6 3
N
5 10 4
Acridine
NH2
N
H ⊕
9–Aminoacridine cation
H2N
N
NH2
H⊕
Proflavine cation
NHR
CH3O
N
Cl
H⊕
R = CH
CH3
(CH2)3N (C2H3)2
Atebrin (Mepacrine, Quinacrine) cation
CH3)2N
N
N(CH3)2
H⊕
Acridine orange cation
R = CH2CH(OH)CH2N(C2H5)2
Acranil cation
NH2
H2
H2
H2
N
H ⊕ H2
9–Amino–1,2, 3, 4,–tetrahydroacridine
cation (THA).
H2N
N
NH2
CH3 ⊕
Acriflavine cation
H2N—
—NH2
N ⊕
C2H3
Ethidium cation

Fig. 8-9 Some proposed intercalative compounds.

discussed in this chapter, aminoacridines and compounds, such as ethidium bromide, have almost identical effects in untwisting and extending the DNA double helices. They can be employed deliberately to modify the shape of DNA molecules.

The interaction of these small molecules with nucleic acids is representative of a wider class of interactions of small molecules with biologic macromolecules; such interactions are likely to be one of the central areas of interest in many biochemical problems, and the methods and ideas developed for this particular system should therefore be of wider import.

B General Binding Properties

The binding curves of proflavine and other aminoacridines on nucleic acids exhibit the following features: The form of the binding curve is that of (A + B) of Figure 8-10, and plots of bound compound per mole of nucleic acid phosphorus, r, over compound concentration, c, (r/c) versus r, are curved even at very high ionic strengths indicating that the heterogeneity is present even when electrostatic effects have been suppressed (Peacocke and Skerrett, 1956). There is general agreement that the binding process can be divided into two categories: (I) a process by which up to 0.1–0.2 molecules of aminoacridine per nucleotide are bound strongly, with a ΔG of -6 to -9 kcal/mole aminoacridine, and (II) a weak process by which further aminoacridines are bound up to the electroneutrality limit of $r = 1.0$ (Peacocke and Skerrett, 1956).

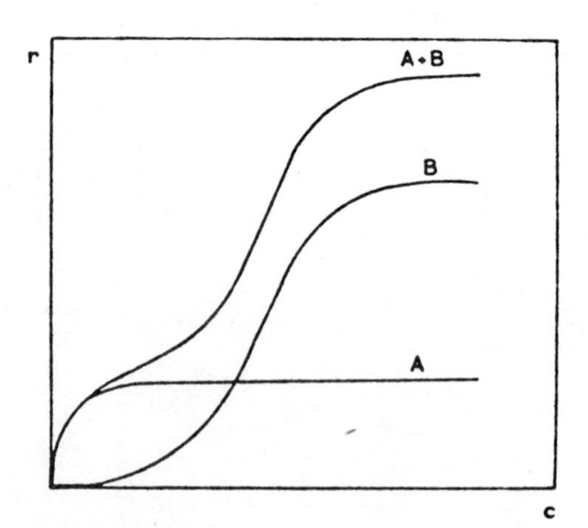

Fig. 8-10 Plots of r vs. c:(a) A = single site binding (b) B = cooperative binding, A + B = sum of A and B. Blake and Peacocke (1967).

The binding curves of ethidium bromide are linear r/c versus r plots, but the corresponding plots for aminoacridines are usually curved. This inherent heterogeneity of the strong binding (I) has been attributed to either a continuous change in the binding constant as a result of continuous modification of the nucleic structure on binding or to a structural heterogeneity in the binding sites dependent on their base composition (Tubbs et al., 1964; Gersch and Jordan, 1965). The form of the binding curves suggests that the weaker binding process II is a cooperative one, whereby the binding of one aminoacridine cation facilitates the binding of the next, especially as it only occurs at values of c at which self-aggregation occurs in proflavine solutions alone. Increase in ionic strength diminishes the extent of overall binding, but the binding by process II is more sensitive to such changes than binding by process I (Peacocke and Skerrett, 1956). Hence it has been concluded that electrostatic forces contribute greatly to both binding processes but are relatively more important in (II) than in I. The effect of temperature on the complex is complicated. Binding of proflavine on DNA decreases with rise in temperature (Chambron et al., 1966*a*; Gersch and Jordan, 1965; Walker, 1965), and this can be explained partly by a decrease in the number of binding sites. However the ΔH of binding is not constant and also decreases with temperature (Chambron et al., 1966*a*). In addition, a cooperative decrease in binding occurs at a temperature that is higher than the actual melting temperature of DNA when binding proflavine at ionic strengths, $\mu = 0.01$ and 0.1 (Chambron et al., 1966*b*).

The effect of disorganization of the double-helical structure of DNA, by heating and cooling rapidly to 0°C, on its ability to bind aminoacridines is of interest in regard to whether the intact double helix is necessary for binding. There is an increase of binding of proflavine by DNA after denaturation at $\mu < 0.01$, which may be underestimated (Drummond et al., 1965). The binding by process II is strongly enhanced relative to binding by I as a result of the denaturation of DNA, as well as by lowering μ. The cooperative binding by II therefore outweighs the strong binding I even at quite low c, when $\mu < 0.01$ and the DNA is denatured. The destruction of the long runs of double helices in the DNA by heating does not reduce the strong binding of proflavine by process I, implying that such helical structures are not a necessary condition of the strong binding I. However, Chambron et al., (1965, 1966*b*) report a marked reduction in binding of proflavine that is coincident, at $\mu = 1.0$, with the breakdown of the double-helical DNA structure. This implies that denatured DNA does not bind proflavine at these high temperatures (>90°C). The effect of denaturation on the relative binding by native and heat-denatured DNA was extended to the binding of 9-aminoacridine, and the evidence shows that denaturation appears to cause some increase in strong binding by process I. The effects on the interaction caused by denaturation with agents other than heat has not been studied extensively. At pH 2.8, DNA still binds proflavine, as has been demonstrated by means of absorption optics (Blake et al., 1967). Thus, acid-denatured DNA at acid pH 2.8 tends to bind proflavine more by II than by I. This observation has been related to the loss, on lowering the pH, of the optical activity that is induced in proflavine when it is bound to DNA (Blake and Peacocke, 1967).

Binding by weak process II may be identified with binding as aggregates. The tendency to bind by process II is apparently enhanced at low ionic strength and by disorganization of the double-helical DNA structure either by heating and shock-cooling or by lowering the pH. Bradley and Wolf (1959) studied in detail the aggregation of dyes like acridine orange on nucleic acids and other polyelectrolytes, and introduced the concept of "stacking," which may be loosely described as the enhanced tendency of a ligand to bind next to a ligand already bound, that is, a cooperative phenomenon. This suggests a model in which the planes of the aminoacridine are parallel to the base rings of the DNA and are stacked on top of each other in the same direction as the DNA axis, with their positive ring nitrogens close to the external phosphate groups of the DNA chains. The forces between bound proflavine molecules are probably weaker than those between DNA and bound proflavine. The tendency of proflavine to interact with proflavine in a direction parallel to the DNA axis should be markedly less than with acridine orange. That this tendency is not absent is shown by the evidence that proflavine on nucleic acids can indeed exhibit an aggregate-type spectrum with the accompanying loss of a good isobestic point. But this tendency is markedly less than with acridine orange. For these reasons it is unwise to assume that all binding by process II, which is "weak" in the energetic sense, is caused by stacking, in Bradley's sense, though it is usually true that all stacked acridines are weakly bound, energetically speaking.

Only those aminoacridines strongly interact with DNA which are in the fully cationic form at the pH (normally 6–8) of the binding experiments, which usually means that all amino groups that may be substituted are present at the 3-, 6-, or 9- positions of the acridine ring (Morthland et al., 1954). It has also been shown that a minimum planar area of 38 $Å^2$ (corresponding to the three rings of acridine) is required in the acridine ring for bacteriostatic activity (Morthland et al., 1954). Hydrogenation of one ring of 9-amino-acridine to form the quinoline derivative, 9-amino-1,2,3,4-tetrahydroacridine (THA), practically eliminates its bacteriostatic activity (Albert, 1966) and reduces its maximum capacity to bind to DNA by process I by more than one-half (Drummond et al., 1965), presumably because of its possession of a bulky, buckled ring in place of one of the aromatic rings of 9-aminoacridine. When long and bulky side chains are attached to the 9-amino group of 9-aminoacridine (as in acranil and atebrin), the binding is not reduced but somewhat enhanced (Drummond et al., 1965). The structure of the complex must be such that the 9-position is free to attach long side chains without detriment to the interaction. The dependence of ability to bind to nucleic acids on the basicity of the acridine has been studied by Lober (1968), who also shows that alkylation of the ring nitrogen (e.g., N-methylation of proflavine to acriflavine) enhances the binding ability of an aminoacridine to an extent not very dependent on the length of the alkyl chain.

It has long been known (Kurnick, 1950; Kurnick and Mirsky, 1950) that changes in secondary structure of the nucleic acid can have considerable influence on the binding process as revealed by the observed metachromasy, the displacement of the absorption spectrum of the bound dye. As discussed earlier, denaturation of DNA causes an increase in the proportion of acridine orange that is bound in the "stacked" form. Denaturation by heating, followed by shock-cooling, causes no decrease in the tendency of DNA to bind proflavine by process I, but enhances the binding by process II, which corresponds to the stacking of acridine orange.

Ribosomal RNA exhibits the same type of binding curve as DNA, although it binds less proflavine, and the inflection point, at $\mu = 0.1$, is at about $r = 0.09$ instead of at $r \simeq 0.2$, as with DNA (Peacocke and Skerrett, 1956; Blake and Peacocke, 1967). This suggests that the maximum binding of proflavine by process I is, for RNA, about half of its value with DNA.

Studies on the polarized fluorescence of atebrin bound to flowing DNA (at $r \simeq 0.015$) show that the plane of the atebrin ring is nearer to that of the bases than to that of the helical axis of DNA, affirming that in the complex the planes of the rings are within $\sim 30°$ of those of the DNA bases (Lerman, 1963). Flow dichroism studies on the complexes of acridine derivatives with DNA show that the DNA itself is slightly more dichroic when the acridine is bound; thus the DNA bases are not tilted from their usual relation to their helical axis (Lerman, 1963; Nagata, 1966). Again, the actual experimental evidence and the incompleteness of orientation of the DNA by flow only allow the assertion that the plane of the acridine ring is within $\pm 30°$ of that of the DNA bases. The change of circular dichroism on flowing complexes of acridine orange and DNA indicates that the acridine orange cations are roughly halfway between perpendicular and parallel to the DNA helix axis. However, the value of r was 0.5; thus at least half the acridine orange was bound by process II, and this evidence could therefore not give an unequivocal indication of the orientation of acridine orange with respect to DNA when bound by process I (Mason and McCaffrey, 1964). The transition from process I to process II binding of proflavine involves a progressive change from an orientation of proflavine approximately perpendicular to the DNA axis to a more disordered state, as demonstrated by the observation of Houssier and Fredericq (1966). When the DNA in nucleohistone is oriented in an electrical field, there is a progressive loss in the negative dichroism of the bound proflavine as the value of r increases.

Although the contour length of DNA is increased as a result of the binding of aminoacridine, the persistence length, which is a measure of the stiffness of the DNA, is in fact decreased from 700 to 300 Å. The absence of a specific increase in the viscosity of denatured DNA on interaction with aminoacridines suggests that the increase in contour length on strong binding of aminoacridines by process I is a specific feature of the interaction with the intact double-helical structure. An increase in contour length has been confirmed by autoradiography, (Cairns, 1962), and light scattering (Mauss et al., 1967) reveals a corresponding increase in the radius of gyration. This expansion is accompanied by a decrease in the mass per unit length along the helical direction, according to low-angle X-ray measurement (Luzzati et al., 1961), and by a loss of the hypochromicity of the DNA at 260 mμ (Walker, 1965; Weill and Calvin, 1963). X-ray diffraction patterns obtained from fibers of the DNA-proflavine complex are (Lerman, 1961) qualitatively consistent with a model in which the two DNA strands both untwist and extend on interacting with DNA, and similar conclusions have been drawn from such studies (Fuller and Waring, 1964) on the complexes of ethidium bromide and DNA discussed in section II of this chapter. The interaction of aminoacridines with native DNA produces not only the configurational changes mentioned above, but it also stabilizes the macromolecule towards thermal denaturation (Chambron, 1965; Gersch and Jordan, 1965; Cairns, 1962; Freifelder et al., 1961; Lerman, 1964) mainly through its electrostatic effect (Chambron et al., 1966*b*).

Pritchard et al. (1966) have proposed that, when acridine cations are bound by process I, the acridine lies between successive nucleotide bases on the same polynucleotide chain, in a plane approximately parallel to the base planes, but at an angle (looking down the polynucleotide chain) such that the positive ring nitrogen is close to the polynucleotide phosphate group (see Figure 8-11). The general picture of process II binding is that of acridine cations attached approximately edgewise and externally to the double helix of DNA, with their positive ring nitrogen atoms close to the phosphate groups. When r is large enough, the acridine rings can stack up on each other in a direction parallel to the helix axis. However, when r is as large as this, the helix will probably be disordered by strong process I binding with intercalation; thus the direction of the helix axis may be ill defined, and no effect on the viscosity of DNA should be anticipated.

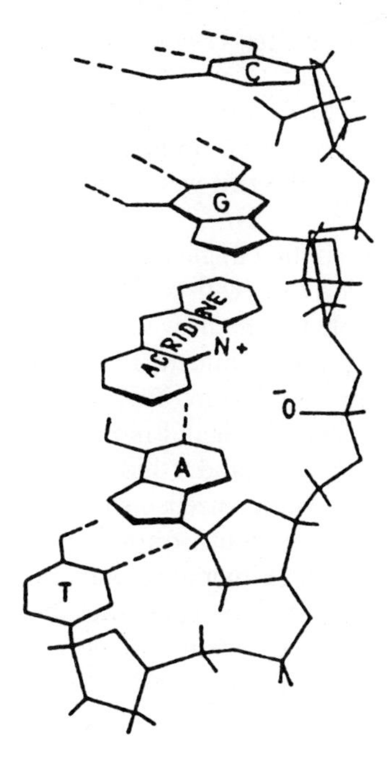

Fig. 8-11 Modified intercalation model for the complex of proflavine with polynucleotide chains. Pritchard et al. (1966).

IV UNWINDING OF DNA BY INTERCALATION AND BINDING INTERACTIONS

Intercalating dyes, especially ethidium, have been used extensively in the determination of the degree of superhelicity of covalently closed, double-stranded DNA (Crawford and Waring, 1967; Bauer and Vinograd, 1968; Wang, 1969, 1971). The binding of increasing amounts of a substance that unwinds the DNA helix, either by intercalation or by some other mechanism, to a DNA containing negative superhelical turns initially causes a reduction in the number of negative superhelical turns. This reduction continues until the DNA molecule contains no superhelical turns. Binding of more of this substance after this "zero turn point" introduces positive superhelical turns. By determining the number of molecules of this substance bound per DNA molecule at the zero turn point, the number of negative superhelical turns per DNA molecule in the absence of this substance can be readily calculated, provided that the unwinding angle per ligand bound is known.

Clearly, such titrations can be done to evaluate the relative winding or unwinding angles of ligands that alter the helix rotation angle of DNA. If Φ_A and Φ_B are the winding angles of ligands A and B, respectively, and n_A

molecules of A, and n_B molecules of B are needed in each case to titrate a given negative superhelical DNA to the zero turn point, then $n_A\Phi_A = n_B\Phi_B$. The relative winding angles Φ_A/Φ_B is simply n_B/n_A. Such titrations were first done by Bauer and Vinograd (1968) for ethidium and 9-aminoacridine. These workers concluded that the winding angles of the two were approximately the same.

The most common technique to determine the winding angle begins with the step in which ligand X is mixed with a cyclic DNA with a few ligase joinable single-chain scissions in a medium suitable for the ligase reaction. The number of ligand bound per nucleotide, v_x, is determined. Ligase is then added to convert the DNA to the covalently closed form. This step is referred to as the closure step. After the closure step, all ligand molecules are removed, and the DNA is titrated with a reference ligand, ethidium, to determine the number of bound ethidium v_E per nucleotide needed to reach the zero turn point. This step is referred to as the ethidium titration step. The number of superhelical turns of the DNA molecule, either at the closure step or at the zero turn point of the titration step, is zero. Therefore, the requirement that the topological winding number be a constant for a closed DNA (Vinograd and Lebowitz, 1966; Glaubiger and Hearst, 1967) can be written as

$$Nv_x\Phi_x + \frac{N}{4\pi}\theta^\circ(\text{c.s.}) = Nv_E\Phi_E + \frac{N}{4\pi}\theta^\circ(\text{t.s.}) \quad (8\text{-}1)$$

$$v_x\Phi_x = v_E\Phi_E + \frac{1}{4\pi}(\theta^\circ(\text{t.s.}) - \theta^\circ(\text{c.s.})) \quad (8\text{-}2)$$

where N is the number of nucleotides per DNA, and θ° is the average helix rotation angle per base pair. The scripts (c.s.) and (t.s.) refer to the quantity at the closure step and the zero turn point of the titration step. All angular quantities in equation 8-1 are in units of radians. Suppose that, for the special case $v_x = 0$, v_E is found to be v_E°. The quantity $(1/4\pi)(\theta^\circ(\text{t.s.}) - \theta^\circ(\text{c.s.}))$ can then be readily evaluated from equation 8–2 as done previously for the determination of the θ° dependence on temperature and ionic media (Wang, 1969). Substituting $(1/4\pi)(\theta^\circ(\text{t.s.}) - \theta^\circ(\text{c.s.})) = -v_E^\circ\Phi_E$ into equation 8-2,

$$v_x\Phi_x = (v_E - v_E^\circ)\Phi_E \quad (8\text{-}3)$$

and

$$\Phi_x\Phi_E = (v_E - v_E^\circ)/v_x \quad (8\text{-}4)$$

Equation 8-4 allows a direct evaluation of the relative winding angles of various ligands.

If the binding of X involves more than one class of sites, with v_x^i moles of x bound per nucleotide to sites of class i, it is easy to show that Φ_x in the equations above is the number average of the winding angle:

$$\Phi_x = \sum v_x^i \Phi_x^i / \sum v_x^i \Phi_x^i / v_x \qquad (8\text{-}5)$$

Therefore, in principle it is possible to compare the change of the torsion of the DNA helix induced by two different intercalating molecules by taking advantage of the physicochemical properties of covalently closed circular DNA.

Unfortunately, a complication arises from the fact that several intercalating molecules are able to bind to DNA on at least two different binding sites—an intercalation site (process I) leading to a change of the torsion of the DNA helix on one hand, and an outside binding site (process II) with an assumed small effect on the torsional angle between DNA base pairs (Saucier et al., 1971). Therefore, when the amount of two different intercalating molecules necessary to relax a given circular DNA are not equal, one does not know whether the proportions of these molecules bound to DNA by intercalation inside the double helix and outside the helix are different, or whether the winding angles are truly different.

Table 8-1 contains a list of compounds that have been shown to intercalate with DNA along with their molecular weights (used as a rough measure of molecular bulk), apparent winding angle, apparent fraction of actually intercalated molecules of the total that interact with DNA, and an estimated 1-octanol/water partition coefficient, P, expressed as $\log(P)$ for convenience. The usefulness of $\log P$ in rationalizing molecular interactions is discussed in detail in Chapter 3, but, succintly, it is a measure of a compound's thermodynamic preference for an aqueous ($\sim$polar) environment or a lipid ($\sim$hydrophobic) medium. The more positive $\log P$, the greater the drive for hydrophobic surroundings.

Using the data in Table 8-1, an attempt to correlate molecular weight or log Ps with either unwinding angle or fraction intercalated is not successful. The magnitudes of the unwinding angles and fraction intercalated, are, as one might expect, directly related. The information in Table 8-1 roughly taken in composite suggests (1) that the sterochemistry and/or environmental thermodynamics of the actual intercalating unit of the molecule, for example, the planar ring moeties, dictates the degree of DNA unwinding, (2) the binding free energy of the nonintercalating portion of the compound coupled with the location of the binding sites competing against the intercalation free energy species the fraction of intercalated compound, (3) that a two-state polar/hydrophobic partitioning is not a realistic model for the nonintercalated/intercalated mediums for DNA helices, and (4) that the intercalation properties are extremely sensitive to small changes in molecular geometry (see the ellipticine analogs in Table 8-1). This last observation might be viewed as a complement to the studies presented in the section

Table 8-1 Some Properties of DNA-intercalating Compounds

Compound	Reference	Molecular Weight	Apparent Unwinding Angle	Apparent Fraction Inter-calated[a]	Calculated Log P^b Neutral	Calculated Log P^b Charged
Ethidium bromide[c]	Waring (1970) Saucier et al. (1971)	344	12°	1.00	3.75–4.35[d]	0.62–1.02
Proflavin	Waring (1970) Saucier et al. (1971)	209	8.4° ± 2.4°	0.7 ± 0.2	0.62–1.02	0.62–1.02
Hycanthone	Waring (1970)	357	6.8° ± 2.7°	0.56 ± 0.22	3.52–3.92	3.06–3.46
Daunomycin	Waring (1970) Saucier et al. (1971)	527	5.2° ± 1.4°	0.44 ± 0.12	(−6.41)–(−5.21)	(−2.46)–(−1.46)
Adriamycin[e]	—	544	—	—	(−6.92)–(−5.12)	(−2.46)–(1.76)
Nogalamycin[f]	Waring (1970)	788	8.1° ± 2.3°	0.68 ± 0.18	—	—
Actinomycin-D	Waring (1970)	1255	11.4° ± 3.°	0.95 ± 0.25	(−11.63)–(−11.03)	0.33–0.93
Quinocrine	Saucier et al.	509	8°	0.66	4.73–5.13	1.55–2.15
Propidium iodide	Waring (1970)	541	12° ± 3.4°	1.0(−) 0.28	7.35–7.75	0.62–1.02
Ellipticine	Le Pecq et al. (1974)	246	9°	0.75	4.80	3.44–3.84
9-methoxy-ellipticine	Saucier et al. (1971) Le Pecq et al. (1974)	276	6.8°	0.56	4.13–4.33	3.33–3.53
N-emthyl-ellipticine	Le Pecq et al. (1974)	260	10.2°	0.85	5.10–5.30	3.40–3.84
N-methyl-9-methoxy-ellipticine	Le Pecq et al. (1974)	290	5°	0.42	4.43–4.53	3.33–3.53
9-hydroxy-ellipticine	Le Pecq et al. (1974)	262	12°	1.0	3.21–3.41	2.41–2.61

[a] The intercalation of one molecule increases the length of the DNA helix by 3.4 Å. Thus a measure of the increased length of the helix by viscosimetry (Cohen and Eisenberg, 1969) can be used to compute the number of intercalated molecules. The ratio of this number to the total number of bound molecules is the apparent fraction intercalated (Saucier et al., 1971).

[b] Estimated from π constants (see Chapter 3 for details). The log *P*s for both the neutral and totally ionized forms of the intercalating compounds were computed.

[c] The accepted reference standard. Unfortunately, there is growing evidence that a value of 12° for the unwinding angle is not correct. Paoletti and LePecq (1971) have suggested a value of 13°, and Pulleyblank and Morgan (1975) have estimated the unwinding angle to be 24°–36°. This range of values is consistent with the crystallographic findings of Tsai et al. (1975) reported in this chapter. They estimate the absolute value of the unwinding angle to be 26°–29°. Obviously, if 12° is not correct, all other apparent unwinding angles in this table must be modified in accordance with equation 8-4.

[d] There is ambiguity as to how one should pick the π fragments of the compounds. This leads to an estimated range of log *P*s. See Chapter 3 for details.

[e] The unwinding angle and fraction intercalated are not measured for this compound. Its overall similarity to duanomycin suggests that it has correspondingly similar intercalation properties.

[f] There is some uncertainty about the molecular structure. Hence log *P*s have not been estimated.

for Act-D interacting with model nucleic acid compounds. In those studies, changes in nucleic acid geometry, for example, choice of base pairs and/or base-pair stacking sequences were shown to result in significant alterations to intercalations. In these studies, changes in the geometry of the probable intercalation unit of the drug compounds alter intercalation behavior. The ellipticines as DNA-intercalating, anticancerous drugs are discussed in some detail with respect to structure–activity relationships in the next section.

V DNA–ANTICANCER DRUG INTERACTIONS

Over 600,000 compounds have been tested for their anticancerous properties. Of these, only about 50 have been officially recognized by the National Institutes of Health as useful in the therapy of human cancer (Wood, 1971). It is very striking that the mechanism of action of over two-thirds of these compounds is related directly or indirectly to DNA. Several of these compounds have been observed to be intercalating agents with DNA. Actinomycin, daunomycin, and adriamycin (Barthelemy-Clavey, et al., 1973; Gellert et al., 1965) (see the previous sections) are examples of DNA intercalating drugs that have strong anticancerous properties.

Nevertheless, there is no a priori reason that DNA intercalating drugs, as any other cytotoxic compound, might have anticancerous properties. One can readily imagine how such molecules might kill a cell, but it is much more difficult to explain how they could selectively destroy cancerous cells. There is no obvious explanation as to what molecular properties cause one DNA intercalating agent to be carcinogenic and another anticancerous (Arcos and Argus, 1974). But it is a fact that cytotoxic compounds can be highly selective for cancerous cells. For instance, the cytotoxic compound BCNU (Bischlorethyl nitroso urea) is able to divide the number of malignant L1210 cells in mice by a factor of the order of 10^6–10^7 without affecting the animal host and leads to the cure of animals injected with massive quantities of tumor cells (Skipper et al., 1965; Schabel et al., 1966). One possible explanation of the high selectivity is that the DNA of cancerous cells is unable to complex with histones or make complexes of low stability. The free DNA would obviously be a preferential target for the intercalating drugs. Moreover, unstable DNA–histone complexes would be dissociated only by drugs that have the largest DNA affinity. In this theory, the ability to intercalate in DNA confers to a drug nothing but a potential cytotoxic property. The anticancerous property is a consequence of the specific ability of the cancerous cells to let the drug reach its target. The experimental evidence favoring this point of view has been reviewed (Werkheiser and Moran, 1973; Golman, 1973).

Extensive physicochemical studies in various systems have shown that, if a set of definite stereochemical conditions are fulfilled, the intercability of a given compound in DNA can be reasonably well predicted Le Pecq, 1971). Among the implicated parameters are the size, shape, planarity, charge, and, more generally, the electronic properties of the molecules. A particular set of compounds that intercalate in DNA with high affinity (Festy, et al., 1971) are derivatives of pyrido- (3-4b) carvozole. Some derivatives of this series (ellipticine) are known to be active both in experimental tumors (Dalton et al., 1967, 1969; Svoboda, et al., 1968; Jewers, et al., 1972) and in man (Mathe, 1970). Figure 8–12 shows the structure of ellipticine and some derivatives whose DNA intercalating properties have been investigated (Le Pecq, et al., 1974). In Table 8-2 are (1) the pks of the different compounds in water, (2) the DNA binding constants, K_{ap}, of the ellipticine derivatives (Le Pecq and Paoletti, 1967), (3) the equilibrium constant $k(E^+)$ for the cationic form of the compounds, (4) the unwinding angle of DNA that is a measure of the extent of conformational change induced by intercalation (Wang, 1969, 1971; Bauer and Vinograd, 1968; Revet et al., 1971; Waring, 1970), and (5) the anticancerous activity of the different ellipticine derivatives as measured using the L1210 leukemia, an experimental tumor, in mice (Le Pecq et al., 1973).

The most striking example of the importance of the pk on the interaction with DNA is seen in a comparison of ellipticine and nor-11-methyl-ellipticine. The removal of a single methyl group from ellipticine in position 11 causes the pk of the molecule to drop by 3 units. The consequence is that DNA affinity of the cations of the two derivatives is almost identical. The biologic activity of ellipticine disappears after demethylation.

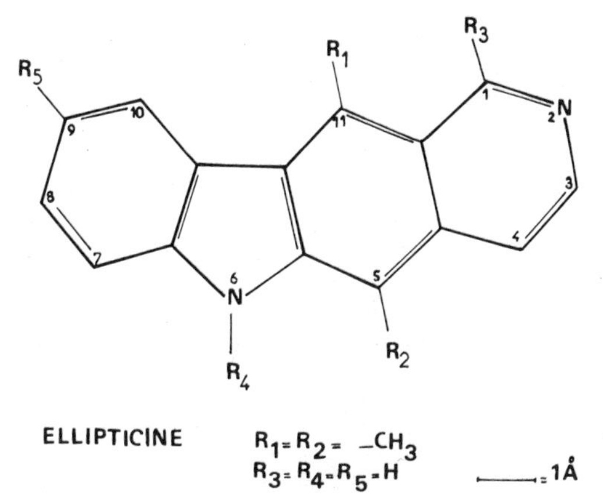

Fig. 8-12 Structure of the ellipticine derivatives. LePecq et al. (1974).

Table 8-2 Physicochemical Properties and Anticancerous Activity of Ellipticine Derivatives. (K_{ap} are measured at 25°C in 0.1 M NaCl, Tris-HCl Buffer, 0.1 M, pH 7.4, With Calf Thymus DNA)

Product	pK	K_{ap}(pH 7.4)	log $K(E^+)$	Unwinding angle	Pharmacological Activity (% of L1210 cells killed) by a Third of the DL50 Dose
N-isopentyl-ellipticine	4.7	$<10^4$ [a]	6.3	8.8[b]	0
N-isopentyl-methoxy-ellipticine	4.5	$<10^4$ [a]	6.7	—	—
nor-5,11-diméthyl-ellipticine	6.35	1.0×10^4	5.08	—	0
nor-11-méthyl-ellipticine	6.30	2.4×10^4	5.52	—	0
9-methoxy-ellipticine	6.8	1.0×10^5	5.7	6.8	90
ellipticine	9.1	1.5×10^5	5.2	9	94
9-bromo-ellipticine	6.1	4.0×10^5	6.92	0[c]	0
N-méthyl-ellipticine	6.1	4.0×10^5	6.92	10.2	92
9-amino-ellipticine	9.8	1.2×10^6	6.08	4	0
N-méthyl-9-methoxy-ellipticine	6.45	2.0×10^6	7.3	5	50
9-hydroxy-ellipticine	9.8	2.0×10^6	6.15	12	99.96

[a] Those compounds are insoluble at pH 7.4. log $K(E^+)$ is deduced from K_{ap} measurement done at pH 5.0 (0.1 *M* NaCl, 0.1 *M* acetate buffer, pH 5.0).

[b] Measurement done at pH 5.0 (NaCl, 0.1 *M*, acetate buffer pH 5.0). All other unwinding angle determinations done in 0.1 *M* NaCl, 0.1 *M* Tris-HCl, pH 7.4 at 25°C.

[c] 9-bromo-ellipticine does not intercalate.

LePecq et al. (1974).

In all cases studied, log $K(E^+)$ is increased by at least 1 unit as soon as the sixth position is substituted by a methyl or isopropyl group. When an electrophilic substituent is placed at position 9, the DNA binding constant increases tenfold (see the 9-amino and 9-hydroxy ellipticine data). It is postulated from stereochemical analysis that the electrophilic substituent is able to directly interact, possibly through a hydrogen bond, with the negatively charged oxygen of the DNA phosphate group, which is close to that position in the intercalated complex.

It is tempting to propose a cause-effect relationship between the pharmacologic activity of intercalating molecules and the type of conformational changes they induce in DNA as measured by the unwinding angle. This is especially so because the extent of conformational change can be quite variable (Saucier et al., 1971; Paoletti and Le Pecq, 1971). However, such a correlation is probably not meaningful. For example, the lack of activity of 9-amino-ellipticine is probably not related to the fact that its intercalation causes an unwinding of the DNA-helix of only 4°, whereas 9-hydroxy-ellipticine unwinds the DNA helix by 12° per bound molecule. This opinion results from the fact that daunomycin, an active anticancerous DNA intercalating agent, also causes only a 4° unwinding of the DNA-helix after intercalation (Saucier, 1971).

REFERENCES

Albert, A. (1951, 1966). *The Acridines*. Arnold, London.

Arcos, J. C. and M. F. Argus (1974). *Chemical Induction of Cancer*. Academic Press, New York, Vols. IIa and IIB.

Arison, B. H. and K. Hoogsteen (1970). *Biochemistry*, **9**, 3976.

Barthelemy-Clavey, V., J. C. Maurizot, and P. J. Sicard (1973). *Biochimie*, **55**, 858.

Bauer, W. and J. Vinograd (1968). *J. Mol. Biol.*, **33**, 141.

Blake, A., H. Fuijita, and A. R. Peacocke (1967). University of Oxford, unpublished observations.

Blake, A. and A. R. Peacocke (1967). *Biopolymers*, **5**, 383.

Bradley, D. F. and M. K. Wolf (1959). *Proc. Natl. Acad. Sci. USA*, **45**, 944.

Brenner, S., L. Barnett, F. H. C. Crick, and A. Prgel (1961). *J. Mol. Biol.*, **3**, 121.

Cairns, J. (1962). *Cold Spring Harbor Symp. Quant. Biol.*, **27**, 579.

Chambron, J. (1965). Thesis, University of Strasbourg.

Chambron, J., M. Daune, and C. Sadron (1966*a*). *Biochim. Biophys. Acta*, **123**, 306.

Chambron, J., M. Daune, and C. Sadron (1966*b*). *Biochim. Biophys. Acta*, **123**, 319.

Cohen, G. and H. Eisenberg (1969). *Biopolymers*, **8**, 45.

Crawford, L. V. and M. J. Waring (1967). *J. Mol. Biol.*, **25**, 23.

Crick, F. H. C., L. Barnett, S. Brenner, and R. J. Watts-Tobin (1961). *Nature*, **192**, 1227.

Dalgeish, D. G., A. R. Peacocke, G. Fey, and C. Harvey (1971). *Biopolymers*, **10**, 1853.

Dalton, L. K., S. Demerac, B. C. Elmes, J. W. Loder, J. M. Swan, and T. Teitei (1967). *Austral. J. Chem.*, **20**, 2715.

Day, R. O., N. C. Seeman, J. M. Rosenberg, and A. Rich (1973). *Proc. Natl. Acad. Sci. USA*, **70**, 849.

Dean, A. C. R. and C. N. Hinshelwood (1966). *Growth, Function and Regulation in Bacterial Cells*. Clarendon Press, Oxford.

Drummond, D. S., V. F. W. Simpson-Gildemeister, and A. R. Peacocke (1965). *Biopolymers*, **3**, 135.

Elliot, W. H. (1963). *Biochem. J.*, **86**, 562.

Festy, B., J. Poisson, and C. Paoletti (1971). *FEBS Letters*, **17**, 321.

Finkelstein, T. and I. B. Weinstein (1967). *J. Biol. Chem.*, **242**, 3663.

Freifelder, D. (1971). *J. Mol. Biol.*, **60**, 401.

Freifelder, D., P. F. Dacison, and E. P. Geiduschek (1961). *Biophys. J.*, **1**, 389.

Fuller, W. and M. J. Waring (1964). *Ber. Bunsenges. Physik. Chem.*, **68**, 805.

Fuller, W., M. H. F. Wilkins, H. R. Wilson, and L. D. Hamilton (1965). *J. Mol. Biol.*, **12**, 60.

Gellert, M., L. E. Smith, D. Neville, and G. Felsenfeld (1965). *J. Mol. Biol.*, **11**, 445.

Gersch, N. F. and D. O. Jordan (1965). *J. Mol. Biol.*, **13**, 138.

Glaubiger, D. and J. E. Hearst (1967). *Biopolymers*, **8**, 691.

Goldberg, I. H. and P. A. Friedman (1971). *Ann. Rev. Biochem.*, **40**, 772.

Golman, I. D. (1973). In *Drug Resistance and Selectivity* (Enrico Mihich, Ed.). Academic Press, New York, p. 299.

Gurskii, G. V. (1966). *Biofizika*, **11**, 737.

Hinshelwood, C. N. (1946). *The Kinetics of Bacterial Change*. Clarendon Press, Oxford.

Houssier, C. and E. Frederieq (1966). *Biochim. Biophys. Acta*, **120**, 434.

Jain, S. C. and H. M. Sobell (1972). *J. Mol. Biol.*, **68**, 1.

Jewers, K., A. H. Manchanda, and H. M. Rose (1972). In *Progress in Medicinal Chemistry* (G. P. Ellis and G. B. West, Eds.). Buttersworths, London, p. 1.

Kersten, W., H. Kersten, and H. M. Rauen (1960). *Nature* (*Lond.*), **187**, 60.

Kirk, J. (1960). *Biochim. Biophys. Acta*, **42**, 167.

Krugh, T. R. (1972). *Proc. Natl. Acad. Sci. USA*, **69**, 1911.

Krugh, T. R. (1974). In *Molecular and Quantum Pharmacology* (E. Bergmann and B. Pullman, Eds.). D. Reidel, Dordrecht-Holland, p. 465.

Krugh, T. R. and J. W. Neely (1973*a*). *Biochemistry*, **12**, 1775.

Krugh, T. R. and J. W. Neely (1973*b*). *Biochemistry*, **12**, 4418.

Krugh, T. R., F. N. Wittlin, S. P. Cramer (1975). *Biopolymers*, **14**, 197.

Kurnick, N. B. (1950). *J. Gen. Physiol.*, **33**, 243.

Kurnick, N. B., A. E. Mirsky (1950). *J. Gen. Physiol.*, **33**, 265.

Le Pecq, J. B. (1971). In *Methods of Biochemical Analysis* (David Glick, Ed.). Wiley-Interscience, New York, Vol. 20, p. 41.

Le Pecq, J. B., C. Gosse, N. Dat-Xuong, and C. Paoletti (1973). *C. R. Acad. Sci.*, **D277**, 2289.

Le Pecq, J. B., M. Le Bret, C. Gosse, C. Paoletti, O. Chalvet, and N. Dat-Xuong (1974). In *Molecular and Quantum Pharmacology* (E. Bergmann and B. Pullman, Eds.). D. Reidel, Dordrecht-Holland, p. 515.

Le Pecq, J. B. and C. Paoletti (1967). *J. Mol. Biol.*, **27**, 87.

Le Pecq, J. B., P. Yot, and C. Paoletti (1964). *C. R. Acad. Sci.*, **259**, 1786.

Lerman, L. S. (1961). *J. Mol. Biol.*, **3**, **18**.

Lerman, L. S. (1963). *Proc. Natl. Acad. Sci. USA*. **49**, 94.

Lerman, L. S. (1964). *J. Mol. Biol.*, **10**, 367.

Lober, G., unpublished data (see Blake & Peacocke, 1967).

Luzzati, V., F. Masson, and L. S. Lerman (1961). *J. Mol. Biol.*, **3**, 634.

Maniatis, T. and M. Ptashne (1973). *Proc. Natl. Acad. Sci. USA*, **70**, 1531.

Mason, S. F. and A. J. McCaffrey (1964). *Nature*, **204**, 468.

Mathe, G., M. Jayat, F. De Vassal, L. Schwarzenberg, M. Schneider, J. R. Schlumberger, C. Jasmin, and C. Rosenfeld (1970). *Rev. Europ. Etudes Clin. Biol.*, **15**, 541.

Mauss, Y., J. Chambron, M. Daune, and H. Benoit (1967). *J. Mol. Biol.*, **27**, 579.

Morthland, F. W., P. P. H. deBruyn, and N. H. Smith (1954). *Expt. Cell Res.*, **7**, 201.

Müller, W. and D. M. Crothers (1968). *J. Mol. Biol.* **35**, 251.

Nagata, C., M. Kodama, Y. Tagashira, and A. Imamura (1966). *Biopolymers*, **4**, 409.

Newton, B. A. (1964). In *Advances in Chemotheraphy* (A. Goldin and F. Hawkins, Eds.). Academic Press, New York, p. 35.

Nicholson, B. H. and A. R. Peacocke (1966). *Biochem. J.*, **100**, 50.

Paoletti, J. and J. B. Le Pecq (1971). *Biochimie*, **53**, 969.

Paoletti, J. and J. B. Le Pecq (1971). *J. Mol. Biol.*, **59**, 43.

Patel, D. J. (1974*a*). *Biochemistry*, **13**, 1476.

Patel, D. J. (1974*b*). *Biochemistry*, **13**, 2388.

Patel, D. J. (1974*c*). *Biochemistry*, **13**, 2396.

Patel, D. J. (1974*d*). In *Peptides, Polypeptides and Proteins* (E. R. Blout, F. A. Bovey, M. Goodman and N. Lotan, Eds.). Wiley-Interscience, New York, p. 459.

Peacocke, A. R. and J. N. H. Skerrett (1956). *Trans. Faraday. Soc.*, **52**, 261.

Pritchard, N. J., A. Blake, and A. R. Peacocke (1966). *Nature*, **212**, 1360.

Pulleyblank, D. E. and A. R. Morgan (1975). *J. Mol. Biol.*, **91**, 1.

Reich, E., R. M. Franklin, and A. J. Shatkin, E. L. Tatum (1961). *Science*, **134**, 556.

Reich, E. and I. H. Goldberg (1964). *Progr. Nucl. Acid Res.*, **3**, 183.

Revet, B., M. Schmir, and J. Vinograd (1971). *Nature* (*New Biol.*), **229**, 10.

Rosenberg, J. M., N. C. Seeman, J. J. P. Kim, F. L. Suddath, H. B. Nicholas, and A. Rich (1973). *Nature*, **243**, 150.

Saucier, J. M., B. Festy, and J. B. Le Pecq (1971). *Biochimie*, **53**, 973.

Saucier, J. M. and J. C. Wang (1972). *Nature* (*New Biol.*), **239**, 167.

Schabel, F. M., H. E. Skipper, W. R. Laster, Jr., M. W. Trader, and S. A. Thompson (1966). *Cancer Chemother. Rep.*, **50**, 55.

Schara, R. and W. Müller (1972). *Eur. J. Biochem.*, **29**, 210.

Skipper, H. E., F. M. Schabel, and W. S. Wilcox (1965). *Cancer Chemother. Rep.*, **45**, 5.

Sobell, H. M. (1973). *Progr. Nucl. Acid Res.*, **13**, 153.

Sobell, H. M. and S. C. Jain (1972). *J. Mol. Biol.*, **68**, 21.

Sobell, H. M., S. C. Jain, T. D. Sakore, and C. E. Nordman (1971). *Nature* (*Lond.*), **231**, 200.

Svoboda, G. H., G. A. Poore, and M. L. Montfort (1968). *J. Pharmacol. Sci.*, **57**, 1720.

Tsai, C., S. C. Jain, C. H. Lam, and H. M. Sobell (1974). American Crystallographic Association Abstracts, March 24–28, University of California, Berkeley, M1.

Tsai, C.-C., S. C. Jain, and H. M. Sobell (1975). *Proc. Natl. Acad. Sci. USA*, **72**, 628.

Tubbs, R. K., W. E. Ditmars, and Q. van Winkle (1964). *J. Mol. Biol.*, **9**, 545.

Vinograd, J. and J. Lebowitz (1966). *J. Gen. Physiol.*, **49**, 103.

Walker, I. O. (1965). *Biochim. Biophys. Acta*, **109**, 585.

Wang, J. C. (1969). *J. Mol. Biol.*, **43**, 25.

Wang, J. C. (1969). *J. Mol. Biol.*, **43**, **263**.

Wang, J. C. (1971). *Biochim. Biophys. Acta*, **232**, 246.

Wang, J. C. (1974). *J. Mol. Biol.*, **89**, 783.

Ward, D. C., E. Reich, and I. H. Goldberg (1965). *Science*, **149**, 1259.

Waring, M. J. (1964). *Biochim. Biophys. Acta*, **87**, 358.

Waring, M. J. (1965). *J. Mol. Biol.*, **13**, 269.

Waring, M. J. (1966). *Biochim. Biophys. Acta*, **114**, 234.

Waring, M. J. (1970). *J. Mol. Biol.*, **54**, 247.

Weill, G. and M. Calvin (1963). *Biopolymers*, **1**, 401.

Wells, R. D. and J. E. Larson (1970). *J. Mol. Biol.*, **49**, 913.

Werkheiser, W. C. and R. G. Moran (1973). In *Drug Resistance and Selectivity* (Enrico Mihich, Ed.). Academic Press, New York, p. 1.

Wood, H. B., Jr. (1971). *Cancer Chemother. Rep.* **2**, 9.

CHAPTER 9

Complexes of DNA with Synthetic Basic Polypeptides: Nucleoprotamine and Nucleohistone Models

The limited amount of structural evidence concerning the interaction of histones with DNA suggests that the complexing linkages occur in the basic residue-rich regions of the histones. Thus several workers have logically chosen to study the interactions between basic synthetic polypeptides with DNA in order to characterize the histone–DNA association structural chemistry. This chapter reports the findings of such modeling studies.

I INTRODUCTION AND GENERAL PROPERTIES OF NUCLEOHISTONES

DNA is found in vivo in association with the basic proteins, protamines, and histones. Protamines are less heterogeneous than histones and are simpler in amino acid composition; they are notable for their very high arginine contents (Bush, 1965). This has led to the view that nucleoprotamines may have relatively simple structures (Feughelman et al., 1955) that may be conveniently modeled by complexes of DNA with synthetic basic polypeptides (Wilkins, 1956). In addition, although simple basic polypeptides like polylysine may not represent the behavior of nuclear proteins that contain tertiary structure, it is reasonable to assume that such basic polypeptides interact with DNA predominantly through a similar mechanism as the segments rich in basic amino acids of histones. This hypothesis is supported by the findings on nucleohistones by Boublik et al. (1971). From their results, a general pattern of behavior emerges in that the basic regions of the polypeptide chains are the primary sites of interaction with DNA, whereas the regions that contain a high proportion of apolar and acidic residues are folded in specific globular structures. Thus the primary sites of interaction between basic proteins and DNA can be simulated by the complex between polylysine and DNA. These complexes have been investigated by different workers by means of X-ray (Wilkins, 1956; Feughelman et al., 1968; Suwalsky and Traub, 1968; 1972) and spectroscopic techniques (Haynes et al., 1970; Leng and Felsenfeld, 1966; Shapiro et al., 1969; Olins et al., 1968; Inoue and Ando, 1968; 1970; Zama and Ichimura, 1971; 1973; Chang et al., 1973), leading to different structural hypotheses. The X-ray results seem to leave little doubt that DNA exists in the high-humidity B-type form, whereas the large changes in optical activity suggest (if asymmetric scattering phenomena may be excluded) a number of hypotheses that involve modifications in the DNA conformation.

II X-RAY STUDIES FOR POLYLYSINE

The most detailed X-ray diffraction pattern of nucleoprotamine was obtained by Suwalsky and Traub (1968; 1972) at 92%, with spacings and intensity distribution closely resembling that reported for Sepia sperm (Feughelman et al., 1955; Wilkins, 1956). The pattern is also similar to that given by semicrystalline NaDNA at the same humidity (Wilkins, 1956). It may therefore be concluded that these features correspond to the DNA moiety of the nucleoprotamine, which conforms to the in vivo B form. However, there are two notable differences between the nucleoprotamine pattern and that of DNA. One is the intensity on the first layer line, which is appreciably stronger in the complex. Second, although both show hexagonal unit cells at 92% relative humidity, NaDNA has three molecules in the cell, whereas nucleoprotamine has only one molecule in a cell with $a = 26.3$ Å and $c = 33.4$ Å.

Decreasing the humidity causes the patterns to become more diffuse, yielding very little structural data. At 0% relative humidity, not a single reflection is observed. Exposing the nucleoprotamine fibers to high humidities leads to the reemergence of the more crystalline patterns. The most detailed pattern of the polylysine–DNA complex is also observed at 92% relative humidity. The observed reflections, about 20 in all, can be indexed on the basis of a hexagonal unit cell with $a = 24.7$ Å and $c = 33.7$ Å. These cell dimensions indicate that there is only one molecule of the complex in the unit cell (Feughelman et al., 1955). It is also evident from the 3.36 Å reflection on the tenth layer line that the DNA molecules in the complex conform to the same helical parameters as DNA in the B form. The polylysine–DNA pattern is very similar to that of nucleoprotamine at the same humidity, which it resembles in intensity distribution as well as in the type of packing. Indeed, no significant differences can be found. The similarity of the patterns suggests that protamine and polylysine molecules may be located on the DNA in a similar manner in the two complexes.

Decreasing the relative humidity causes the patterns of polylysine–DNA to become increasingly diffuse. The X-ray diagram of a fiber photographed at 0% relative humidity indicates a helix having about nine nucleotide pairs per turn in a pitch of approximately 30 Å. These are the helical parameters of the C form of DNA (Marvin et al., 1961). This conformation can be maintained up to about 75% relative humidity. At 85% relative humidity, there is a reversible C to B transformation. Neither the A form nor a pattern showing the simultaneous existence of more than one conformation has been observed at any humidity.

III STEREOCHEMICAL MODELS FOR POLYLYSINE

Suwalsky and Traub (1972) have used the X-ray structural data to suggest DNA–polylysine complexing models. Their studies were based on the assumption that the negatively charged phosphate groups of DNA were completely neutralized by the positively charged amino of lysine. This conforms to the approximately equimolar amounts of these groups found by analyses of the complexes. The interactions were assumed to be made through specific hydrogen bonds. Models were considered with the polypeptide wound around either the small or the large groove of DNA, with all the side chains bound to oxygens of DNA bases or forming cross links to neighboring DNA molecules (Suwalsky et al., 1969). Approximate atomic coordinates for the various models were measured or calculated. These model studies show that it is stereochemically possible for the polypeptide chains to bind in either of the DNA grooves. However, the Fourier transforms of the two types of models for the polylysine–DNA complex, having the polypeptide in the trans configuration with the side chains pointing alternately up and down, show significant differences that suggest that polylysine—and also protamine—lies in the small groove of DNA. Nevertheless, Suwalsky and Traub (1972) were not able to find any one satisfactory model that could account for all the observed intensities. It may be that this complex, like spermine-DNA (Suwalsky et al., 1969), has both intramolecular and intermolecular binding.

De Santis et al. (1973) have investigated the stereochemical conditions of a regular interaction between DNA and polylysine on the basis of theoretical conformational analysis coupled with available experimental structural data. This approach has been used rather well for gramicidin S, actinomysin, and in the study of the complex between DNA and actinomycin (de Santis, 1973). Specifically, de Santis et al. (1973) studied the interactions of polylysine with the DNA B form by fitting the polypeptide chain into the two grooves and localizing the deepest minima of the van der Waals interaction energy. Steric interactions leave a limited mobility to the mutual arrangement of polynucleotide and polypeptide chains in the narrow groove of DNA, where X-ray data indicate that the polypeptide chain is localized (Wilkins, 1956; Feughelman et al., 1968; Suwalsky and Traub, 1968; 1972; Subivana and Puigjaner, 1973). De Santis et al. (1973) conclude that the β conformation of the polypeptide is not specific for the large or the narrow groove of the double helix, in contrast to some chemical evidence and X-ray results (Wilkins, 1956; Feughelman et al., 1968; Subirama and Puigjaner, 1973). This is because the β-type structures are characterized by large (10 Å) radial cylindrical coordinates, which result directly from the geometrical condition that the polypeptide chain have the same symmetry as DNA.

De Santis et al. (1973) suggest that the adoption of an α-helical conformation can result in an intimate fitting of the polylysine chain into the narrow groove of DNA that accounts for the quasi-invariance of the X-ray equatorial spacings of DNA–polylysine fibers (Feughelman et al., 1968) with respect to DNA. An additionally stable complementary fit between the polypeptide and DNA can be achieved by slight changes of the tilting and the distance from the helical axis of the base pairs. These features can form the basis of an explanation of the changes observed in the CD spectra, together with the possible perturbation due to the hydrogen bonds of the n $-$ π^* optically active transitions. It should be noted that a regular sequence of right-handed and left-handed α-helical conformations can be adopted by poly(L-lysine) as well as by poly(D-lysine) in agreement with the similar affinity for DNA of these two polymers (Shapiro et al., 1969). Figure 9-1 shows a three-dimensional illustration of the DNA–polylysine complex for the β polypeptide conformation. Figure 9-2 is the same illustration for the α_R-helical polypeptide conformation. These are the structures tested by de Santis et al. (1973).

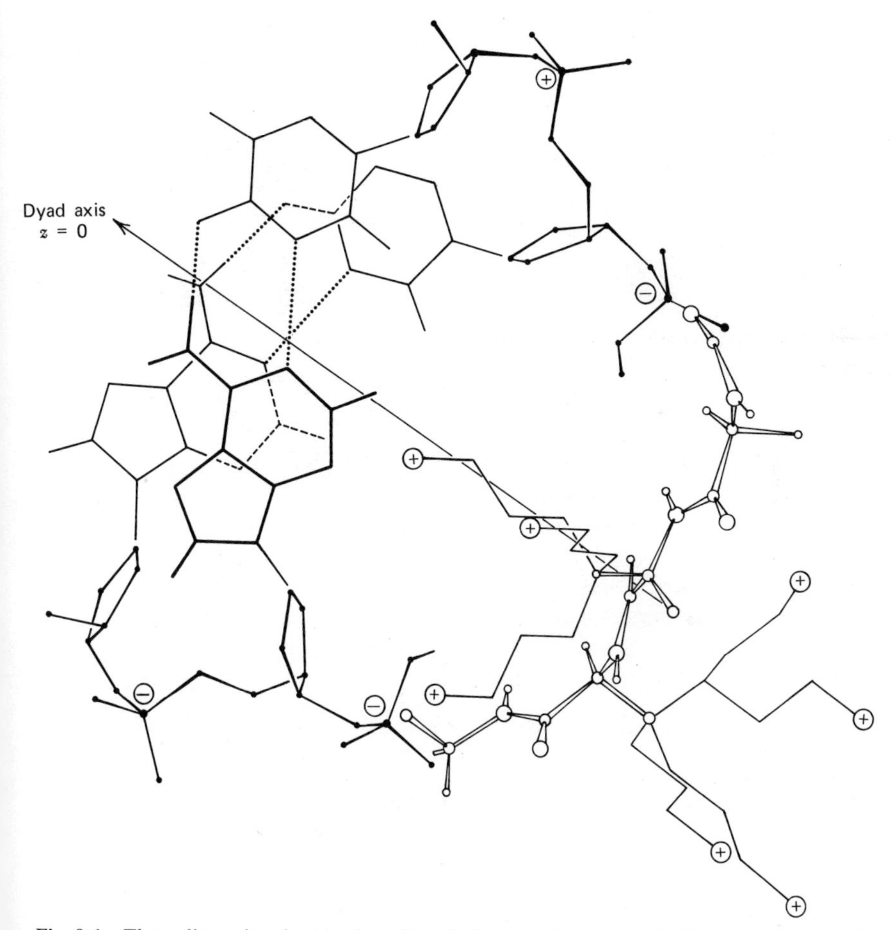

Fig. 9-1 Three-dimensional projection of a cylindrical section perpendicular to the helical axis of DNA poly(L-lysine) complex. The polypeptide chain is in a β conformation, and the possible positions of the charged amino groups are indicated. DeSantis et al. (1973).

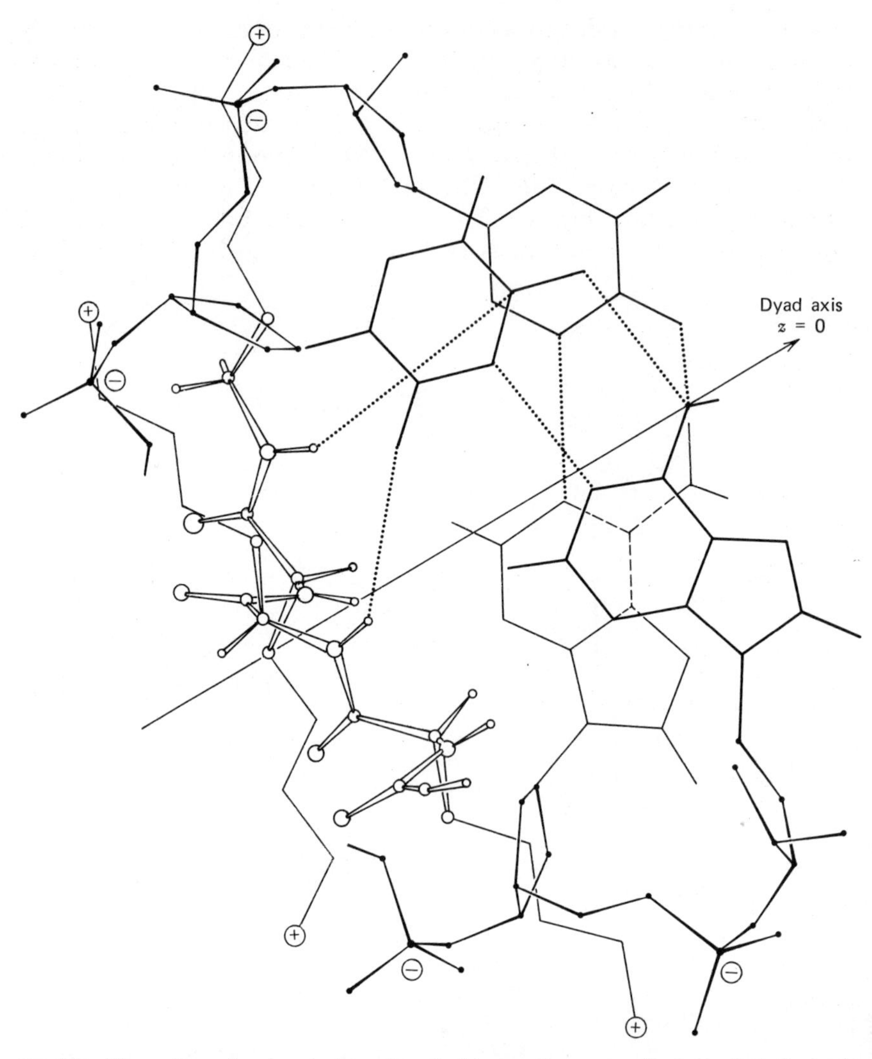

Fig. 9-2 Three-dimensional projection of a cylindrical section perpendicular to the helical axis of DNA poly(L-lysine) complex. The polypeptide chain is in a right-handed-left-handed α helix alternate conformation; the favorable positions of the charged ζ-amino groups are indicated. De Santis et al. (1973).

IV STUDIES OF POLYARGININE–DNA COMPLEXES

Fibers of the polyarginine–DNA complex yield X-ray patterns that differ in some respects from those of polylysine (Suwalsky and Traub, 1972). At 92% relative humidity, a well-oriented fiber shows a pattern with pronounced streaks, as well as spots, indicating some disorder along and about the fiber axis (Marvin et al., 1961). Fibers of polyarginine–DNA photographed at lower humidities also show X-ray patterns conforming to the B form of DNA. The mode of packing remains the same, although the molecules are closer together at low humidities. Polyarginine models show much less flexibility in binding to DNA than those for polylysine (Suwalsky and Traub, 1972). One of these has the polypeptide in the small groove with its side chains up and down. Each branched guanidyl groups is hydrogen bonded through the terminal amino groups to the O_3 of two neighboring phosphates. Such "clamping" of the polynucleotide chains by polyarginine may explain the lack of transformation from the B form of DNA to the C form.

Both in terms of intermolecular packing and enhancement of the first layer line intensity, polylysine–DNA resembles nucleoprotamine more closely than does polyarginine–DNA (Suwalsky and Traub, 1972). In view of the very high arginine content of protamines, this is a rather surprising result and apparently implies that the nonbasic amino acids play an essential role in facilitating an optimal steric interaction between protamine and DNA.

Model studies indicate that, in contrast to polylysine, polyarginine cannot wind smoothly along long stretches of the DNA small groove and achieve maximum hydrogen bonding without conformational distortions. It apparently forms a very tight complex with DNA fixing its B conformation. Nonbasic residues, however, allow more flexibility, which may explain why the nucleoprotamine structure needs a high water content to maintain it. The frequent occurrence of groups of four successive arginine residues in protamines (Ando and Suzuki, 1967; Ando and Watanabe, 1969) suggests that such short sequences bind particularly well to DNA. Perhaps more complex models consisting of polymers of hexapeptides with four arginine and two nonbasic amino acid residues may better simulate both the location and the mode of binding of protamine in the nucleoprotamine complex.

V SEQUENTIAL POLYPEPTIDE HISTONE MODELS

Work in this direction has begun with the recent synthesis of several sequential polypeptides containing lysine (lys) and glycine (gly), I, II, and III (Brown et al., 1974) so that studies can be made

$$\begin{aligned} &\text{I poly(L-lysgly)}_n &&= \text{lys-gly-lys-gly-lys-gly}\\ &\text{II poly(L-lysglygly)}_n &&= \text{lys-gly-gly-lys-gly-gly}\\ &\text{III poly(gly-L-lysglygly)}_n &&= \text{lys-gly-gly-gly-lys-gly} \end{aligned}$$

of the influence of peptide sequence on the structure of DNA in complexes (Williams and Kielland, 1975).

The results indicate that (in the case of each of the polypeptides) changes in the original solution CD spectrum of DNA occur upon increasing the lysine/phosphate (L/P) ratio. The large-magnitude changes in the CD spectra at L/P ratios above 0.5 are thought to be the result of at least three different, but not unrelated, causes: (1) long-range order has been introduced into the DNA (Carroll, 1972; Fasman et al., 1970; Sponar and Eric, 1972); (2) solvent surrounding the DNA has been excluded (Carroll, 1972; Fasman et al., 1970; Sponar and Eric, 1972); (3) an increase in interhelix interactions caused by aggregation of the DNA has taken place (Fasman et al., 1970; Sponar and Eric, 1972; Jordan et al., 1972). Small-magnitude changes in the CD spectra at all L/P ratios could be the result again of at least three different, but not unrelated, causes: (1) the conformation of DNA has changed from the B type to the C type (Tunis-Schneider and Maestre, 1970; Fasman et al., 1970); (2) the charge shielding of the DNA sugar–phosphate backbone has been altered (Studdert et al., 1972; Ivanov et al., 1972); (3) the interhelix interactions in the aggregated state of the DNA have been reduced (Fasman et al., 1970; Jordan et al., 1972). The differences are unlikely to be caused by differences in the molecular weight of the polypeptides, since each is very similar in regard to molecular weight and distributions of molecular weights (Brown et al., 1974).

Differences noted in the L/P ratio where maximal distortion of the DNA CD spectrum occurs (Figure 9-3) can probably be related to one or more of three interrelated effects: (1) the poly (L-lysine) and the lysine-glycine polypeptides adopt different conformations in the final complex; (2) the state of aggregation of the DNA in the two types of complexes differs; and (3) the spectral differences might also be related to differences in the abilities of the polypeptides to adopt discrete conformations. This is consistent with the finding that poly(L-lysine), as assessed by changes in its solution optical rotatory dispersion (ORD) and CD spectra, has been shown to have the ability to adopt several discrete conformations, dependent on conditions of pH, ionic strength, and temperature (Holworth and Doty, 1965; Sarkar and Doty, 1965; Davidson et al., 1966; Townend et al., 1966; Davidson and Fasman, 1967; Dearborn and Wetlaufer, 1970; Tiffany and Krimm, 1969; 1972). In contrast, the constancy of the CD spectrum of the lysine-glycine

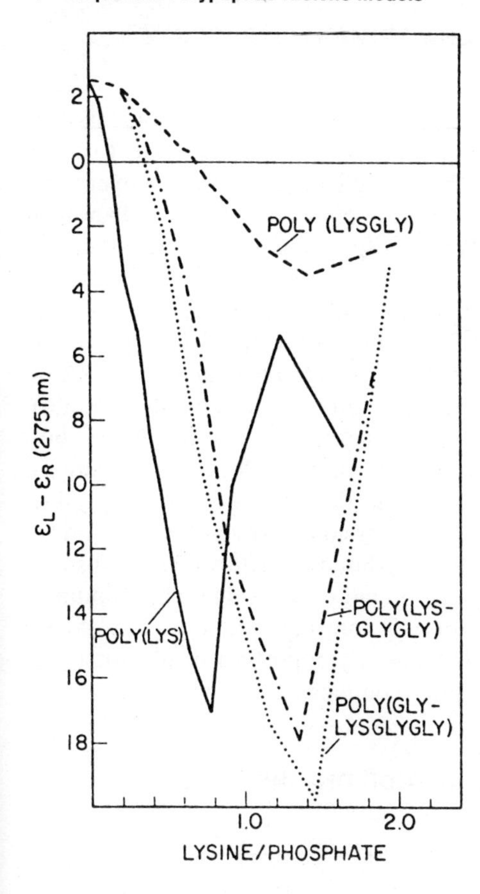

Fig. 9-3 Variation of molar ellipticity ($\varepsilon_L - \varepsilon_R$) at 275 nm with lysine/DNA-phosphate ratio in complexes between the lysine and lysine/glycine polypeptides and DNA. No corrections for sample turbidity have been made. Experimental conditions: DNA-phosphate concentration: 1.76×10^{-4} *M*; 0.15 *M* sodium chloride, sodium cacodylate, 0.001 *M*, pH.7.0. Williams and Kielland (1975).

polypeptides upon changes in pH and ionic strength (Brown et al., 1974), as well as the presence of the helix-breaking amino acid, glycine (Ananthanarayanan et al., 1972), strongly suggest that they are unable to sustain significant populations of radically different conformations in solution.

It is clear that differences in the polypeptide sequences can result in differences in the juxtaposition of charged groups within the complexes. Based on the predominance of simple electrostatic interactions between the polypeptides and the DNA, it has been suggested (Brown et al., 1974), after modification of an earlier proposal (Wilkins, 1956), that juxtaposition of the positively charged side chain of the lysine-containing peptides and the negatively charged phosphates of one strand of the DNA could account for the differences observed in the abilities of the polypeptides to precipitate DNA from solution (Brown et al., 1974).

VI MODIFICATIONS IN DNA PRIMARY STRUCTURE

Interaction between polylysine and DNAs of varied G + C contents has been studied using thermal denaturation and circular dichroism (Jei Li et al., 1975). For each complex, there is one melting band at a lower temperature T_m, corresponding to the helix-coil transition of free base pairs, and another band at a higher temperature T'_m, corresponding to the transition of polylysine-bound base pairs. For free base pairs, with natural DNAs and poly(dA-dT), a linear relation is observed between the T_m and the G + C content of the particular DNA used. This is not true with poly(dG)·poly(dC), which has a T_m about 20°C lower than the extrapolated value for DNA of 100% G + C. For polylysine-bound base pairs, a linear relation is also observed between the T'_m and the G + C content of natural DNAs, but neither poly(dA-dT) nor poly(dG)·poly(dC) complexes follow this relationship. There is also a noticeable similarity in the CD spectra of polylysine and polyarginine–DNA complexes, except for complexes with poly(dA-dT). The calculated CD spectrum of polylysine-bound poly(dA-dT) is substantially different from that of polyarginine-bound poly(dA-dT). These findings suggest that base pair content may not only control thermal stability of free DNA (Marmur and Doty, 1962), but also be important in dictating the overall strength of the DNA–polypeptide complexing interaction.

VII STRUCTURAL FEATURES OF DNA–HISTONE INTERACTIONS

Limited structural information is available concerning the intermolecular organization of histones and DNAs. The most well-suited spectroscopic probes, CD/ORD, fluorescence, and calorimetric techniques, to study these complexes unfortunately yield little direct structural data. The molecular geometry associated with DNA–histone complexes must be indirectly inferred from spectral differences observed in various interacting systems involving known fragmental or congeneric molecular modifications.

Still, some general features of DNA–histone complexes seem to be:

1. Thermal denaturation studies indicate multiple melting bands characteristic of free DNA at low temperatures and bound histone–DNA structures at high temperatures. Hwan et al. (1975) report that histone V (f2c) from chick erythrocytes shows three melting bands (Figure 9-4) when complexed with chick or calf thymus DNA. Adler et al. (1975) also report a threefold melt transition behavior for histone IV (f2a1).

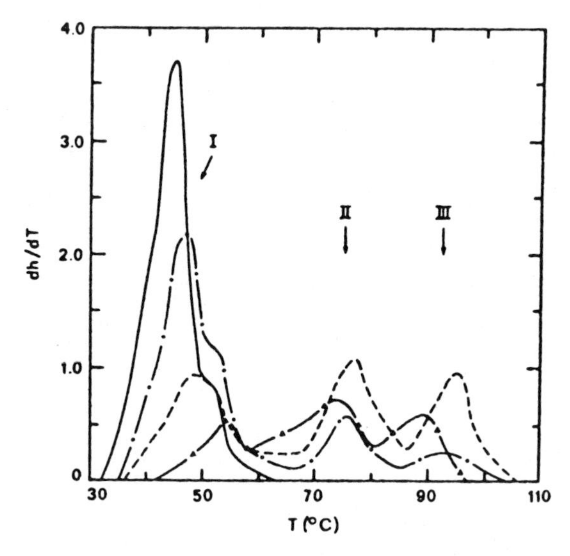

Fig. 9-4 Derivative melting profiles of histone V-calf thymus DNA complexes. r = 0(—), 0.5(–·–), 1.0(– – –), and 1.5(–▲–). "r is the ratio of amino acid residues per mucleotide." Hwan et al. (1975).

2. Preferential DNA binding sites are located along the primary structure of histones. The binding sites are probably α-helical when complexed to the DNA. Bradbury et al. (1975), using PMR spectroscopy as well as other spectroscopic probes, have found that histone H1(F1) does not contain β structures, but does contain a certain amount of α helix. The likely locations of the helical segments are residues 42–45 and 58–75. These are probably the binding sites to DNA and the amount of α-helical content increases slightly with complexing. There is some evidence that both DNA binding sites complex on a single DNA. Adler et al. (1975) were able to cleave histone IV (102 amino acids) at residue 84. The large basic N-terminal fragment (1–83) was bound as well to DNA as the whole intact histone, but did not produce the characteristic C.D. distortion of the complete histone–DNA complex. The small C-terminal fragment (85–102) was found to bind weakly to DNA. Similar findings for histone IV have been reported by Ziccardi and Schumaker (1973).

3. Each histone has a characteristic binding potential to DNA. The base pair composition of the DNA can have a slight to marginal effect on the strength of the interaction. Hwan et al. (1975) report that in the binding regions there are 1.5 amino acid residues/nucleotide for nucleohistone V, whereas Li (1973) found 2.9 to 3.3 residues/nucleotide for nucleohistone I (f1). Hwan et al. (1975) also observed that, when two bacterial DNAs of

varied A + T (adenine + thymine) content simultaneously compete for binding of histone V, the more (A + T)-rich DNA is selectively favored.
4. The structural stability of each histone is unique with respect to salt concentration and pH. Isenberg and co-workers (D'anna and Isenberg 1972; 1974; Wickett et al., 1972) have found that histone LAK (IIb1) is more sensitive to salt and pH changes than either histone GRK (IV; f2a1) or histone KAS (II2; f2b).

Interhistone interactions and chromatin organization are discussed in Chapter 13, Section VI. Some discussion of nucleohistone solvation behavior is given in Chapter 7.

REFERENCES

Adler, A. J., A. W. Fulmer, and G. D. Fasman (1975). *Biochemistry*, **14**, 1446.

Ananthanarayanah, V. S., R. H. Andretta, D. Poland, and H. A. Sheraga (1972). *Macromolecules*, **4**, 417.

Ando, T. and K. Suzuki (1967). *Biochim. Biophys. Acta*, **140**, 377.

Ando, T. and S. Watanabe (1969). *Int. J. Protein Res.*, **1**, 221.

Boublik, M., E. M. Bradbury, C. Crane-Robinson, and H. W. E. Rattle (1971). *Nature* (*Lond.*), **229**, 149.

Bradbury, E. M., P. D. Cary, G. E. Chapman, C. Crane-Robinson, S. E. Denby, H. W. Rattle, M. Boublik, J. Palau, and F. J. Aviles (1975). *Eur. J. Biochem.*, **52**, 605.

Brown, J., J. R. Langlois, D. R. Lauren, B. K. Stoochnoff, and R. E. Williams (1974). *Can. J. Chem.*, **42**, 3140.

Bush, H. (1965). *Histones and Other Nuclear Proteins*. Academic Press, New York.

Carroll, D. (1972). *Biochemistry*, **11**, 421.

Chang, C., M. Weiskopf, and H. J. Li (1973). *Biochemistry*, **12**, 3028.

D'Anna, J. A., Jr. and I. Isenberg (1972). *Biochemistry*, **11**, 4017.

D'Anna, J. A., Jr. and I. Isenberg (1974). *Biochemistry*, **13**, 2093.

Davidson, B. and G. D. Fasman (1967). *Biochemistry*, **6**, 1616.

Davidson, B., N. Tooney, and G. D. Fasman (1966). *Biochem. Biophys. Res. Commun.*, **23**, 156.

Dearborn, D. G. and D. B. Wetlaufer (1970). *Biochem. Biophys. Res. Commun.*, **39**, 314.

De Santis, P. (1973). In *Conformation of Biological Molecules and Polymers* (E. Bergmann and B. Pullman Eds.). Academic Press, New York, p. 493.

De Santis, P., R. Rizzo, and M. Savino (1973). *Macromolecules*, **6**, 520.

Fasman, G. D., B. Schaffhausen, L. Goldsmith, and A. Adler (1970). *Biochemistry*, **9**, 2814.

Feughelman, M., R. Langridge, W. E. Seeds, A. R. Stokes, H. R. Wilson, C. W. Hooper, M. H. F. Wilkins, R. K. Barcly, and L. D. Hamilton (1955). *Nature*, **175**, 834.

Feughelman, M., Y. Mitsui, Y. Iitaka, and M. Tsuboi (1968). *J. Mol. Biol.*, **38**, 129.

Haynes, R. A., M. Garrett, and W. B. Gratzer (1970). *Biochemistry*, **9**, 4410.

Holzworth, G. and P. Doty (1965). *J. Am. Chem. Soc.*, **87**, 218.

Hwan, J. C., I. M. Leffak, H. J. Li, P. C. Huang, and C. Mura (1975). *Biochemistry*, **14**, 1390.

Inoue, S. and T. Ando (1968). *Biochem. Biophys. Res. Commun.*, **32**, 501.

Inoue, S. and T. Ando (1970). *Biochemistry* **9**, 395.

Ivanov, V. I., L. E. Minchenkova, A. K. Schyolkina, and A. J. Poletayev (1973). *Biopolymers*, **12**, 89.

Jei Li, H., L. Herlands, R. Santella, and P. Epstein (1975). *Biopolymers*, **14**, 2401.

Jordan, C. F., L. S. Lerman, and J. H. Venable (1972). *Nature* (*New Biol.*), **236**, 67.

Leng, M. and G. Felsenfeld (1966). *Proc. Natl. Acad. Sci. USA*, **56**, 1325.

Marmur, J. and P. Doty (1962). *J. Mol. Biol.*, **5**, 109.

Marvin, D. A., M. Spencer, M. H. F. Wilkins, and L. D. Hamiltin (1961). *J. Mol. Biol.*, **3**, 547.

Olins, D. E., A. L. Olins, and P. H. von Hippel (1968). *J. Mol. Biol.*, **33**, 265.

Sarkar, P. and P. Doty (1965). *Proc. Natl. Acad. Sci. USA*, **55**, 981.

Shapiro, J. I., M. Leng, and G. Felsenfeld (1969). *Biochemistry*, **8**, 3219.

Sponar, J. and I. Fric (1972). *Biopolymers*, **11**, 2317.

Studdert, D. S., M. Patroni, and R. C. Davis (1972). *Biopolymers*, **11**, 761.

Subirana, J. A. and L. C. Puigjaner (1973). In *Conformation of Biological Molecules and Polymers* (E. Bergmann and B. Pullman, Eds). Academic Press, New York, p. 645.

Suwalsky, M. and W. Traub (1968). *Israel J. Chem.*, **6**, Vp.

Suwalsky, M. and W. Traub (1972). *Biopolymers*, **11**, 2223.

Suwalsky, M., W. Traub, U. Shmueli, and J. A. Subirana (1969). *J. Mol. Biol.*, **42**, 363.

Tiffany, M. L. and S. Krimm (1969). *Biopolymers*, **8**, 347.

Tiffany, M. L. and S. Krimm (1972). *Biopolymers*, **11**, 2309.

Townend, R., T. F. Kumoskinski, S. N. Timasheff, G. D. Fasman, and B. Davidson (1966). *Biochem. Biophys. Res. Commun.*, **23**, 163.

Tunis-Schneider, M. J. B. and M. F. Maestre (1970). *J. Mol. Biol.*, **52**, 521.

Wickett, R. R., H. J. Li, and I. Isenberg (1972). *Biochemistry*, **11**, 2952.

Wilkins, M. H. F. (1956). *Cold Spring Harbor Symp. Quant. Biol.*, **21**, 75.

Wilkins, M. H. F. (1966). *Cold Spring Harbor Symp. Quant. Biol.*, **21**, 75.

Williams, R. E. and S. L. Kielland (1975). *Can. J. Chem.*, **52**, 542.

Zama, M. and S. Ichimura (1971). *Biochem. Biophys. Res. Commun.*, **44**, 936.

Zama, M. and S. Ichimura (1973). *Biochim. Biophys. Acta*, **294**, 214.

Ziccardi, R. and V. Schumaker (1973). *Biopolymers*, **12**, 3231.

Zimmer, C. and G. Luck (1973). *Biochim. Biophys. Acta*, **312**, 215.

CHAPTER 10

Glycosaminoglycan and Amylose Structural Organization

Polysaccharides in general, and glycosaminoglycans in particular, were long considered to be devoid of a uniformily repeating primary structure and, consequently, any appreciable secondary structure. This viewpoint was further enhanced for glycosaminoglycans in the tissue state, since they are found covalently complexed to a protein factor yielding, in composite, a gel-like material. However, recent studies reported in this and the next chapter indicate that these biopolymers do possess some primary and secondary structure that is involved in some yet unknown, but organized, interaction with the protein components of tissues. The extensive studies of amylose and some of its analogs also bring home this finding of secondary structure and consequential interpolymeric organization involving the polysaccharides.

I GENERAL ROLE OF GLYCOSAMINOGLYCANS IN CONNECTIVE TISSUE

In Chapter 11, glycosaminoglycan–polypeptide interactions are discussed within the framework of connective tissue models. Although the primary and secondary structures of polypeptides are relatively well known and understood, such is not the case for the glycosaminoglycans. The glycosaminoglycans (mucopolysaccharides) form a group of polysaccharides that are widely distributed throughout mammalian connective tissues. These tissues are characterized by a high proportion of extracellular material, and include bone, tendon, cartilage, skin, and ligament. With the exception of bone, the extracellular matrix of connective tissue consists largely of fibers (e.g., collagen and elastin) imbedded in an apparently formless gel-like material known as ground substance. This ground substance is composed predominantly of protein–polysaccharide covalent complexes known as proteoglycans, where the glycosaminoglycan chains are attached as side chains to a protein core.

In addition to proteoglycan, the ground substance contains amino acids, sugars, lipids, and electrolytes (e.g., sodium, potassium, and chloride ions) that function in the interstitial fluid. All substances going to and from the cells must pass through the ground substance; thus changes in its state and composition can be expected to exert a significant influence on the life of individual cells and tissues.

II PROTEOGLYCANS

The proteoglycans are heterogeneous in composition and molecular weight; Mathews and Lozaityte (1958) have proposed a comb-like model for the structure of proteoglycans in which the polysaccharide chains are arranged in pairs along the protein core. This model has since been revised by Hascall and Heinegard (1974), who consider the polysaccharide chains to be distributed in a regular manner along the protein core. Tsiganos et al. (1971) have also shown that the proteoglycans from pig laryngeal cartilage have several core proteins differing in length, distribution, and number of glycosaminoglycans attached. However, the average molecular weight of the bound glycosaminoglycan chains does not vary from one core protein to another. The major polysaccharide components of the proteoglycans are chondroitin 4-sulfate, chondroitin 6-sulfate, dermatan sulfate, heparan sulfate, and keratan sulfate. The structural formulas for these acid glycosaminoglycans, as well as hyaluronic acid and chondroitin, are shown in Figure 10-1. All of

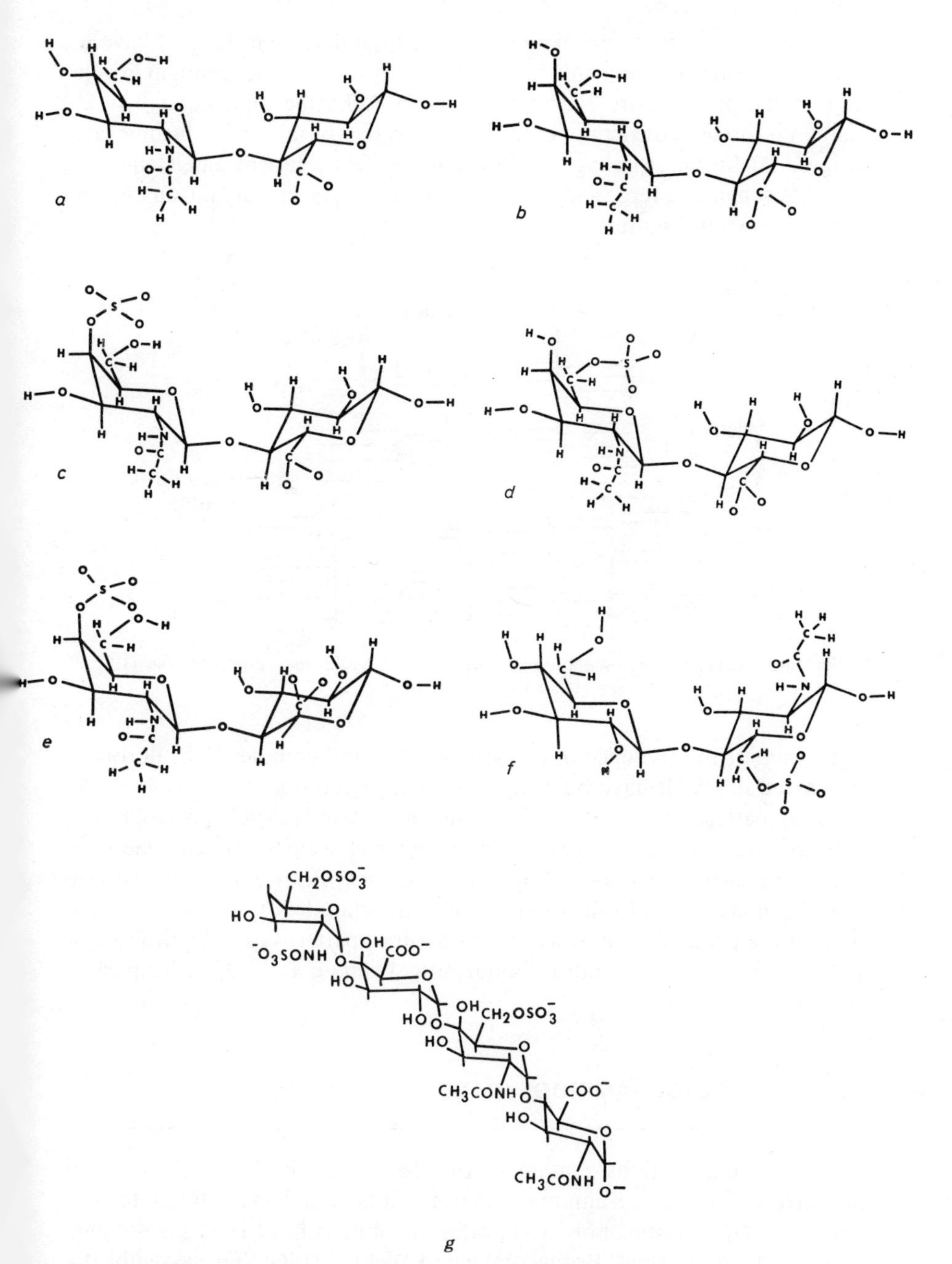

Fig. 10-1 Glycosaminoglycans: (*a*) hyaluronic acid, (*b*) chondroitin, (*c*) chondroitin 4-sulfate, (*d*) chondroitin 6-sulfate, (*e*) dermatan sulfate, (*f*) keratan sulfate, (*g*) heparan sulfate.

these polysaccharides approximate to repeating disaccharides and have the same backbone, differing only in the type and position of the pendant groups. Part of the protein core does not have polysaccharide chains attached, but functions in noncovalently binding the proteoglycan to hyaluronic acid (Figure 10-1*a*) via a linkage or "locking" protein (Hascall and Heinegard, 1974). A schematic representation of such a proteoglycan–hyaluronate complex is shown in Figure 10-2.

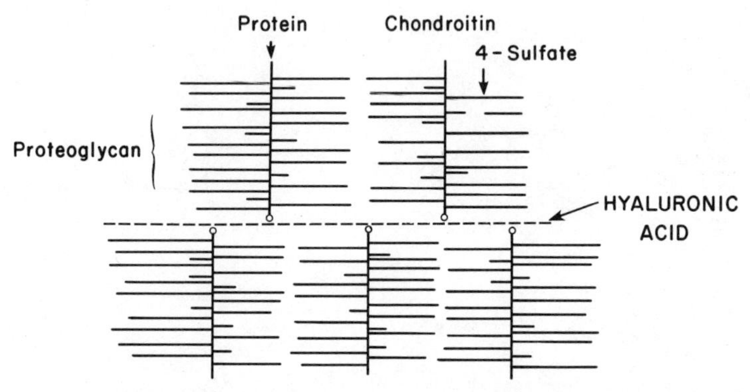

Fig. 10-2 Schematic representation of a proteoglycan/hyaluronate complex. Cael (1975).

In spite of the presence of the protein core and complexed hyaluronate, Atkins et al. (1974) have been able to obtain good quality X-ray fiber diffraction patterns from oriented, chondroitin 4-sulfate-rich proteoglycan–hyaluronate complexes. Similar fiber patterns were obtained when the hyaluronic acid was removed from the complex. Both diffraction patterns have been attributed to the orientation of the chondroitin 4-sulfate component of the proteoglycan, since the X-ray fiber patterns are identical to the pattern obtained from pure chondroitin 4-sulfate in a 2_1 helical chain conformation.

III GLYCOSAMINOGLYCANS

The succeeding sections contain summaries of the physical and structural properties of the glycosaminoglycans in solution and the solid state. The size and shape of mucopolysaccharides in solution have been investigated to a very limited extent. Brimacombe and Webber (1964) have assembled a review of this area, and a summary of the physiochemical properties of several glycosaminoglycans is included in a review of acid polysaccharide–protein complexes by Mathews (1967). Considerable work is being done to determine the solid-state structures of glycosaminoglycans, especially by Arnott and co-workers and Atkins and co-workers. The studies are referenced in the appropriate places below.

A Hyaluronic Acid

Hyaluronic acid is a nonsulfated, high-molecular-weight (10^5–10^6) glycosaminoglycan found, for example, in vitreous humor, skin, umbilical cord, cartilage, and synovial fluid. It is an unbranced, alternating copolymer of β-(1,4)-D-glucuronic acid and β-(1,3)-*N*-acetyl-D-glucosamine (Figure 10-1*a*), and is thought to be involved in binding water within the interstitial spaces, holding cells together, and aggregating proteoglycans.

In solution, hyaluronic acid has been shown to form an extended three-dimensional network that retards the movement of other solutes through the solution to an extent dependent on the size of the latter (Ogston and Phelps, 1960; Ogston and Preston, 1966; Laurent and Pietruszkiewicz, 1961). The same continuous network sieve was obtained from hyaluronic acids of molecular weights 1.7×10^6–1.4×10^5. A fivefold increase in concentration for the smaller molecular weight sample was seen, and all network formation was independent of electrostatic effects (Ogston and Phelps, 1960; Laurent and Pietruszkiewicz, 1961). Hydrodynamic studies have been carried out on several hyaluronic acid samples. A summary of the findings is given in Table 10-1. The general conclusion is that hyaluronic acid in solution is more rigid than a random coil model. Moreover, the more pure the preparation, that is, less protein bound, the more extended the biopolymer and the lower its molecular weight.

Oriented films and fibers of hyaluronate salts (e.g., those of Na^+, K^+, and Ca^{2+}) have been shown to adopt a variety of chain conformations and crystalline packing arrangements as a function of the counterion, pH, ionic strength, and degree of hydration. For example, sodium and calcium hyaluronate are found to exist as lefthanded threefold helices with an axially projected rise per disaccharide (h) of 9.5 Å (Atkins and Sheehan, 1971, 1972; Atkins and Mackie, 1972; Atkins et al., 1972*a*; Atkins et al., 1974; Sheehan et al., 1974; Winter et al., 1975), indicating a very extended chain conformation. In addition, both sodium and potassium hyaluronate can exist in two different fourfold conformations, namely, an extended conformation with a fiber repeat of 36.0 Å and a more compact structure with a fiber repeat of 33.6 Å (Atkins et al., 1974; Arnott, 1973; Atkins and Sheehan, 1973; Arnott et al., 1973*a*; Guss et al., 1975). Because of its compactness, the latter structure was considered originally to exist as antiparallel double helices; however,

Table 10-1 Molecular Weight and Hydrodynamic Properties of Some Glycosaminoglycans in Solution

Glycosaminoglycan	Molecular Weight	Hydrodynamic Property	References
Hyaluronic acid	13×10^6	RG = 3750 A	Ogoton & Phelps (1960)
	$2\text{–}10 \times 10^6$	EED = 4600 A	Ogoton & Stanier (1950)
			Balazs (1958)
	3.4×10^6	RG = 2400 A	Laurent & Gergely (1955)
	3.4×10^6	EED = 10,000 A	Blix & Snellman (1945)
	0.5×10^6	EED = 7000 A	Stone (1969)
	0.2×10^6	EED = 4800 A	Stone (1969)
	11×10^6	RG = 2020 A	Ogston & Preston (1966)
		RG = 2030 A	Ogston & Preston (1966)
Heparin	12,000 – 16,000	FR = (1.93–2.18)	Barlow et al. (1961)
			Barlow et al. (1961)
			Helbert & Marini (1963)
	13,970	FR = 2.3	
	5,540	FR = 1.9	Lasker (1965)
Chondroitin	$17\text{–}50 \times 10^{3\,a,\,b}$	$[\eta] = 3.1 \times 10^{-4}\ M_w^{0.74}$	Mathews (1959)
Sulfates	$22 \times 10^{3\,c}$	— —	Mathews (1956)
Keratin	$9\text{–}10 \times 10^{3\,d}$	— —	Wortman (1964)
Sulfates	$15\text{–}20 \times 10^{3\,e}$	— —	Wortman (1964)

RG, radius of gyration.
* EED, end-to-end distance.
FR, frictional ratio.
[a] For chondroitin-4-sulfate, chondroitin-6-sulfate and chondroitin sulfate B.
[b] Osmotic pressure and light-scattering methods.
[c] Highest molecular weight fraction of chondroitin sulfate B.
[d] Sample from the cornea.
[e] Sample from the mucleus pulposus.

rigid body least-squares refinement of the X-ray intensities by Guss et al. (1975) indicate that the chains are left-handed single helices. Figure 10-3 shows two views of the refined tetragonal structure of sodium hyaluronate, as well as an enlarged view of the sodium ion environment as reported by Guss et al. (1975).

When oriented filaments of hyaluronate are soaked in solution consisting of 80% ethanol and 20% 0.2 *N* HCl, the ionized carboxyl groups of the glucuronic residues are converted into the "free acid" form. During this conversion, an untwisting of the 3_{-1} or 4_{-1} helices occurs with the formation of a more extended 2_1 helix with a fiber repeat of 19.6 Å (Atkins et al., 1972*a*).

B Chondroitin Sulfates

The chondroitin sulfates are a group of chemically similar glycosaminoglycans that are alternating copolymers of β-(1,4)-D-glucuronic acid and β-(1,3)-*N*-acetyl-D-galactosamine. The two main members of this group are chondroitin 4-sulfate and chondroitin 6-sulfate, which have sulfate groups bound to the C-4 and C-6 positions, respectively (Figure 10-1). Both occur covalently bound to a protein core in proteoglycan subunits and are the major constituents of bone and cartilage connective tissues.

Molecular weight and hydrodynamic data on three chondroitins are given in Table 10-1. The molecular weight distributions of chondroitin sulfates have varied. M_w/M_m for chondroitin-4-sulfate range from 1.25, close to monodisperse, to 1.95 (from pressure and light-scattering measurements in 0.1 *M* NaCl) (Mathews, 1959), while ratios for chondroitin sulfate B have been estimated to be 1.42 by the same procedures and 1.2 from sedimentation equilibrium at 1.5 *M* NaCl (Tanford et al., 1964). Molecular weights of the order of 17,000–50,000 for these two chondroitins as well as chondroitin-6-sulfate indicate 34–100 disaccharide units per molecule as compared with about 2000 for hyaluronic acid.

Viscosity–molecular weight studies on the three chondroitin sulfates discussed above (Mathews, 1959; 1962) indicate that they are all roughly linear, flexible chains. A comparison of sulfated chondroitin-6-sulfate and mammalian chondroitin-6-sulfate points out that the extra sulfate group renders the chain more extendable in aqueous solutions without added counterions than either chondroitin-6-sulfate or chondroitin-4-sulfate. The chondroitin sulfates are also considered to be flexible chains when bound covalently to proteins in the proteoglycan complex (Anderson and Meyer, 1962; Anderson et al., 1963; Lindahl and Roden, 1961). The protein–chondroitin sulfate unit of shark cartilage is reported to be a rod-like structure roughly 3500 Å long with a molecular weight of 4×10^6. These rod-like structures self-aggregrate in solution.

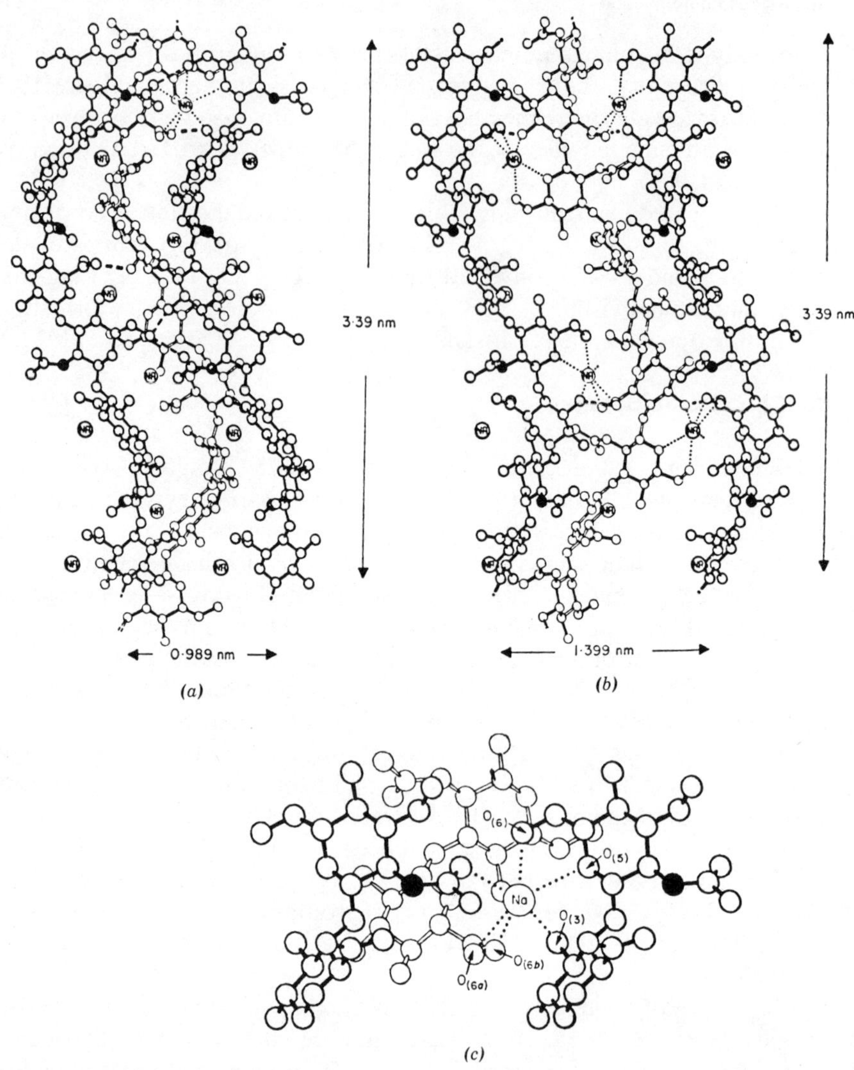

Fig. 10-3 Views of the refined tetragonal structure of sodium hyaluronate. Chains at the cell corners are shown emphasized. Hydrogen bonds are shown by dashed lines and $Na^+ \ldots O$ contacts by dotted lines. (For clarity only some of these interactions have been shown.) (*a*) View along the (100) direction. The center chain (open bonds) is translated a/2 behind the corner chains. Only the contacts to the sodium ion NA, which binds to all three chains in this figure, are shown. The *a* and *c* dimensions are indicated. (*b*) View along the (110) direction. The two corner chains are at diagonally opposite corners of the unit cell. The helix axes of the three chains shown are coplanar. Contacts to sodium ions that interact with two chains are shown. The sixth coordination positions (to chains not in this figure) are indicated. Gus et al. (1975). (*c*) Details of the sodium ion environment in the tetragonal structure. This is an enlargement of a section of Figure 10-3*a* showing the labels of atoms bound to the sodium ion. Gus et al. (1975).

X-ray diffraction studies of sodium chondroitin 6-sulfate (Atkins et al., 1974; Arnott, 1973; Arnott et al., 1973*a*,*b*; Atkins et al., 1972*b*) show that this polydisaccharide can adopt a variety of helical conformations in the solid state. Like the hyaluronate, both threefold and twofold helical conformations are observed for the salt and "free acid" forms, respectively. In the threefold structure, the fiber repeat is 28.7 Å (h = 9.6 Å), and the helix is considered to be lefthanded. The 2_1 conformation of the "free acid" form is surprisingly less extended than the 3_{-1} helix and has a fiber repeat of 18.6 Å (Dea et al., 1973) with h = 9.3 Å.

Unlike hyaluronate, however, sodium chondroitin 6-sulfate has not been observed to form a fourfold conformation, although it can be crystallized in what has been interpreted as a left-hand 8_{-3} helix that is intermediate between the 3_{-1} and 2_1 conformations (Arnott, 1973; Arnott et al., 1973*a*,*b*). The fiber repeat is 78.2 Å with a rise per disaccharide of 9.8 Å.

Similar to chondroitin 6-sulfate, the sodium salt of chondroitin 4-sulfate (Atkins et al., 1974; Isaac and Atkins, 1973; Atkins and Laurent, 1973) adopts a threefold conformation (fiber repeat = 28.8 Å; h = 9.6 Å), whereas the free acid form is an extended 2_1 helix (fiber repeat = 19.6 Å; h = 9.8 Å).

C Dermatan Sulfate

Dermatan sulfate is the predominant glycosaminoglycan of skin, and is found to a lesser extent in tendon and aorta, where its linkage to the protein core of proteoglycans is similar to that of the chondroitin sulfate proteoglycans. It is distinguished from the other connective tissue polysaccharides by the replacement of the usual D-glucuronic acid, common to the other members of the group, by its C-5 epimer, L-iduronic acid. In this situation the equatorial COOH group of D-glucuronic acid is transformed to an axial configuration at C-5, provided the ring is in the C1 conformation. Diagrams of the two uronic acid epimers are shown in Figure 10-4. Dermatan sulfate is therefore an alternating copolymer of β(1,4)-L-iduronic acid and β(1,3)-*N*-acetyl-D-galactosamine-4-sulfate. Generally it is found that dermatan sulfate has more sulfate groups than the idealized formula of Figure 10-1*e*, and the extra sulfate groups (up to 0.4 per disaccharide) are probably attached at the C-6 position of the galactosamine ring.

X-ray diffraction studies (Atkins et al., 1974; Arnott et al., 1973; Atkins and Laurent, 1973; Atkins and Isaac, 1973) of sodium dermatan sulfate indicate that this polydisaccharide can adopt an 8_{-3} helical conformation (fiber repeat = 74.4 Å; h = 9.3 Å) similar to that of chondroitin-6-sulfate. Atkins and Isaac (1973) report that conversion from the salt to a "free acid" form results initially in helix extension to a 3_{-1} conformation (h = 9.5 Å),

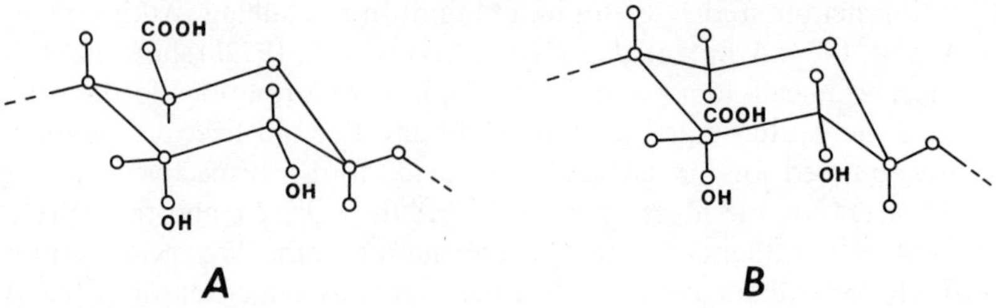

Fig. 10-4 Structural formula for uronic acid and its C-5 epimer, L-iduronic acid. Both structures have the C-1 chair conformation. Cael (1975).

which is transformed by annealing or prolonged standing in air to a more extended 2_1 structure with a rise per disaccharide of 9.7 Å. However, the results of Arnott et al. (1973*b*) indicate that the 3_{-1} conformation of dermatan sulfate can be obtained directly from the sodium salt, with conversion to the 8_{-3} and 2_1 conformations occurring from prolonged standing in air at 84% relative humidity.

D Heparin and Keratin Sulfates

The few available hydrodynamic properties of heparin are reported in Table 10-1. These values indicate that the heparin chain becomes relatively more extended at lower degrees of polymerization, which is reminiscent of the anomalous effect of amylose fractions (Cowie, 1961). Lasker (1965) has shown that heparin behaves hydrodynamically more like an extended coil than an ellipsoid of revolution. Heparin does not associate to an appreciable degree as compared to hyaluronic acid. It has been observed that mild acid treatment causes a collapse of the heparin preparation with a fall in frictional ratio, but without a significant loss of sulfate content (Jensen et al., 1948).

Virtually no physical or structural properties of the keratin sulfates are reported in the literature. Table 10-1 contains estimates of the molecular weights of keratin sulfate.

IV STRUCTURE OF AMYLOSE IN SOLUTION AND SOLID STATES

Amylose serves as food-storage carbohydrate in plants, where it occurs in starch granules. The linear biopolymer is composed of α-(1 → 4)-linked D-glucose residues (see Figure 10-5). Starch is believed to occur in three crystal-

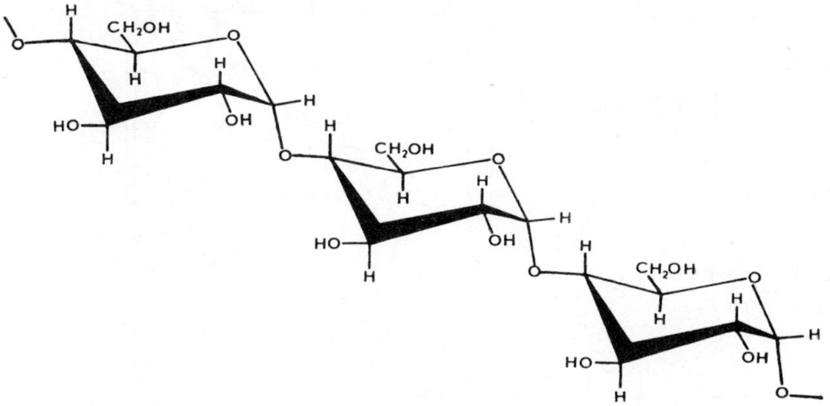

Fig. 10-5 Structural formula for amylose. Cael et al. (1973).

line forms, differentiated by their X-ray diffraction patterns (Katz and Van Itallie, 1930). The A form is found in cereal starch, whereas the B form occurs in tuber starches. The C form is rarer; it is observed in, for example, tapioca and banana starches. All three X-ray patterns can also be obtained with isolated amylose. It seems that the A, B, and C structures are very similar, and are probably different hydrates with the same chain conformation (Bear and French, 1941).

Amylose can be isolated from starch by precipitation from solution with various organic solvents such as butyl alcohol. The precipitated amylose is generally complexed with solvent molecules and exhibits the X-ray pattern of the so-called V form. Removal of the complexing agent by using aqueous methanol yields the V_a structure. The subscript was originally used to indicate what was thought to be an anhydrous structure, but this form almost certainly contains some water molecules within the crystal lattice. The V_a structure has been shown by X-ray diffraction (Rundle and French, 1943; Zobel et al., 1967) to consist of amylose helices having six glucose residues per turn, repeating in 7.90 Å. A cylindrial projection (Blackwell et al., 1969) of this helix is shown in Figure 10-6*a*. Model building suggests that, of the three hydroxyl groups, two are involved in intramolecular hydrogen bonding, and the third could be hydrogen bonded to a water molecule within the central cavity of the helix.

Winter and Sarko (1974*a,b*) have more recently investigated the crystal and molecular structure of V_a amylose. The intrachain hydrogen bonds are substantially in agreement with previous studies. A revised model for the packing of the chains in the unit cell and the presence of crystallographic water is proposed. A projection in the *ab*-plane is shown in Figure 10-6*b*.

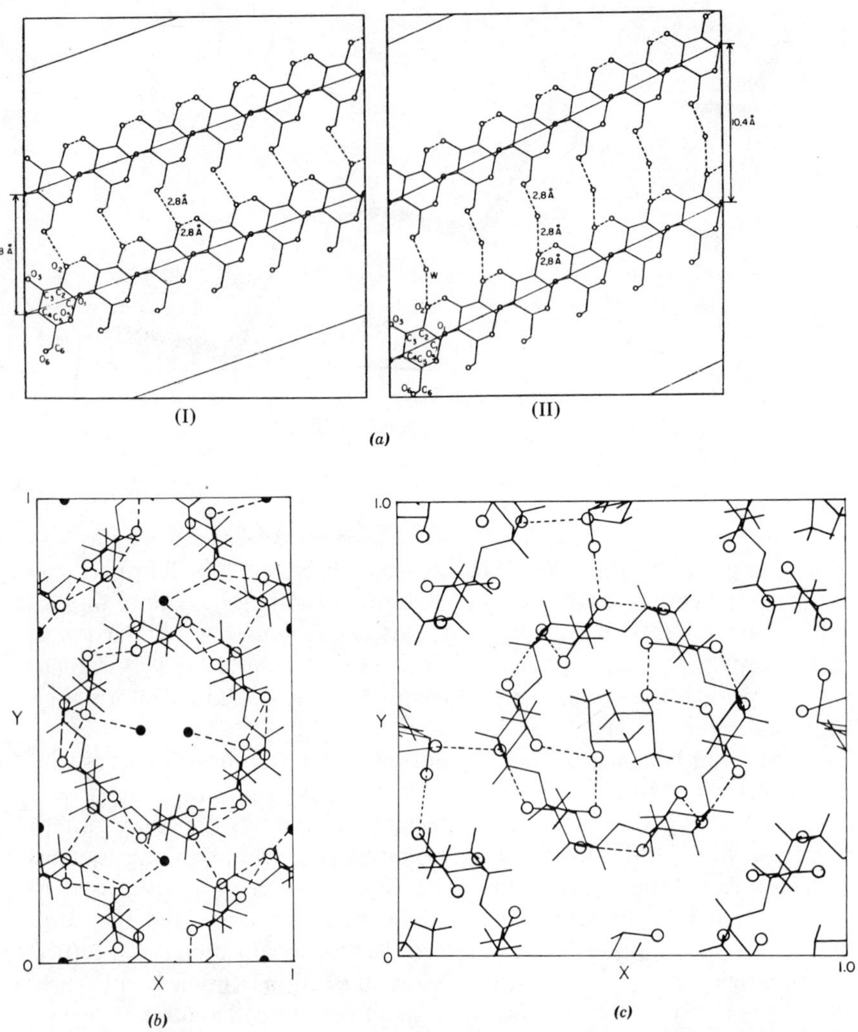

Fig. 10-6 (*a*) (I) Cylindrical projection of the V-amylose helix. (II) Cylindrical projection of the proposed B-amylose helix, showing the inclusion of a water molecule, W, between successive turns. Blackwell et al. (1969). (*b*) V*a* = amylose in *ab* projection. Hydroxyl oxygens are denoted by ○, water by ●, and possible hydrogen bonds by ---. (*c*) V-DMSO structure in *ab* projection. Hydroxyl and DMSO oxygens are denoted by circles and possible hydrogen bonds by dashed lines. Winter and Sarko (1974 a,b).

Packing appears to be stabilized by corner-to-center chain 0(2)—0(2) hydrogen bonds. Winter and Sarko (1974*a,b*) also investigated the nature of the transition from the amylose-DMSO complex to V_a amylose. The complex structure in *ab*-projection is shown in Figure 10-6*c*. They conclude that the transition involves translation of the amylose chains parallel to the *a* and *b* unit cell axes with only slight changes in the orientation of the helix. They observed no significant change in the chain conformation as a result of the transition. The V_a structure is converted by humidification first into a higher hydrate known as the V_h form, which is thought (Zobel et al., 1967) to have the same 6_1 helical conformation, but the interhelix separation is increased from 13.0 to 13.7 Å because of the inclusion of water molecules between the chains. Further humidification, followed by soaking in water, leads to formation of the B structure, with the A and C forms sometimes observed as intermediates. The structure proposed for B-amylose (Blackwell et al., 1969) has chains with six residues per turn, repeating in 10.4 Å. It has also been suggested (Blackwell et al., 1969) that the mechanism for conversion from the V into the B form involves stretching out of the helix, thereby breaking the interturn hydrogen bond, which is reformed by insertion of a water molecule between the two hydroxyl groups, as shown in Figure 10-6*b*. Other structures suggested for B-amylose are 3 and 4 single helices (Kreger, 1951, Sundararajan and Rao, 1969) and a double helix of 6 chains (Kainuma and French, 1972).

The conformation of amylose in aqueous systems has not been established unambiguously. The solubilization of amylose in water and the tendency of solutions to precipitate on standing, retrogradation, are the main experimental problems that have impeded the determination of the solution structure of this biopolymer. The bulk of solution–structure studies of amylose and some of its derivatives have been based on hydrodynamic properties. Most investigations of this nature have also considered the effect of complexation for the long-known amylose–iodine complex (Rundle and Baldwin, 1943; Rundle and French, 1943; Foster and Zucker, 1952). Three general types of conformations for amylose in solution have been proposed: (1) the tight helix model, (2) the random coil model, and (3) the extended helix model. Table 10-2 summarizes the investigations that have led to the various solution models of amylose; Figure 10-7 schematically illustrates these three models. The extended helix model, which might be viewed as a compromise structure between the tight helix and random coil, seems to account for much of the experimental data. The tight helix model is not consistent with the most recent intrinsic viscosity measurements made for both noniodine complex and iodine complex forms of amylose (Senior and Hamori, 1973). The random coil model is consistent with the intrinsic viscosity data, but is

Table 10-2 Proposed Solution Structures of Amylose and Some of its Analogs

Compounds	Solvent Environment	Source	Proposed Structure
Amylose	H_2O	Banks & Greenwood (1969) Banks et al. (1971) Banks & Greenwood (1971) Banks & Greenwood (1972a)	In neutral, aqueous KCl solution a random coil at the θ-temperature.
	H_2O–0.78 *M* NaCl DP[513-10,800][a]	Goebel & Brant (1970) Brant & Dimpfl (1970)	$C_x^b = 5.1$ and independent of DP.
	Neutral aqueous solution	Rao & Foster (1963a) Hollo & Szejtli (1957)	A stiff coil with essentially helical backbone contour. At high pH there is a partial breakdown of the helical structure, leading to increasing flexibility.
	Aqueous-based solvents	Banks & Greenwood (1972*b*)	Expanded coil in alkaline solution Random coil in neutral aqueous salt solution. Helical form in A neutral solution + complexing agent. An alkaline solution + complexing agent. 0.01 *M* KOH + 0.3 KCl.

H_2O/DMSO mixed solvent	Dintzis & Tobin (1969)	Optical rotation cannot distinguish between random coil and the partially extended helix.
95% H_2O–5% DMSO	Pfannemüller et al. (1971)	DP[50-80], well-ordered, stiff chains. DP $>$ 80, worm-like helical chains.
Dilute aqueous + 8.67 mM KI + 1% ethanol	Senior & Hanori (1973)	Loose extended helical regions interrupted by short disordered regions. Complexing with iodine produces a helical contraction of the loose helical regions in order to yield ion entrapment.
Formamide, H_2O, DMSO, acetone, and $\frac{1}{2}$ N NaOH–H_2O DP[1,13,000]	Burchard (1963)	A coil that is drained to different solvents. Radius of gyration and θ point decrease rapidly with temperature.
Formamide, DMSO, and 0.15 M KOH	Banks & Greenwood (1963, 1968*a,b*, 1971)	A random coil that undergoes solvent expansion, but maintains the same basic backbone structure, i.e. long-range interactions increase while short-range do not change.

Table 10-2 (Continued)

Compounds	Solvent Environment	Source	Proposed Structure
	Ethylenediamine Formamide	Dintzis et al. (1970)	Random coil-like structure with amylose aggregate particles relatively spherical. Random coil-like structure with aged aggregate particles asymmetric.
	Theoretical calculation	Rao et al. (1969)	$C_\infty = 6.6 + 0.2$ exp., $C_\infty = 6.9$ theor. A random coil with short helical sequences in neutral aqueous solution.
	DMSO	Cael et al. (1973)	Amylose does not retain, even locally, the V conformation in DMSO. The chains are probably extended.
	DMSO	Casu et al. (1966)	The CH_2OH hydroxyl groups are bonded to Me_2SO molecules. The 2-OH and 3′-OH groups spend some time intramolecularly hydrogen bonded.

Carboxy-methylamylose	H_2O–0.78 *M* NaCl DP[232-1710]	Goebels & Brant (1970) Brant & Dimpfl (1970)	C_x = 5.3, which is independent of DP. A statistical coil with no identifiable helical character.
	H_2O–0.5 *M* NaCl pH = 8 T = 35°C	Brant and Min (1969)	C_∞ = 10.
	H_2O–0.15 *M* NaCl	Patel et al. (1967) Patel and Patel (1971)	Statistical coil C_∞ = 7.9, with no helical character over all pHs studied.
Amylose Acetate	Dilute nitromethane	Cowie (1961*a*,*b*)	Partially extended helix.
Amylose Triacetate	Nitromethane	Banks & Greenwood (1968*a*)	C_∞ = 13.4; partially extended structure.
Diethylamino-ethylamylose hydrochloride	H_2O–0.78 *M* NaCl DP[232-1710]	Goebels & Brant (1970) Brant & Dimpfl (1970)	C_x = 6.4, independent of DP. Statistical coil with no helical character.

[a] DP, degree of polymerization.
[b] C_x, characteristic ratio for DP = x.

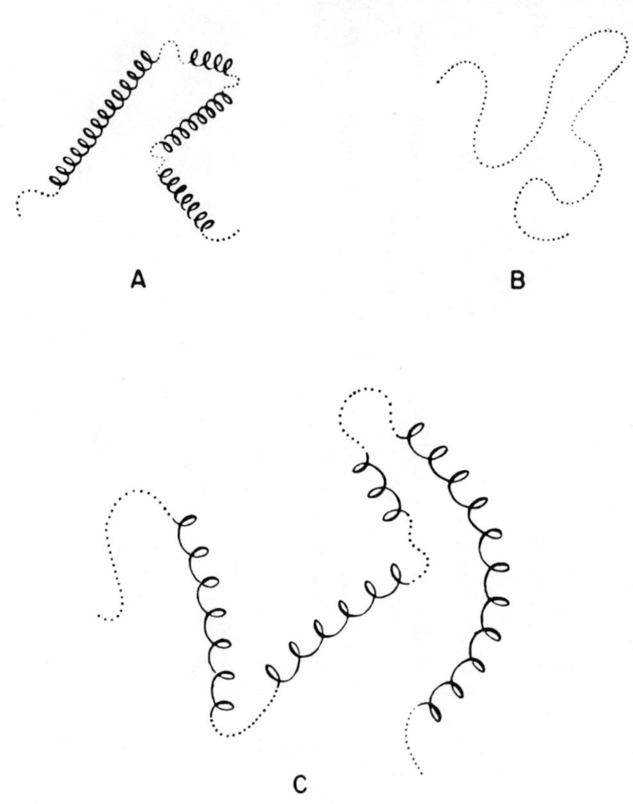

Fig. 10-7 Schematic representation of the different models proposed for the conformation of an amylose macromolecule in aqueous solution (only part of the molecule is represented): (A) the Holló-Szejtli tight-helix model, (B) the Banks-Greenwood random-coil model, and (C) the extended-helix model suggested by Senior and Hamori. (The dotted segments designate disordered regions. Drawings (A) and (C) are not intended to specify the chirality of the helical segments.) Senior and Hamori (1973).

not consistent with the kinetics of iodine–amylose complex formation (Thompson and Hamori, 1969, 1971; Thompson, 1970).

The extended helix model attributes the drop in intrinsic viscosity upon iodine complexation to the entrapment of iodine ions in the loose helical regions that contract into tight helices of the "V" amylose type. From the original work of Hanes (1937), it was believed that the iodine ions are complexed within the helical cavity of the polysaccharide (Freudenberg et al., 1939). Bates et al. (1943) confirmed the Hanes model for the solid amylose–iodine complex by X-ray diffraction and, from titration studies, concluded that there is one iodine ion per six residues. Baldwin et al. (1944) reached the conclusion that the ion saturation level depends largely on the iodide concentration of the solvent system. Robin (1964) has proposed, from observation on the benzamide–iodine complex, that the iodine in the helical cavity of amylose is composed of a linear chain of triiodide ions bent with respect to each other. Various iodine ion species are thought to be involved in the complexes. Gibert and Marriot (1948) have suggested that the blue color of the solid amylose–iodine complex might contain $3I_2 \cdot 2I^-$ ions. Yamagishi et al. (1972) have determined the rate constant of the amylose unit–I_3 complex formation to be $7 \times 10^8\ M^{-1}\ sec^{-1}$, which is close to that of a diffusion-controlled process.

Nicolson et al. (1966) estimated the hydration enthalpy of amylose to be 10.4 kcal/mole. Water bound in crystalline systems generally forms hydrogen bonds with enthalpies of approximately, 3–4 kcal/g bond (Pimentel and McClellan, 1960). Hence a value of 10.4 kcal may imply that water molecules in the carbohydrate matrix have three (or even four) points of attachment.

Around pH 13, aqueous amylose solutions display a major change in their intrinsic viscosity and optical rotation (see Figure 10-8) (Rao and Foster, 1963*a*; Foster, 1965; Doppert and Staverman, 1966*a,b*). Reeves (1954) suggested that some of the glucose units undergo a change in ring conformation at pH 13 upon ionization of certain hydroxyl groups. Rao and Foster (1963*b*) demonstrated no change in ring conformation in maltose for increasing pH and concluded that this is also the case for amylose. Senior and Hamori (1973) attribute these effects to the breakdown of the helical structures of the molecule because of the electrostatic charge repulsions on the dissociating hydroxyl groups of the glucose residues.

Carboxymethyl amylose (CMA) shows intrinsic viscosity and optical rotation properties quite different from amylose at low pH (see Figure 10-9). The viscosity of an aqueous solution of CMA increases with increasing pH from 2 to 5, whereas the viscosity of an aqueous amylose solution remains constant from 2 to 12 (Rao and Foster, 1965; Patel and Patel, 1971). The rapid change in viscosity and optical rotation for a relatively small change in pH is indicative of a helix-coil transition in some systems such as poly(L-glutamic acid), PGA (Doty, 1959). In other systems such as poly (acrylic acid), PAA, a rapid change in viscosity and/or optical rotation has been attributed to molecular aggregration/dissociation. Kano and Ikegami (1966) have defined an apparent binding constant as a function of the degree of ionization, α. Figure 10-10 shows plots of the apparent binding constant versus α for CMA, PAA, and PGA. It can be seen that the CMA curve is very similar to PAA and shows no discontinuity, indicative of a helix-coil transition, as does the PGA curve. Thus some type of intermolecular aggregration process probably takes place between pH 2 and 5 for CMA.

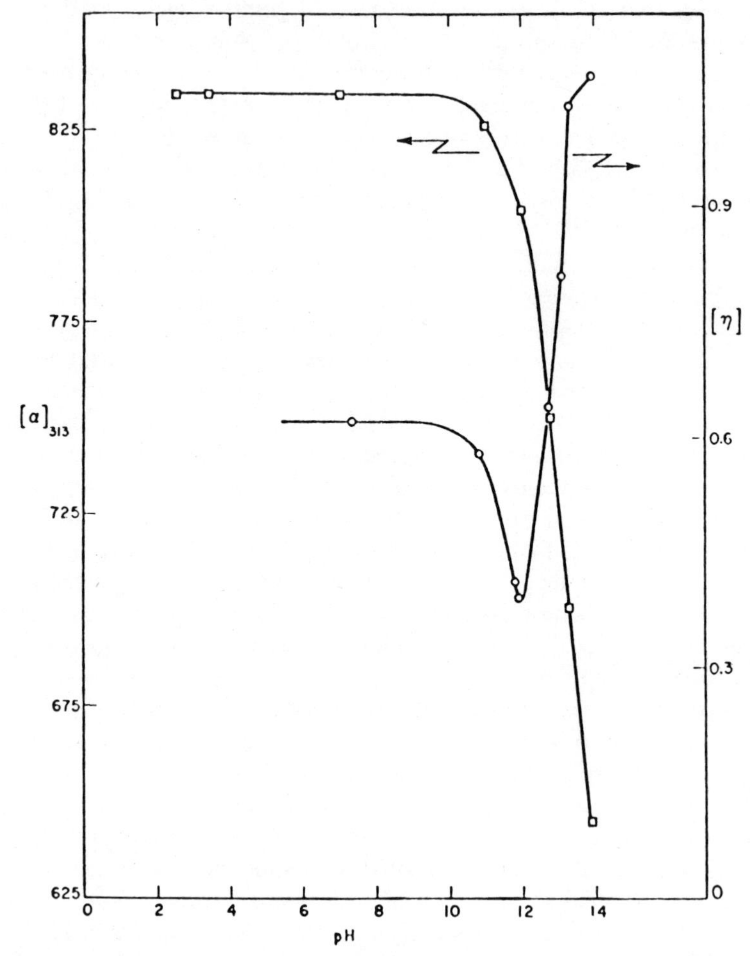

Fig. 10-8 Dependence on pH of the specific optical rotation (at 313 mμ) and intrinsic viscosity of aqueous solutions of an amylose fraction. Rao and Foster (1963a).

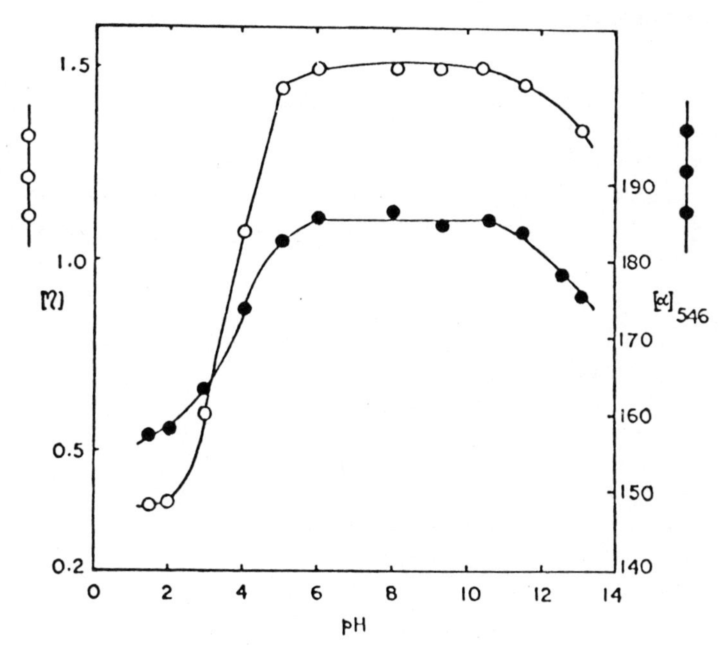

Fig. 10-9 Plot of specific optical rotation, $[\alpha]_{546}$ and viscosity vs. pH for CMA at 35°C. Patel and Patel (1971).

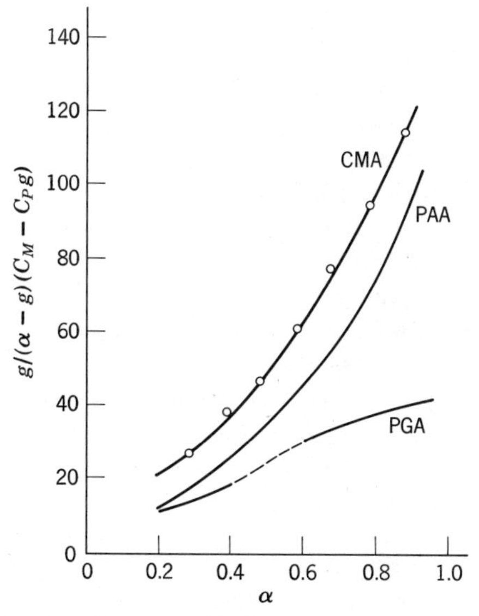

Fig. 10-10 Comparative plot of apparent binding constant $g/[(\alpha - g)(c_M - c_P g)]$ and α for CMA, PGA, and PAA. Patel and Patel (1971).

REFERENCES

Anderson, B., P. Hoffman, and K. Meyer (1963). *Biochem. Biophys. Res. Commun.*, **5**, 430.

Anderson, B. and K. Meyer (1962). *Fed. Proc.*, **21**, 171.

Arnott, S. (1973). *Trans. Am. Cryst. Assoc.*, **9**, 31.

Arnott, S., J. M. Guss, D. W. L. Hukins, and M. B. Mathews (1973*a*). *Science*, **180**, 743.

Arnott, S., J. M. Guss, D. W. L. Hukins, and M. B. Mathews (1973*b*). *Biochem. Biophys. Res. Commun.*, **54**, 1377.

Atkins, E. D. T., R. Gaussen, D. H. Issac, V. Nandanwar, and J. R. Sheehan (1972*b*). *J. Polym. Sci.* (*B*), **10**, 863.

Atkins, E. D. T., T. E. Hardingham, D. H. Isaac, and H. Muir (1974). *Biochem. J.*, **141**, 919.

Atkins, E. D. T. and D. H. Isaac (1973). *J. Mol. Biol.*, **80**, 773.

Atkins, E. D. T., D. H. Issac, I. A. Nieduszynski, C. F. Phelps, and J. F. Sheehan (1974). *Polymer*, **15**, 263.

Atkins, E. D. T. and T. C. Laurent (1973). *Biochem. J.*, **133**, 605.

Atkins, E. D. T. and W. Mackie (1972). *Biopolymers*, **11**, 1685.

Atkins, E. D. T., C. F. Phelps, and J. K. Sheehan (1972*a*). *Biochem. J.*, **128**, 1255.

Atkins, E. D. T. and J. K. Sheehan (1971). *Biochem. J.*, **125**, 92P.

Atkins, E. D. T. and J. K. Sheehan (1972). *Nature* (*New Biol.*), **235**, 253.

Atkins, E. D. T. and J. K. Sheehan (1973). *Science*, **179**, 562.

Balazs, E. A. (1958). *Fed. Proc.*, **17**, 1086.

Baldwin, R. R., R. S. Bear, and R. E. Rundle (1944). *J. Am. Chem. Soc.*, **66**, 111.

Banks, W. and C. T. Greenwood (1963). *Makromol. Chem.*, **67**, 49.

Banks, W. and C. T. Greenwood (1968*a*). *Carbohydr. Res.*, **7**, 349.

Banks, W. and C. T. Greenwood (1968*b*). *Carbohydr. Res.*, **7**, 414.

Banks, W. and C. T. Greenwood (1969). *Eur. Polymer J.*, **5**, 649.

Banks, W. and C. T. Greenwood (1971). *Biopolymers*, **12**, 141.

Banks, W. and C. T. Greenwood (1972*a*). *Biopolymers*, **11**, 315.

Banks, W. and C. T. Greenwood (1972*b*). *Carbohydr. Res.*, **21**, 229.

Banks, W., C. T. Greenwood, D. J. Housston, and A. R. Proctor (1971). *Polymer*, **12**, 452.

Barlow, G. H., L. J. Coen, and M. M. Mozen (1964). *Biochim. Biophys. Acta*, **83**, 272.

Barlow, G. H., N. D. Sanderson, and P. D. McNeil (1961). *Arch. Biochem. Biophys.*, **94**, 581.

Bates, F. L., D. French, and R. E. Rundle (1943). *J. Am. Chem. Soc.*, **65**, 142.

Bear, R. and D. French (1941). *J. Am. Chem. Soc.*, **63**, 2298.

Blackwell, J., A. Sarko, and R. H. Marchessault (1969). *J. Mol. Biol.*, **42**, 379.

Blix, G. and O. Snellman (1945). *Arkivkemi Minerol. Geol.*, **A19** (**32**), 1.

Brant, D. A. and W. L. Dimpfl (1970). *Macromolecules*, **3**, 655.

Brant, D. A. and B. K. Min (1969). *Macromolecules*, **2**, 1.

Brimacombe, J. S. and J. M. Webber (1964). *Mucopolysaccharides*. BBA Library 6, Elsevier, Amsterdam.

Burchard, W. (1963). *Makromol. Chem.*, **64**, 110.

Cael, J. J. (1975). Ph.D. Thesis, Dept. of Macromolecular Science, Case Western Reserve University, Cleveland, Ohio.

Cael, J. J., J. L. Koenig, and J. Blackwell (1973). *Carbohydr. Res.*, **29**, 123.

Casu, B., M. Reggiani, G. G. Gallo, and A. Vigevani (1966). *Tetrahedron*, **22**, 3061.

Cowie, J. M. G. (1961*a*). *Makromol. Chem.*, **42**, 230.

Cowie, J. M. G. (1961*b*). *J. Polymer Sci.*, **19**, 455.

Dea, I. C. M., R. Moorhouse, D. A. Rees, S. Arnott, J. M. Guss, and E. A. Balazs (1973). *Science*, **179**, 560.

Dintzis, F. R., G. E. Babcock, and R. Tobin (1970). *Carbohydr. Res.*, **17**, 216.

Dintzis, F. R. and R. Tobin (1969). *Biopolymers*, **7**, 581.

Doppert, H. L. and A. J. Staverman (1966*a,b*). *J. Polymer Sci. A-1*, **4**, 2367a, 2373b.

Doty, P. (1959). *Rev. Mod. Phys.*, **31**, 107.

Foster, J. F. (1965). In *Starch: Chemistry and Technology* (R. L. Whistler and E. F. Paschall, Eds.). Academic Press, New York, p. 349.

Foster, J. F. and D. Zucker (1952). *J. Phys. Chem.*, **56**, 170.

Freudenberg, K., E. Schaaf, G. Dumpert, and T. Ploetz (1939). *Naturwissenschaftern*, **27**, 850.

Gilbert, G. A. and J. V. R. Marriot (1948). *Trans. Faraday Soc.*, **44**, 84.

Goebel, K. D. and D. A. Brant (1970). *Macromolecules*, **3**, 634.

Guss, J. M., D. W. L. Hukins, P. J. C. Smith, W. T. Winter, S. Arnott, R. Moorhouse, and D. A. Reese (1975). *J. Mol. Biol.*, **95**, 359.

Hanes, C. S. (1937). *New Phytologist.*, **36**, 189.

Hascall, V. C. and D. Heinegard (1974). *J. Biol. Chem.*, **249**, 4232.

Helbert, J. R. and M. A. Marini (1963). *Biochemistry*, **2**, 1101.

Hollo, H. and J. Szejtli (1957). *Periodica Polytechnica, Chem. Eng.*, **1**, 223.

Isaac, D. H. and E. D. T. Atkins (1973). *Nature* (*New Biol.*), **244**, 252.

Jensen, R., O. Smellman, and Sylvén (1948). *J. Biol. Chem.*, **174**, 265.

Kainuma, K. and D. French (1972). *Biopolymers*, **11**, 2241.

Kano, N. and A. Ikegami (1966). *Biopolymers*, **4**, 823.

Katz, J. R. and T. B. Van Itallie (1930). *Z. Physik. Chem.*, **A150**, 90.

Kreger, D. R. (1951). *Biochim. Biophys. Acta*, **6**, 406.

Lasker, S. E. (1965). Dissertation Abstr., 65, *12*, 579: S. E. Lasker, *Some Solution Properties of Fractionated Bovine Heparin*. Univ. Microfilms, Ann Arbor, Mich.

Laurent, T. C. and J. Gergely (1955). *J. Biol. Chem.*, **215**, 325.

Laurent, T. C. and A. Pietruszkiewicz (1961). *Biochim. Biophys. Acta*, **49**, 258.

Lindahl, V. and L. Rodén (1961). *Biochem. Biophys. Res. Commun.*, **5**, 430.

Mathews, M. B. (1956). *Arch. Biochem. Biophys.*, **61**, 367.

Mathews, M. B. (1959). *Biochim. Biophys. Acta*, **35**, 9.

Mathews, M. B. (1962). *Biochim. Biophys. Acta*, **58**, 92.

Mathews, M. B. (1967). *The Connective Tissue*, Internat. Acad. of Pathology, Monograph No. 7, pp. 304. Williams & Williams, Baltimore.

Mathews, M. B. and I. Lozaityte (1958). *Arch. Biochem. Biophys.*, **74**, 158.

Nicholson, P. C., G. U. Yuen, and B. Zaslow (1966). *Biopolymers*, **4**, 677.

Ogston, A. G. and C. F. Phelps (1960). *Biochem. J.*, **78**, 827.

Ogston, A. G. and B. N. Preston (1966). *J. Biol. Chem.*, **241**, 17.

Ogston, A. G. and J. E. Stanier (1950). *Biochem. J.*, **46**, 364.

Patel, J. R. and R. D. Patel (1971). *Biopolymers*, **10**, 839.

Patel, J. R., C. K. Patel, and R. D. Patel (1967). *Starke*, **19**, 330.

Pfannemüller, B., H. Mayerhöfer, and R. C. Schulz (1971). *Biopolymers*, **10**, 243.

Pimentel, G. C. and A. L. McClellan (1960). *The Hydrogen Bond*. Freeman, San Francisco, p. 254.

Rao, V. S. R. and J. F. Foster (1963*a*). *Biopolymers*, **1**, 527.

Rao, V. S. R. and J. F. Foster (1963*b*). *J. Phys. Chem.*, **67**, 951.

Rao, V. S. R. and J. F. Foster (1965). *Biopolymers*, **3**, 185.

Rao, V. S. R., N. Yathindra, and P. R. Sundararajan (1969). *Biopolymers*, **8**, 325.

Reeves, R. E. (1954). *J. Am. Chem. Soc.*, **76**, 4595.

Robin, M. B. (1964). *J. Chem. Phys.*, **40**, 3369.

Rundle, R. E. and R. R. Baldwin (1943). *J. Am. Chem. Soc.*, **65**, 554.

Rundle, R. E. and D. French (1943). *J. Am. Chem. Soc.*, **65**, 558.

Senior, M. B. and E. Hamori (1973). *Biopolymers*, **12**, 65.

Sheehan, J. K., E. D. T. Atkins, and I. A. Nieduszynski (1974). *J. Mol. Biol.*, **95**, 153.

Stone, A. L. (1969). In *Structure and Stability of Biological Macromolecules*. (S. N. Trimasheff and G. O. Fasman, Eds.). Marcel Dekker, New York, Vol. 2, p. 353.

Sundararajan, P. R. and V. S. R. Rao (1969). *Biopolymers*, **8**, 313.

Tanford, C., E. Marler, E. Jury, and E. A. Davidson (1964). *J. Biol. Chem.*, **239**, 4034.

Thompson, J. C. (1970). Ph.D. Dissertation, University of Delaware.

Thompson, J. C. and E. Hamori (1969). *Biopolymers*, **8**, 689.

Thompson, J. C. and E. Hamori (1971). *J. Phys. Chem.*, **75**, 272.

Tsiganos, C. P., T. E. Hardingham, and H. Muir (1971). *Biochim. Biophys. Acta*, **229**, 529.

Winter, W. T. and A. Sarko, (1974). *Biopolymers*, **13**, 1447a; 1461b.

Winter, W. T., P. J. C. Smith, and S. Arnott (1975). *J. Mol. Biol.* (in press).

Wortman, B. (1964). *Biochim. Biophys. Acta*, **83**, 288.

Yamagishi, A., T. Imamura, and M. Fujimoto (1972). *Bull Chem. Soc. Jap.*, **45**, 2304.

Zobel, H. F., A. D. French, and M. E. Hinkle (1967). *Biopolymers*, **5**, 837.

CHAPTER 11

Interaction of Polypeptides and Collagen with Glycosaminoglycans and Proteoglycans

The characterization of the molecular structure of connective tissue is a most difficult task. The limited extent of sequential primary structure in the component classes of macromolecular species, as reflected by the amorphous nature of this composite material, is a major source of the obstacles to structural studies. However, Blackwell and co-workers have demonstrated the utility of studying dilute solution mixtures of connective tissue components as structural models for connective tissue. This chapter is a summary of their novel work.

I INTRODUCTION

Glycosaminoglycans occur as components of connective tissue in the amorphous "ground substance" that interacts with the strong protein fibers of these composite materials. In cartilage, for example, the predominant glycosaminoglycans are chondroitin 4- and 6-sulfate, which form a gel-like matrix between the collagen fibers. Most, if not all, of the polysaccharide chains in connective tissue are attached to proteins by covalent bonds (Lindahl and Rodén, 1964). Still, knowledge of the predominant noncovalent interactions between glycosaminoglycans and proteins is essential for an understanding of the structure and function of connective tissue.

II CHARACTERIZATION OF THE GLYCOSAMINOGLYCAN–POLYPEPTIDE INTERACTION

As an initial model for connective tissue structure, Gelman, Blackwell, and co-workers (see references for Table 11-1) have used difference circular dichroism (CD) spectroscopy to study the interactions between glycosaminoglycans and some cationic homopolypeptides, as well as collagen, in dilute aqueous solutions at neutral pHs. Figure 10-1 lists the structural geometries of six of the glycosaminoglycans with simple repeating disaccharide units. For the structural properties of the remaining glycosaminoglycans that have been studied, the reader is referred to the appropriate references reported in Table 11-1. The side chain units of each of the three cationic homopolypeptides are shown in Figure 11-1. For a discussion of the primary structure of collagen, the reader should consult the references given in Section 2 of Chapter 15.

Three classes of structural information have been identified in the interactions of polypeptides and glycosaminoglycans: (1) the type and extent of induced polypeptide conformation resulting from the interaction, (2) the ratio of the relative number of disaccharide units to peptide residues in solution producing the "strongest" polypeptide–glycosaminoglycan interaction, and (3) the temperature at which the induced polypeptide conformation melts out for the optimum disaccharide–peptide ratio. Table 11-1 summarizes the results of the work of Gelman, Blackwell, and co-workers. Figure 11-2 contains the CD spectra of mixtures of poly(L-arginine) (PLA) and chondroitin 4-sulfate (C4-S) in water at neutral pHs at various disaccharide–peptide ratios and the same spectra with the chondroitin 4-sulfate spectral contribution removed. The difference CD spectra yield direct information concerning changes in polypeptide conformation. The

...Polypeptide–Glycosaminoglycan (P + G) Aqueous Solutions at Neutral pHs.

Glycosaminoglycan	Polypeptide Solution Conformation				Disaccharide:Peptide Ratio	T_m °C (P + G)	Polypeptide
	Polypeptide Only		(P + G)				
	Type	Percent	Type	Percent			
Chondroitin 6-sulfate (ref. 1, 2, 3, 4, 5)	C.C.[a]	100	α[b]	>80	1:1	47.0 ± 1.0	Poly(L-lysine)
	C.C.	100	α	>80	1:2	76.0 ± 1.0	Poly(L-arginine)
	C.C.	100	C.C.	100	No preference	None	Poly(L-ornithine)
	C.T.H.[c]	100	C.T.H.	100~	5.5:100	46~	Collagen[d]
Hyaluronic acid (ref. 7, 8)	C.C.	100	R[e]	—	1:1	None	Poly(L-lysine)
	C.C.	100	α	35–40	1:1	35.0	Poly(L-arginine)
	C.T.H.	100	C.T.H.	100~	10:100	46~	Collagen[d]
Chondroitin 4-sulfate (ref. 2, 5, 8)	C.C.	100	α	20~	1:1	25.0	Poly(L-lysine)
	C.C.	100	α	>80	1:2	54.5	Poly(L-arginine)
	C.C.	100	C.C.	100	No preference	None	Poly(L-ornithine)
	C.T.H.	100	C.T.H.	100~	14.0:100	46~	Collagen[d]
Heparitin sulfate (ref. 7)	C.C.	100	R	—	1:2	None	Poly(L-lysine)
	C.C.	100	α	>80	1:1	65.5	Poly(L-arginine)
Keratan sulfate-1 (ref. 7, 8)	C.C.	100	R	—	1:1.2	None	Poly(L-lysine)
	C.C.	100	α	>80	1:1.2	>90	Poly(L-arginine)
	C.T.H.	100	C.T.H.	100~	10:100	46~	Collagen[d]
	C.C.	100	C.C.	100	No preference	None	Poly(L-ornithine)
Dermatan sulfate (ref. 5, 8)	C.C.	100	α	60	1:1.4	74.5	Poly(L-lysine)
	C.C.	100	α	>80	1:1.4	>90	Poly(L-arginine)
	C.C.	100	C.C.	100	No preference	None	Poly(L-ornithine)
	C.T.H.	100	C.T.H.	100~	7:100	46	Collagen[d]

Table 11.1 (continued)

Glycosaminoglycan	Polypeptide Solution Conformation				Disaccharide:Peptide Ratio	T_m°C (P + G)	Polypeptide
	Polypeptide Only		(P + G)				
	Type	Percent	Type	Percent			
Heparin (ref. 9)	C.C.	100	α	>80	1:2.3 ± 0.1	>90	Poly(L-lysine)
	C.C.	100	α	>80	1:3.3 ± 0.1	>90	Poly(L-arginine)
	C.C.	100	α	>80	1:2.3 + 0.1	56.0 + 1.0	Poly(L-ornithine)
Intact proteoglycan[f] (ref. 10)	C.C.	100	α	Low[g]	1:1.0 ± 0.1	N.A.	Poly(L-lysine)
	C.C.	100	α	High[h]	1:0.60 ± 0.05	56.0 ± 1.0	Poly(L-arginine)
	C.C.	100	C.C.	100	No preference	None	Poly(L-ornithine)

[a] C.C., "charged coil" conformation (Tiffany and Krimm, 1969).
[b] α, alpha helical conformation.
[c] C.T.H., collagen triple helical structure.
[d] Soluble calf skin collagen (Gelman and Blackwell, 1973*c*).
[e] R, random chain organization.
[f] Sample from bovine nasal septum cartilage (Gelman et al., 1974).
[g] The proteoglycan CD spectra were not constant. Hence helical content could only be qualitatively estimated.
[h] Same as [g] except helix content is estimated to be extensive.

Reference code
1. Gelman et al. (1972)
2. Gelman and Blackwell (1973*a*)
3. Gelman et al. (1973*a*)
4. Gelman et al. (1973*b*)
5. Gelman and Blackwell (1973*b*)
6. Gelman and Blackwell (1973*c*)
7. Gelman and Blackwell (1974*a*)
8. Gelman and Blackwell (1974*b*)
9. Gelman and Blackwell (1973*d*)
10. Gelman et al. (1974)

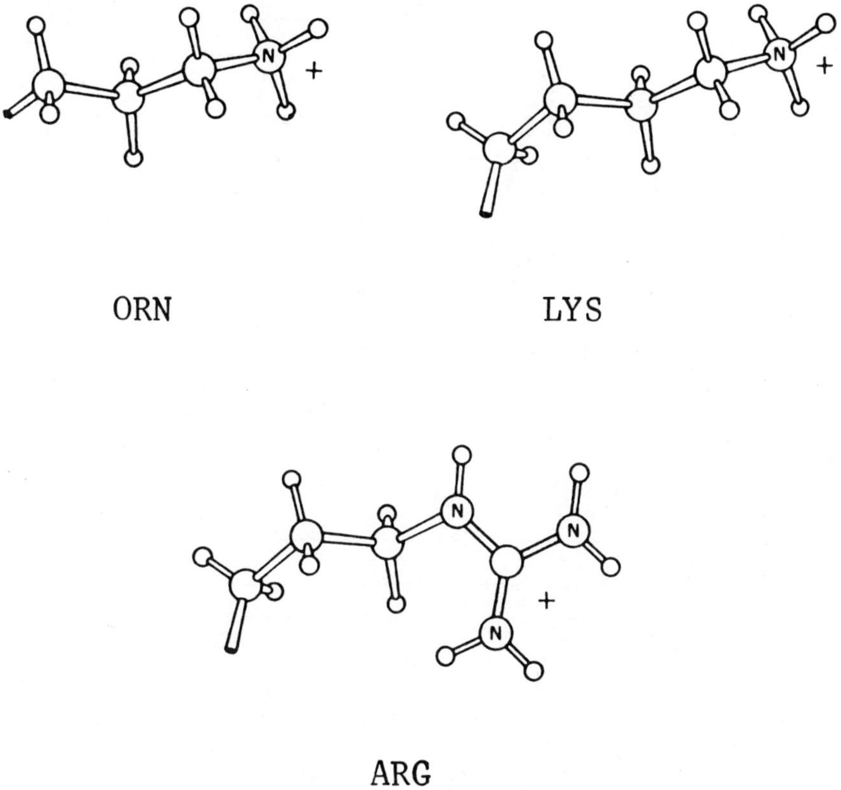

ORN LYS

ARG

Fig. 11-1 Ornithine, lysine and arginine cationic polypeptide side chains.

spectra in Figure 11-2 are typical for the systems studied. Figure 11-3, which is again representative of the polypeptide–glycosaminoglycan systems studied, contains plots of the ellipticity at 221 nm, $[\theta]_{221}$, as a function of temperature for mixtures of collagen and keratin sulfate-1 at several disaccharide to 100 amino acid residues ratios. Note that for some ratios there is biphasic melting out of the constituent collagen and keratin sulfate-1 structures.

Several general conclusions can be drawn from the results seen for the interactions between acid glycosaminoglycans and cationic homopolypeptides. The interactions are seen to be dependent on the following:

1. The length of the polypeptide side chain: the strength of the interaction increases across the series: ornithine $<$ lysine $<$ arginine.

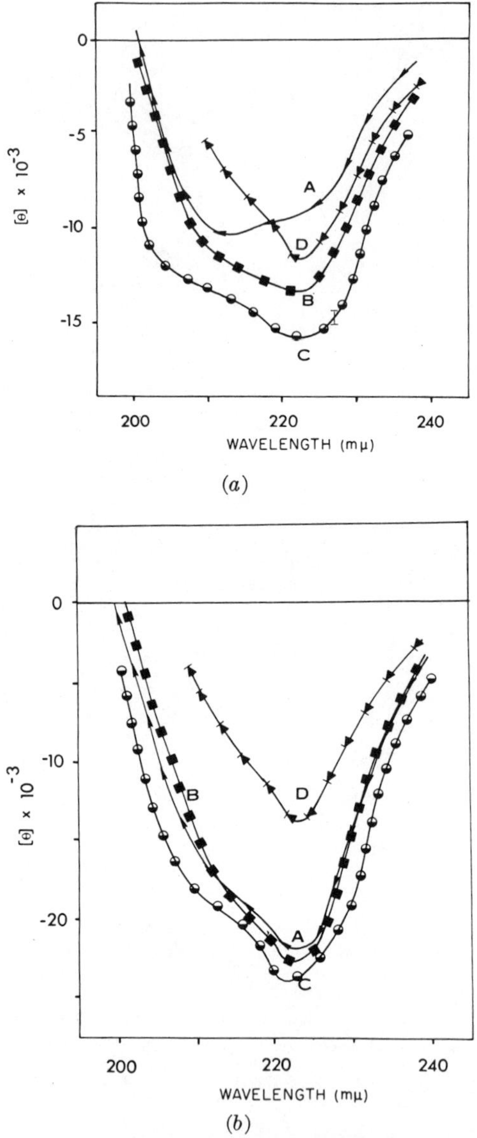

Fig. 11-2 (*a*) The CD spectra of mixtures of PLA and C4-S, in water at neutral pH, at various ratios. (A) 0.5:1, (—◀—); (B) 1:1, (—■—); (C) 2:1, (—◒—); (D) 3:1, (—|◀—). (*b*) The data presented in Figure 3*a* after subtraction of the C 4-S contribution. (A) 0.5:1, (—◀—); (B) 1:1, (—■—); (C) 2:1, (—◒—); (D) 3:1, (—|◀—). Gelman and Blackwell, (1973*b*).

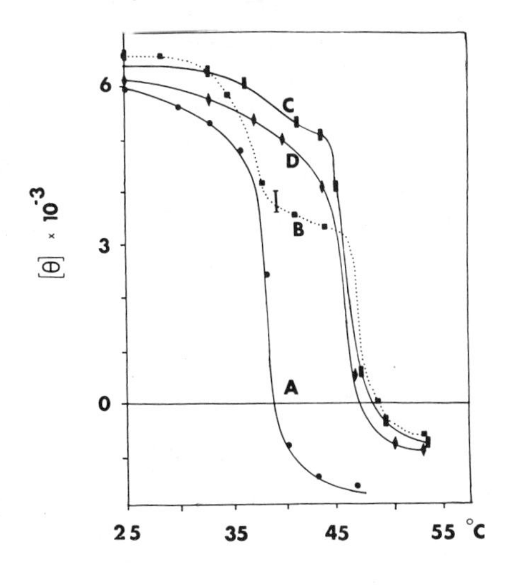

Fig. 11-3 Plot of ellipticity at 221 nm, $[\theta]_{221}$, as a function of temperature of mixtures of collagen and keratan sulfate-1 at various values of r': A, $r' = 0$, (● — ●); B, $r' = 6.6$, (■ ∘ ■); C, $r' = 9.6$, (▮ — ▮); D, $r' = 10$, (◆ — ◆). Gelman and Blackwell (1974*b*).

2. The position of the sulfate groups: a stronger interaction results when the sulfate is on the 6 side chain in chondroitin 6-sulfate than directly on the galactose ring as in chondroitin 4-sulfate.
3. The degree of sulfation: this changes the stoichiometry of the interaction.
4. The position of the carboxyl groups: the equatorial carboxyls of the chondroitin sulfates interact with arginine side chains, but the axial groups of dermatan sulfate do not. However, this shielding of the carboxyls may be partly responsible for the strength of the dermatan sulfate interactions, since the interactions with keratan sulfate, which has no carboxyl groups and less sulfates than dermatan sulfate, show an equally high thermal stability.
5. The glycosidic linkages: this may affect the interactions by controlling the chain conformation of the glycosaminoglycans.

If the observed T_m for the α-helix-induced conformations are indicative of the strength of glycosaminoglycan–polypeptide interaction, six of the glycosaminoglycans that induce α-helical conformations can be ranked in the order of increasing strength of interaction: hyaluronic acid $<$ chondroitin 4-sulfate $<$ heparitin sulfate $<$ chondroitin 6-sulfate $<$ keratan sulfate $\leq$ dermatan sulfate. It is not possible to distinguish between keratan sulfate and dermatan sulfate on the basis of T_m, since both systems, when mixed with poly(L-arginine), melt above 90°C. Keratan sulfate, however, induces the random rather than the α helix for poly(L-lysine) and is considered to show weaker interactions than dermatan sulfate on this basis.

Studies on the binding of cations to these polysaccharides (Scott et al., 1964; Scott and Dorling, 1965) yield the same general trends as observed by

Gelman, Blackwell, and co-workers. That is, the critical electrolyte concentration was low for those polysaccharides having only carboxyl groups (hyaluronic acid); next came those with both sulfate and carboxyl groups (chondroitin 4-sulfate), and the highest was observed for those glycosaminoglycans having only sulfate groups (keratin sulfate).

Blackwell and co-workers (Schodt et al., 1976) have given some attention to the effects of changes in salt concentration and pH on the interactions between the basic polypeptides and connective tissue glycosaminoglycans in dilute aqueous solution. They find that interactions with poly(L-lysine) undergo transitions as the ionic strength is increased. The polysaccharides can be placed in order of increasing strength of interaction: chondroitin 6-sulfate, dermatan sulfate, heparin, using the ionic strength at which the interactive conformational effect on the polypeptide is disrupted. Scattering effects are observed prior to the transition, which indicates increased aggregation. No such transitions are observed for the poly(L-arginine)-glycosaminoglycan mixtures, suggesting again that this polypeptide interacts more strongly with glycosaminoglycans than poly(L-lysine).

Similar pH-induced conformational transitions occur in the pH range 2.5–3.8 and are thought to be a result of the deionization of carboxyl groups. For poly(L-lysine), the conformational effect is disrupted at low pH. In the case of poly(L-arginine), similar behavior is seen for hyaluronic acid mixtures. The transitions for the mixtures containing the sulfated polysaccharides, however, appear to correspond to an increase in scattering, and the stronger interactions are maintained via the sulfate groups.

The conformation-directing effect of poly(L-lysine) appears to follow a natural structure division; only chondroitin 6-sulfate, chondroitin 4-sulfate, and dermatan sulfate force this cationic homopolypeptide into the α helix. These three glycosaminoglycans are very similar chemically and are important to collagen fibril formation. In contrast, poly(L-lysine) adopts a random conformation in the presence of hyaluronic acid, heparitin sulfate, and keratan sulfate, which have markedly different structures from the previous group.

It is surprising to note that the T_m of the glycosaminoglycan collagen mixtures are roughly the same, $\sim 46°C$, regardless of the choice of glycosaminoglycan. However, the disaccharide–peptide ratio does vary markedly from one glycosaminoglycan to another. If the disaccharide–peptide ratios are indicative of the strength of the collagen–glycosaminoglycan interaction, the glycosaminoglycans may be ranked in decreasing order of interaction as follows: chondroitin 6-sulfate > dermatan sulfate > hyaluronic acid > keratan sulfate > chondroitin 5-sulfate.

The interactions involving heparin are analogous to those observed for the first six glycosaminoglycans listed in Table 11-1, with the exception that heparin exhibits a significant conformation-directing influence on poly(L-ornithine). Moreover, the high melt temperatures of all three glycosaminoglycans–polypeptide mixtures reported in Table 11-1 are indicative of stronger stabilizing interactions for heparin than any other glycosaminoglycan investigated. Gelman and Blackwell (1973*d*) attribute this high thermal stability to the high sulfate content of heparin.

III PROTEOGLYCAN–POLYPEPTIDE MIXTURES

Proteoglycans exist as aggregates in connective tissue, (Sajdera and Hascall, 1969) which, like their component polysaccharides, undergo ionic interactions with tropocollagen (DiSalvo and Schubert, 1966; Toole and Lowther, 1967, 1968; Mathews and Decker, 1968; Stevens et al., 1969; Mathews, 1970; Obrink, 1973). Greenfield and Fasman (1969) found that in poly(L-lysine)-proteoglycan mixtures the polysaccharide possessed weak conformation-directing effects on the polypeptide, with the α-helix content estimated at 20–25%. These findings are consistent with those of Gelman et al. (1974) reported in Table 11-1. For bovine nasal septum proteoglycan, the interactions are similar to those for chondroitin 4-sulfate, which comprises approximately 63% of the total polysaccharide. Gelman et al. (1974) also found that the interactions produce a conformational change in the protein core, which, based on analysis of the Smith degradation product, shows that the protein core can adopt a substantial α-helical content when interacting with poly(L-arginine).

In order to investigate the possible effects of the arrangement of glycosaminoglycan chains in pairs along the protein core of the proteoglycan, Gelman et al. (1974) also studied the interactions of chondroitin sulfate "doublets." The doublet sample consisted of pairs of chondroitin sulfate chains connected by a short peptide chain, a portion of the original protein core. The interactions for the doublets are significantly different from those for the separated chains, indicating that the arrangement of the polysaccharide side chains in pairs (and presumable larger groupings) along the protein backbone contributes to the interaction properties of the intact proteoglycan.

In a recently reported investigation, Schodt and Blackwell (1976) studied the interactions of five proteoglycans (PG-I–PG-IV and PG-B) possessing different protein and glycosaminoglycan contents with poly(L-arginine), $(Arg)_n$. The properties of the proteoglycans along with the characteristic arginine/disaccharide residue ratio for maximum interaction as determined by the melt temperature, T_m, of the mixture are given in Table 11-2. Each proteoglycan "induces" an α-helical conformation rather than the extended coil form in $(Arg)_n$ that is normally observed at neutral pH. The arginine/disaccharide residue ratio at maximum interaction seems to be relatively

Table 11-2 Source, Composition, and Molecular Weight of Proteoglycan Samples

Sample	Source	MW(10^6)	% Protein	Glycan	Glycan MW(10^3)
PG-I	Lamprey notochord	0.5	50	C4S	6
PG-II	Sturgeon notochord	1.3	26	C4S	7
PG-III	Sturgeon cartilage	2.7	8	C6S	25
PG-IV	Rat chondrosarcoma		7	C4S	20
PG-B	Bovine nasal septum cartilage	2.5	9	C4S	
				C6S	25
				KS2	

Table 11-3 Comparison of Residue Ratios and Melting Temperatures at Maximum Interaction for Various $(Arg)_n$–PG Mixtures

		PG-I	PG-II	PG-III	PG-IV	PG-B
$(Arg)_n$	Ratio	2:1	2:1	1.3:1	2.2:1	1:1
	T_m (°C)	53	51	>75	58	56

fixed for varying protein content of the proteoglycans that contain chondroitin 4-sulfate. The thermal stability of the proteoglycan interaction is the same as for the component polysaccharide. Consequently, using the strength of interaction with $(Arg)_n$ as a "benchmark," the properties of proteoglycan and the constituent glycosaminoglycans are the same. Schodt and Blackwell (1976) suggest that this situation very likely extends to collagen–proteoglycan systems.

Some work has also been done on the interactions of keratan sulfate-2-rich proteoglycan (Schodt and Blackwell, 1976). A much lower thermal stability is seen for mixtures containing these glycosaminoglycans than for those involving corneal keratan sulfate-1.

REFERENCES

Di Salvo, J. and M. Schubert (1966). *Biopolymers*, **4**, 247.

Gelman, R. A. and J. Blackwell (1973*a*). *Biochim. Biophys. Acta*, **279**, **452**.

Gelman, R. A. and J. Blackwell (1973*b*). *Biopolymers*, **12**, 1959.

Gelman, R. A. and J. Blackwell (1973*c*). *Connective Tissue Res.*, **2**, 31.

Gelman, R. A. and J. Blackwell (1973*d*). *Arch. Biochem. Biophys.*, **159**, 427.

Gelman, R. A. and J. Blackwell (1974*a*). *Biopolymers*, **13**, 139.

Gelman, R. A. and J. Blackwell (1974*b*). *Biochim. Biophys. Acta*, **342**, 254.

Gelman, R. A., J. Blackwell, and M. B. Mathews (1974). *Biochem. J.*, **141**, 445.

Gelman, R. A., D. N. Glaser, and J. Blackwell (1973*b*). *Biopolymers*, **12**, 1223.

Gelman, R. A., W. B. Rippon, and J. Blackwell (1972). *Biochem. Biophys. Res. Commun.*, **48**, 708.

Gelman, R. A., W. B. Rippon, and J. Blackwell (1973*a*). *Biopolymers*, **12**, 541.

Greenfield, N. and G. D. Fasman (1969). *Biochemistry*, **8**, 4108.

Lindahl, V. and L. Roden (1964). *J. Biol. Chem.*, **241**, 2113.

Mathews, M. B. (1970). In *Chemistry and Molecular Biology of the Intercellular Matrix* (E. A. Balazs, Ed.). Academic Press, New York, p. 1155.

Mathews, M. B. and L. Decker (1968). *Biochem. J.*, **109**, 517.

Obrink, B. (1973). *Eur. J. Biochem.*, **33**, 387.

Sajdera, S. W. and V. C. Hascall (1969). *J. Biol. Chem.*, **244**, 2384.

Schodt, K. P. and J. Blackwell (1976). *Biopolymers*, (in press).

Schodt, K. P., R. A. Gelman, and J. Blackwell (1976). *Biopolymers*, (in press).

Scott, J. E. and J. Dorling (1965) *Histochemie*, **5**, 221.

Scott, J. E., G. Quintarelli, and M. C. Dellovo (1964). *Histochemie*, **4**, 13.

Stevens, F. S., J. Knott, D. S. Jackson, and V. Podrazky (1969). *Biochim. Biophys. Acta*, **188**, 307.

Tiffany, M. L. and S. Krimm (1969). *Biopolymers*, **8**, 347.

Toole, B. P. and D. A. Lowther (1967). *Biochem. Biophys. Res. Commun.*, **29**, 515.

Toole, B. P. and D. A. Lowther (1968). *Biochem. J.*, **109**, 857.

CHAPTER 12

Interaction of Ions with Biologic Macromolecules

The addition of ions to an aqueous solution containing soluble organic molecules is generally thought of as a concentration-dependent process in which intra- and intermolecular structure involving the organic molecules decreases as ion content increases. When biologic macromolecules are the organic component, this is only half of the story, and even then it is incomplete. Such nonspecific ion interactions can add structure to the organic component, as well as delete it, depending on the electrostatic potential inherent to the macromolecules. In addition, some biologic macromolecules contain specific and selective ion-binding sites leading to relatively concentration-independent interactions. Moreover, the intermolecular structuring of water molecules by ions can lead to indirect interactions with the macromolecules.

I INTRODUCTION

It is useful in a discussion of ion interactions to categorize the perturbants into those which interact directly with the biopolymer, that is, bind directly at specific sites, and those which do not. Site binding interactions are characterized by large association constants. Hence these interactions demonstrate both saturation at low additive concentration and a strong dependence of the observed effect on macromolecule (site) concentration. In the limit of low levels of affinity, site binding would become operationally indistinguishable from indirect effects of the additive on the structure of the solvent. In such solvent structure-mediated effects, macromolecular conformations are modified indirectly as a consequence of additive-induced alterations in solvent–macromolecule interactions and thus in the relative stability of various macromolecular conformations.

For the class of indirect ion interactions, we are generally dealing with solutions containing additives ranging in concentration up to several molar. Hence no water molecule will be more than two or three molecular diameters removed from the "surface" of an ion. Generally a water molecule located at the solvent–macromolecule interface, and thus strongly influenced in its interactions by this proximity, will also be within the competitive "sphere of influence" of at least one ion as well. Thus, in the most general sense of the word, we are always speaking of "binding" interactions. Whether it is thought to act "directly" or "indirectly" on the structural organization of the macromolecules, an affector molecule or ion must be acting at distances of no more than a few water molecule diameters from a macromolecular "surface". Moreover, interactions involving ions must be classified as short range.

II INDIRECT INTERACTIONS OF IONS OF NEUTRAL SALTS WITH MACROMOLECULES

A large portion of the first part of this section on indirect ion interactions is a summary of the excellent review by Von Hippel and Scheich (1969).

Indirect ion interactions of neutral salts exert specific effects on the conformational stability of proteins and nucleic acids, on association/dissociation equilibria between macromolecules, on enzyme activity, on the stability of macroscopic fibrous structures made up of proteins or nucleic acids, and on the rates of macromolecular transconformation reactions. The effectiveness of a given ion in carrying out these processes transcends details of macromolecular composition or conformation (von Hippel and Wong, 1964). For example, ions such as ClO_4^-, SCN^- and Ca^{2+}, which are particularly effective in decreasing the stability of the "native" conformation of a fibrous protein in water, are also effective destabilizers of globular proteins and nucleic acids. The same ions are also very effective dissociating agents. The effectiveness of these functions seems to be independent of charge sign and magnitude. These observations are illustrated in Figure 12-1 by ranking ions according to increasing effectiveness in disordering the "native" conformations of four dissimilar macromolecules.

A particularly effective means of investigating neutral salt effects on the stability of macromolecular conformation is to examine thermally induced transitions in macromolecules that are components of solutions also containing the salt reagents.

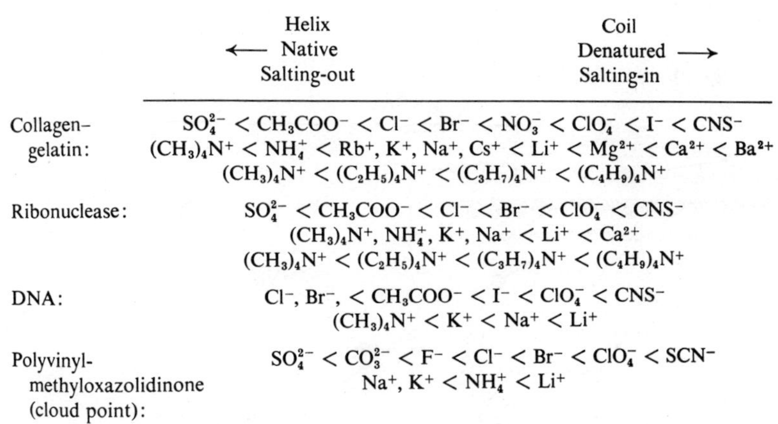

Helix
← Native
Salting-out
Coil
Denatured →
Salting-in
Collagen–gelatin:
$SO_4^{2-} < CH_3COO^- < Cl^- < Br^- < NO_3^- < ClO_4^- < I^- < CNS^-$
$(CH_3)_4N^+ < NH_4^+ < Rb^+, K^+, Na^+, Cs^+ < Li^+ < Mg^{2+} < Ca^{2+} < Ba^{2+}$
$(CH_3)_4N^+ < (C_2H_5)_4N^+ < (C_3H_7)_4N^+ < (C_4H_9)_4N^+$
Ribonuclease:
$SO_4^{2-} < CH_3COO^- < Cl^- < Br^- < ClO_4^- < CNS^-$
$(CH_3)_4N^+, NH_4^+, K^+, Na^+ < Li^+ < Ca^{2+}$
$(CH_3)_4N^+ < (C_2H_5)_4N^+ < (C_3H_7)_4N^+ < (C_4H_9)_4N^+$
DNA:
$Cl^-, Br^-, < CH_3COO^- < I^- < ClO_4^- < CNS^-$
$(CH_3)_4N^+ < K^+ < Na^+ < Li^+$
Polyvinyl-methyloxazolidinone (cloud point):
$SO_4^{2-} < CO_3^{2-} < F^- < Cl^- < Br^- < ClO_4^- < SCN^-$
$Na^+, K^+ < NH_4^+ < Li^+$

Fig. 12-1 Relative effectiveness of various ions in stabilizing or destabilizing the "native" form of collagen, ribonuclease, DNA, and polyvinylmethyl-oxazolidinone in aqueous solution. Von Hippel and Scheich (1969).

A Ribonuclease

Perhaps the protein most thoroughly analyzed using this experimental design is ribonuclease. The ribonuclease transition was first observed by Harrington and Schellman (1956), who included the effects of a neutral salt, LiBr, as part of their investigation. Mandelkern and Roberts (1961) reexamined the transition of ribonuclease in LiBr. Both groups noted that the specific optical rotation became more dextrorotatory as the LiBr concentration was increased. Mandelkern and Roberts also noted that the transition temperature decreased for increasing concentration of LiBr, demonstrating the salt destabilization potential on protein conformation.

Von Hippel and Wong (1965) carried out an extensive study of the effects of various neutral salts on the thermal unfolding transition in ribonuclease. Figure 12-2 depicts the transition temperature as monitored by optical rotation for ribonuclease in a variety of neutral salt solutions at different concentrations. Although it is not clearly brought out in Figure 12-2, the principle of ion additivity (von Hippel and Schleich, 1969) applies; for example, the

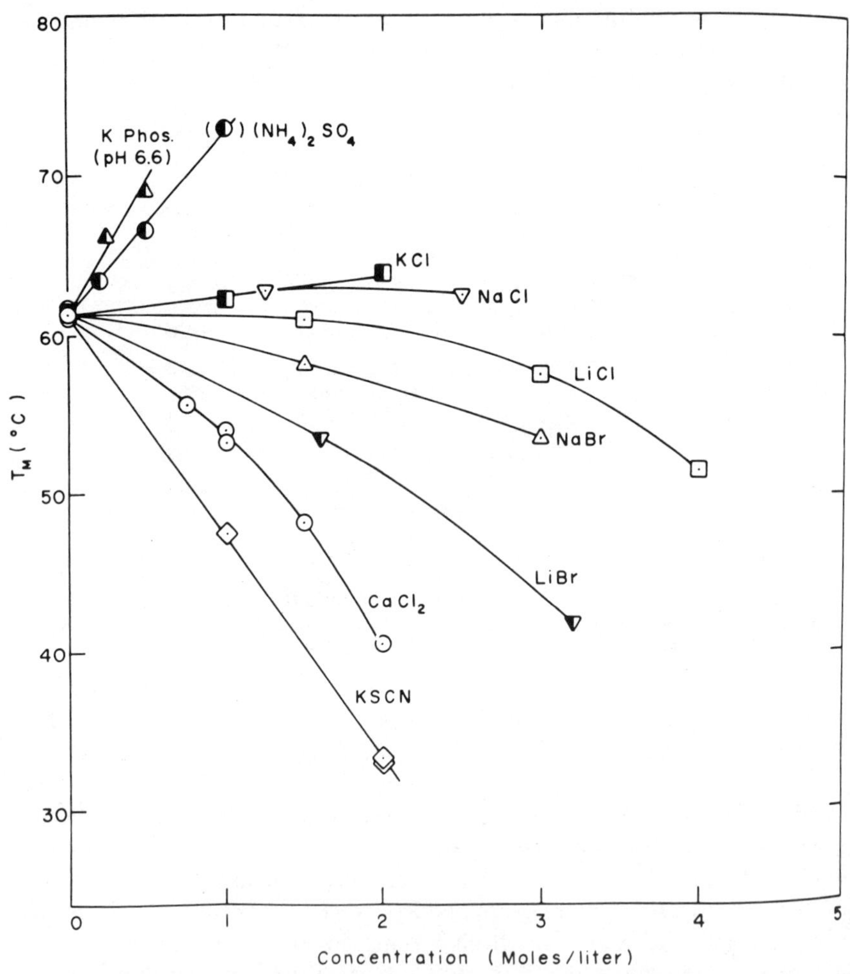

Fig. 12-2 Transition temperatures of ribonuclease as a function of concentration of various added salts. All the solutions were adjusted to pH 7.0 and also contained 0.15 *M* KCl and 0.013 *M* sodium cacodylate; ribonuclease concentration, $\simeq$ 5mg/ml. (Reproduced by permission of the American Society of Biological Chemists; Von Hippel and Wong, 1965).

effect of 2 *M* LiBr on the T_m of ribonuclease is roughly calculable from the sum of the effects of 2 *M* LiCl plus 2 *M* NaBr, minus the effect of 2 *M* NaCl.

The plots of T_m versus salt concentration for ribonuclease are not all linear; most of them show some concave downward shape at low salt concentration. By analogy to DNA (see below), this may reflect an electrostatic component of the ribonuclease transition that tends to stabilize the native form of the molecules relative to the denatured. Another factor that might also affect the shape of the curves in Figure 12-2 is the progressive conversion of the pre- or post-transition forms of the molecule (or both) into a high salt form.

B Collagen

The fibrous protein collagen undergoes a reversible, cooperative thermal transition from a rigid, three-chain ordered structure to a random form called gelatin (von Hippel, 1967). As mentioned in Chapters 4 and 5, several synthetic polytripeptides and a few synthetic polyhexapeptides also undergo triple-helix to random-chain thermal transitions in dilute solutions (Brown, Hopfinger, and Blout, 1972; Traub and Piez, 1971). The T_m of these transitions is dependent on the polarity of the solvent and the primary structure of the polypeptides (Brown, Hopfinger, and Blout, 1972).

Figure 12-3 shows plots of melting temperature, T_m, versus concentration of $CaCl_2$ in the aqueous collagen solution. Note that increasing salt concentration decreases the T_m, suggesting that ions of neutral salts generally destabilize protein structure. Second, different species of collagen have significantly different T_m's, which remain at constant differences as a function of increasing salt concentration. Presumably, different collagen species have slightly different primary structures whose effect on T_m is independent of ion interactions, but is an inherent property of the macromolecular structure in pure aqueous solution. Each of the T_m versus concentration curves in Figure 12-3 is linear, and the curves have nearly identical slopes. This suggests that the interactions of ions with the collagen molecules are very nearly identical for all three collagen species, and there are probably no specific concentration (of $CaCl_2$)-dependent electrostatic interactions that drastically alter T_m for small changes in $CaCl_2$ concentration.

C DNA

DNA is an example of a helical macromolecule that can exist in the form of a charged polyelectrolyte in which all of the charges have the same sign, thus repelling one another. Consequently, a strong conformational driving force in such macromolecules is the repulsive electrostatic interactions that work to

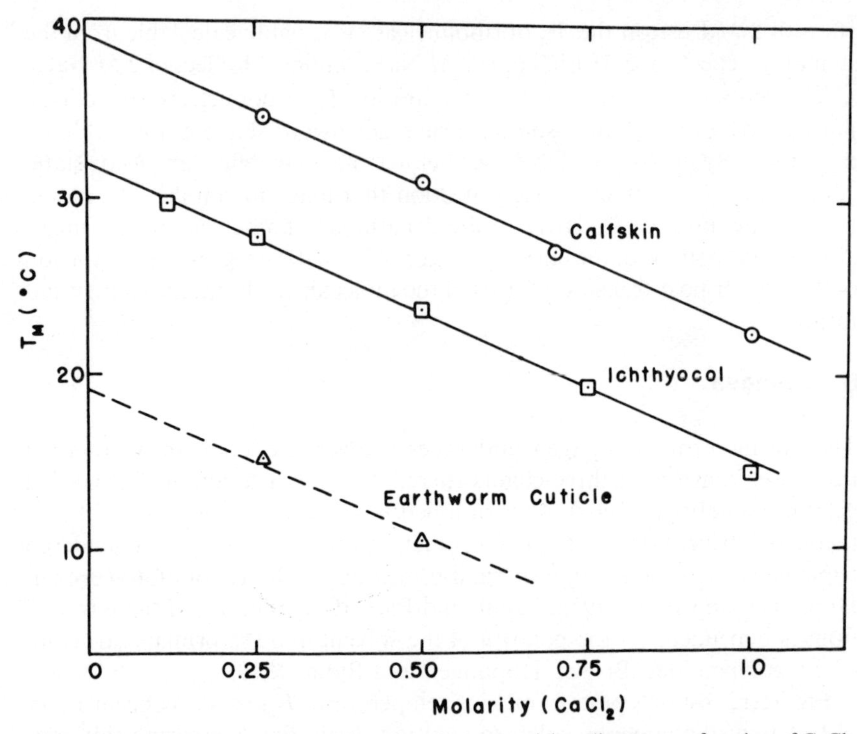

Fig. 12-3 T_m for calfskin, ichthyocol, and earthworm cuticle collagens as a function of $CaCl_2$ concentrations. (Reproduced by permission of the American Chemical Society; von Hippel and Wong, 1963).

unfold the molecule so that the like charges are as far apart as possible. DNA contains one negatively charged phosphate group per nucleotide unit and thus is subject to considerable intramolecular electrostatic destabilization. The effects of uni-univalent electrolytes on the helix-coil T_m of various DNAs and synthetic polynucleotides have been investigated by a number of workers (Dove and Davidson, 1962; Schildkraut and Lifson, 1965; Marmur and Doty, 1962; Hamaguchi and Geiduschek, 1962). Figure 12-4 shows the T_m versus the logarithm of the salt concentration for various DNA solutions. The most striking observation, in view of the previous discussions, is that the T_m increases with increasing salt concentration. This is just the opposite of solutions of nearly *neutral* macromolecules like collagen and ribonuclease. In addition, plots of T_m versus concentration are linear over the log of the concentration. Kotin (1963) and Schildkraut and Lifson (1965) have attempted to provide a theoretical basis for the logarithmic relationship through an electrostatic component that measures the difference in free

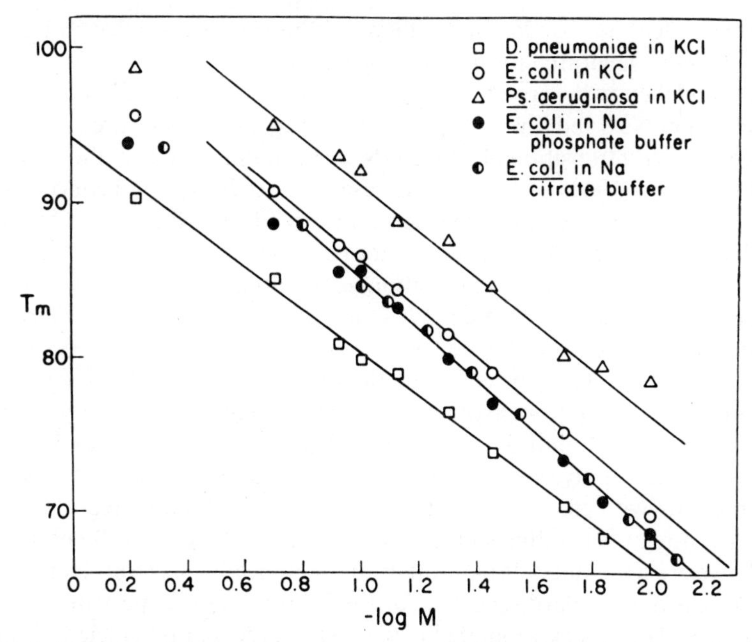

Fig. 12-4 Variation of the transition temperature of DNA with the logarithm of the salt concentration (molar). The sodium citrate buffer was prepared by diluting a stock solution of 1.50 *M* NaCl + 0.15 *M* sodium citrate to give the desired sodium ion concentration. The sodium phosphate buffer was prepared by diluting a stock solution 0.15 *M* Na_2HPO_4 + 0.30 *M* NaH_2PO_4 to give the desired sodium ion concentration. The final pH of both types of buffer solutions was approximately 7. [Reproduced by permission of John Wiley & Sons (Interscience); Schildkraut and Lifson, 1965.]

energy due to interphosphate electrostatic repulsions at low ionic strength. The correct logarithmic relationship was achieved largely through the use of a Debye-Hückel representation of ionic interactions. However, counterion binding may negate the reasonableness of a Debye-Hückel approximation, and the results achieved may be fortuitous, giving the right answer but the wrong explanation.

Ross and Scruggs (1964*a,b*) were able to estimate the relative binding affinities to DNA of some univalent cations to be: $Li^+ > Na^+ > K^+ > (CH_3)_4N^+$ and some divalent cations to be: $Mn^{2+} > Mg^{2+} > Ca^{2+}$. Ross and Scruggs conclude that the binding sites for both classes of ions are the phosphate groups.

Manning (1972) proposed that Mg^{2+} should condense onto DNA in preference to monovalent ions, and that the sole function of monovalent ions in DNA solutions should be Debye-Hückel screening. Since, as in the case of monovalent ion condensation, there is a higher residual electrostatic free energy in the denatured form, screening stabilizes this form relative to native DNA and hence lowers the T_m. Record (1975) has studied the effects of monovalent Na^+ and divalent Mg^{2+} cations on the temperature and breadth of the helix-coil transition of phage DNA. His findings confirm those of Dove and Davidson (1962) for the limiting cases of zero divalent ion concentration and saturating levels of divalent ion and extend their findings to the intermediate region of Mg^{2+} concentrations. Record (1975) also develops a theory for the dependence of transition temperature on the ion concentration using the approach of Wyman (1964), modified to account for electrostatic nonideality of the polyelectrolytes.

Hanlon et al. (1975) believe that the ordered transconformational reactions of DNA in solution involve two factors, aqueous hydration and cationic binding, which cannot be completely separated from one another (see Chapter 7). Succintly, the structural conversions appear to depend on specific interactions with the cation in the medium, since, even at high hydration levels and relatively dilute solutions of these ions, the percent B character of DNA varies from one salt to another. If the ionically unperturbed spectrum is taken as the standard reference for the B form (i.e., the spectrum in 0.2 *M* $(CH_3)_4$ NCl or that extrapolated to 0 *M* NaCl), the relative efficiencies of the ions in inducing the structural transformation to a more C-like state at dilute concentrations of salt will vary from 0 effectiveness in the case of Na^+ to a maximal effectiveness in the case of NH_4^+ and Cs^+. Other ions fall in the order: $Na^+ < K^+ \leq Li^+ < Cs^+ \leq NH_4^+$. The nature of these differences is puzzling, since the order bears no relationship to ionic size and hydration (Pauling, 1960; Robinson and Stokes, 1959; Harned and Owen, 1958), relative affinities of the ion for the phosphate groups of DNA, and the fraction of phosphate sites occupied at a given ionic concentration (Ross and Scruggs, 1964*a,b*; Strauss et al., 1967), or any other single ionic property in aqueous solutions at room temperature. Clearly, the interaction leading to the transformation of the secondary structure of DNA by these specific ion effects involves a combination of two or more ionic properties that dictates where in the macromolecular structure they bind and, once bound, how they affect the other important factor in these reactions, the hydration of the DNA macromolecule in solution (see Chapter seven).

III AN ION-INDUCED POLYPEPTIDE CONFORMATION

One of the most recent, and still not completely resolved, conformational findings to come from ion–polypeptide interaction studies is the possible existence of a unique polyelectrolyte-ordered coil conformation. Indirect experimental evidence and theoretical calculations suggest that this new conformation is approximately a left-handed 10_4 helix (for L-residues) that is realized only when there is a uniform distribution of like charges along the polypeptide chain.

The first experimental evidence for this ordered polyelectrolyte conformation was presented by Tiffany and Krimm (1969), who observed that the charged form of poly (L-glutamic) acid in aqueous solution gave CD/ORD curves similar to those of poly (L-proline) in the left-handed, 3_1 helical conformation, but shifted to a lower wavelength. Treatment of the poly(L-glutamic) acid solution with $LiClO_4$ changed the CD/ORD curves to the form normally associated with denatured proteins. It was also noted (Tiffany and Krimm, 1968) that poly(L-proline) in aqueous solution under the influence of concentrated $CaCl_2$ shows a CD spectrum similar to that of poly(L-glutamic) acid in solution with $LiClO_4$. Although there are similarities between the poly(L-glutamic) acid and poly(L-proline) systems in terms of the effect of concentrated electrolytes, the nature of the effect has been disputed. Some workers have suggested that divalent ions coordinate to the carbonyl oxygen in the peptide bond and consequently modify the nature of the active chromophore (von Hippel and Schleich, 1969; Urry et al., 1971; Balasubramanian and Shaikh, 1973). The Raman spectrum of the collapsed form of poly(L-proline) seems to suggest that a mixture of cis and trans peptide bonds is present, seemingly providing support for the carbonyl coordination hypothesis (Rippon et al., 1970). However, no similar evidence is available for poly(L-glutamic) acid where it seems reasonable that the negatively charged polyelectrolyte establishes a charged cloud of counterion including lithium (Baddiel et al., 1967).

In retrospect, the classic pH versus viscosity studies of poly(L-glutamic) acid (Doty et al., 1957) can be taken to support the existence of the ordered polyelectrolyte conformation. As one would expect, the high-pH form exhibits a much lower viscosity than the α helix (low pH). However, as shown in Figure 12-5*a*, the high-pH viscosity is not the minimum value. This indicates that the completely ionized form of poly(L-glutamic) acid is less flexible (and presumably exhibits less variation in backbone conformation, or less "randomness") than some intermediate, partially ionized state. In addition, Iizuka and Jaug (1965) have demonstrated, as shown in Figure 12-5*b*, that the intrinsic viscosity of ionized poly(L-glutamic) acid in LiBr solution decreases as the molarity of the salt increases. This relationship is also noted in other salts (Iizuka and Jaug, 1965). Although this effect is typical of a polyelectrolyte (Flory, 1969), such a dramatic change in the viscosity must indicate a large increase in chain flexibility (and conformational freedom), implying that charged poly(L-glutamic) acid at zero salt

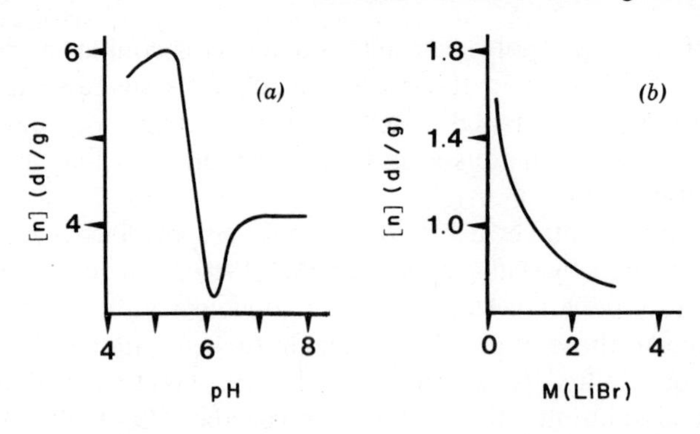

Fig. 12-5 *a*. Intrinsic viscosity [η] vs. pH for poly(Glu) in 0.2 *M* NaCl/dioxane (2:1). *b*. Intrinsic viscosity [η] vs. salt concentration for poly(Glu) at pH 7.3. Hiltner (1972).

concentration assumes a conformation of some order—the ordered polyelectrolyte conformation. For less than fully ionized conditions, unbalanced electrostatic interactions prevent the polypeptide chain from assuming a reasonably ordered conformation (more random components) and, consequently, a low viscosity, as observed for partially ionized poly(L-glutamic) acid.

The disruption of the α-helix of poly(L-glutamic) acid and poly(L-lysine) as the completely ionized form develops suggests that the predominant conformation-directing influence is the electrostatic repulsion between side chains. Krimm and Mark (1968) have performed a simplified theoretical calculation in which the conformation of the charged poly(amino acid) is assumed to be dictated solely by electrostatic forces. The lowest free space energy conformation was found to be a left-handed helix possessing 2.5 residues per turn and a repeat of about 3.4 Å. Subsequently, Hilter et al. (1972) carried out extensive conformational analyses of poly(L-glutamic acid) and found a left-handed helix with 2.4 residues per turn and a rise per residue of 3.2 Å. The structural similarity of these preferred charged conformations found theoretically and the left-handed 3_1 helices of poly(L-proline) and polyglycine II has been inferred to account for the similar shapes of the CD spectra of both poly(L-glutamic acid) and poly(L-lysine) in the charged form to that of poly(L-proline). The differences in spectral shapes has been, in part at least, attributed to the intrinsic rotatory property differences of imino and amino acids. Rippon and Walton (1971, 1972) have shown that poly(ala-gly-gly) adopts a left-handed 3_1 helical conformation. The corresponding CD spectrum is very similar to that of charged poly(L-glutamic acid), supporting the existence of the new polyelectrolyte conformation.

Remember that polyglycine does not yield a CD spectrum; thus there was no CD spectrum for a left-handed 3_1-helix of amino acids until that of poly(ala-gly-gly) was reported.

A number of investigators studying relatively exotic polymer–solvent systems have reported CD spectra similar to that of charged poly(L-glutamic acid). A list of these studies is given by Balasubramanian and Roche (1970) and some particular examples are: poly(γ-ethyl-L-glutamate) in sulfuric acid–water mixtures (Steigman et al., 1969), poly(γ-methyl-L-glutamate) in fluoro gem diols (Balasubramanian and Roche, 1970) and poly(phenylalanine) in methane sulfonic acid–water mixtures (Peggion et al., 1970).

If the CD spectra are indicative of the ordered polyelectrolyte conformation and if this conformation is forced on the polypeptide through like-charge electrostatic repulsions, an ion-polypeptide binding mechanism is probably involved in the realization of the conformation. The polypeptide examples given above do not contain the ionizable groups needed to supply electrostatic ion-pair repulsions. However, if the acid components of the solvent systems bind in some uniform fashion to the peptide residues and are ionized at some time (either pre- or postbinding), the bound solvent ions could provide the needed electrostatic interactions. Presumably, the favorable electrostatic interactions between the bound solvent ions and the peptide binding sites hold the ions firmly enough to survive the like-ion electrostatic repulsions leading, overall, to the ordered polyelectrolyte conformation.

Although the effects of salts on the conformation of polyamino acids and proteins have been extensively studied in aqueous solutions, nonaqueous solvents, on the other hand, have received only limited attention (Frazen et al., 1966; Lotan et al., 1967; Bradbury and Fenn, 1968; Shiraki and Imahori, 1966). Shiraki and Imahori (1966) have shown that lithium chloride (2–5 *M*) affects the UV and IR spectra, as well as the optical rotary properties of poly(L-tyrosine) in methanol. These workers have interpreted these findings to indicate an α helix to random coil transition, although no evidence has been presented for the actual formation of the random-coil state. Poly-N^5-(4-hydroxybutyl)-L-glutamine (PHBG) has been shown to assume the α-helical conformation in methanol (Lotan et al., 1966; Joubert et al., 1970). The addition of LiCl salt to the methanol–PHBG system leads to a helix-coil transition at about 3 *M* LiCl on the basis of optical rotary dispersion (ORD) measurements (Lotan, 1973). Intrinsic viscosity measurements, on the other hand, suggest two transitions for increasing salt concentrations. Figure 12-6 shows the ORD and viscosity curves for PHBG as a function of LiCl concentration. The viscosity curve is nearly identical to that observed for poly(L-glutamic) acid as a function of solution pH shown in Figure 12-5*a*. Lotan suggests that the first transition seen for increasing LiCl concentration is an α helix to random state transition, and the second transition is a random

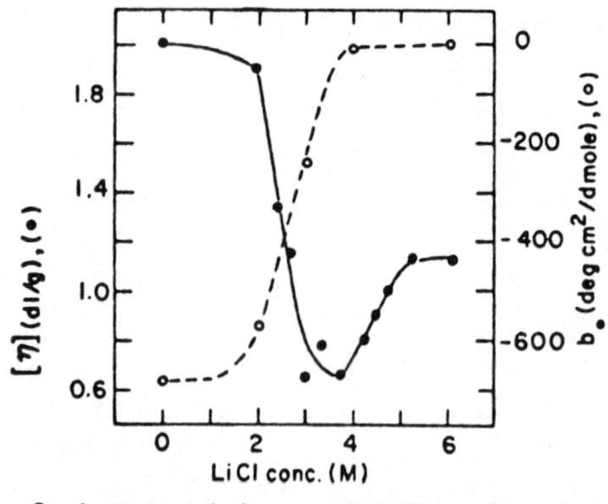

Conformational changes of PHBG in the LiCl–methanol solvent system, as followed by (O) ORD and (●) viscosity.

Fig. 12-6 Conformational changes of PHBG in the LiCl-methanol solvent system as followed by (○) ORD and (●) viscosity. Lotan (1973).

state to the ordered polyelectrolyte conformation transition. Lotan further suggests that the polyelectrolyte conformation is induced as a result of Li^+ ions binding to the polypeptide just as in the mixed water-containing polymer solvent systems discussed above, and solvation of the Cl^- ions by methanol. Lotan (1974) has shown that this pseudopolyelectrolyte effect is even more pronounced in trifluoroethanol (TFE), as could be predicted from the known properties of the LiCl–TFE solvent system. That is, in the presence of LiCl, polypeptides behave as polyelectrolytes and that conformational changes occur at solvent compositions that are related to the anion binding ability of the alcohol.

IV SPECIFIC BINDING OF METAL IONS TO PROTEINS

Proteins that undergo specific association with metal ions have been termed metalloproteins. Since many enzymes complex with metal ions, this protein subset has been given the particular title of metalloenzymes. Vallee and Wacker (1970), and Friedman (1974) have reviewed the field of metalloproteins. Table 12-1 (constructed from data in their review) is a summary of the known metalloproteins along with the associated complexing metal ions. The metal atoms of metalloproteins are bound so firmly that they are not removed from the protein by the isolation procedure. The role of metals in the tertiary and quaternary structure of proteins is slowly, but steadily, being unraveled (Drum et al., 1967). It has proved possible with increasing frequency to replace the "native" metal atom of metalloenzymes with others not

Table 12-1 A List of Many Metalloproteins and Some of Their General Properties (Compiled from Vallee and Wacker, 1970)*

Protein	Metal	MW	Stoichiometry of metal	Function
VANADIUM				
Hemovanadin V	—	—	—	Unknown
IRON				
Transferrin	Fe(Mn, Cu)	80,000	2	Iron transport
	Fe	67,000	2	Iron transport
Conalbumin	Fe(Mn, Cu)	77,000	2	Iron transport
Lactotransferrin	Fe	80,000	2	Iron transport
	Fe	—	2	Iron transport
Ferritin	Fe	Apoferritin 460,000	$(FeOH)_8FeOPO_3H_2)$ 75% of weight	Iron
				Respiratory protein
Hemerythrin	Fe	107,000	16	
Ferredoxin (clostridial)	Fe	6,000	7	
Spinach ferredoxin	Fe	11,600	2	
Chromatium ferredoxin	Fe	5,600	3	Electron carrier
Alfalfa ferredoxin	Fe	11,500	2	
Methanobacterium ferrodoxin	Fe	6,000	2	
Adrenodoxin (adrenal ferrodoxin)	Fe	13,000	2	
Testodoxin	Fe	Unknown	Unknown	
Putida redoxin	Fe	12,000	2	
Rubredoxin	Fe	5,900–7,000	1	
	Fe	—	1	
Metapyrocatechase	Fe	140,000	3	
Pyrocatechase	Fe	95,000	2	Cleavage of aromatic compound

Table 12-1 *(continued)*

Protein	Metal	MW	Stoichiometry of metal	Function
Protocatechuate 3,4-dioxygenase	Fe	700,000	8	Cleavage of aromatic compounds
Agavain	Fe	52,000	1	Proteolytic action
Malate vitamin K reductase	Fe	Unknown	Unknown	Specific to malate
NADH-dehydrogenase	Fe	80,000	4	
Succinate dehydrogenase	Fe	200,000	4	
	Fe	Unknown	8Fe/flavine	Respiratory protein
	Fe	200,000	4Fe, 1 flavine	
Dihydroorotate dehydrogenase	Fe	115,000	4 2FAD	
Aldehyde oxidase	Fe Mo	300,000	8Fe, 2Mo, 2FAD	
Xanthine oxidase	Fe Mo	300,000	8Fe, 2Mo, 2FAD	Electron carrier
	Fe Mo	Unknown	4Fe, 2Mo, perFAD	
	Fe, Mo	300,000	8Fe, 2Mo, 2FAD	
	Fe, Mo	250,000	8Fe, 2Mo, 2FAD	
ZINC AND CADMIUM				
Metallothionein	Zn Cd	6,600	6 (Zn + Cd)	Unknown
Procarboxypeptidase A	Zn	90,000	1	Precursor of carboxypeptidase A
	Zn	52,500	1	Precursor of carboxypeptidase A
Procarboxypeptidase B	Zn	67,400	1	Precursor of carboxypeptidase B

Carboxypeptidase A	Zn	34,300	1	Hydrolysis of C-terminal amino acid residues
	Zn	34,800	1	
	Zn	36,500	1	
Carboxypeptidase B	Zn	34,300	1	Hydrolysis of C-terminal amino-acid residues
	Zn	34,300	1	Protein hydrolysis
Neutral protease	Zn	44,700	1–2	
Alcohol dehydrogenase	Zn	150,000	4	Oxidation of ethanol or reduction of acetal dehyde
	Zn	80,000	4	
	Zn	87,000	> 2Zn	
Glutamic dehydrogenase	Zn	1,000,000	2–6Zn	
D-Glyceraldehyde-3-phosphate dehydrogenase	Zn	137,000	2–3Zn	Oxidation Processes
Lactic dehydrogenase	Zn	Unknown	Unknown	
Malic dehydrogenase	Zn	40,000	1	
D-Lactic cycochrome reductase	Zn	50,000	4–6Zn	
Alkaline phosphatase	Zn	89,000	4	Hydrolyzes monophosphate esters
	Zn	Unknown	Unknown	
	Zn	Unknown	Unknown	
Aldolase	Zn	65,000–75,000	1	Conversion of fructose-1-diphosphate
	Zn	50,000	1	
Proteinase	Zn, Ca	26,000	1Zn, 2Ca	Protein hydrolysis
Phospholipase	Zn	Unknown	Unknown	Hydrolysis of phospholipids
Dipeptidase	Zn	47,200	1	Hydrolysis of dipeptides
Leucine aminopeptidase	Zn	326,000	12[a]	Leucine specific Hydrolysis
Leucine aminopeptidase	Zn	300,000	4–6	
Thermolysin	Zn	37,500	1,800–2,200[b]	
Carbonic anhydrase	Zn	30,000	1	Hydration of CO_2 and dehydration of bicarbonate
	Zn	28,000	1	
	Zn	29,000	1	

Table 12-1 (*Continued*)

Protein	Metal	MW	Stoichiometry of metal	Function
COPPER				
Hemocyanin	Cu	25,000–75,000 (subunit)	2	Respiratory protein
Ceruloplasmin	Cu	151,000	8	Unknown
Erthrocuprein	Cu	35,000	2	Unknown
Hepatocuprein	Cu	~30,000	2	Unknown
Cerebrocuprein	Cu	~30,000	2	Unknown
Neonatal hepatic mitochondrocuprein	Cu	Insoluble	4.2–5.5% 2.5–4.4%	Unknown Unknown
Plastocyanin	Cu	21,000	2	Photosynthetic electron transport
Pseudomonas blue protein	Cu	15,000–17,000	1	Evidence for electron transport
Azurin	Cu	14,600	1	Evidence for electron transport
Mung bean blue protein	Cu	22,000	1	Evidence for electron transport
Ascorbic acid oxidase	Cu	140,000	8	Oxidation processes
Rhus blue protein (stellacyanin)	Cu	16,800	1	Possible electron carrier
Monoamine oxidase	Cu	170,000	1	
Amine oxidase	Cu	195,000	3	
Diamine oxidase	Cu	96,000	1	
	Cu	185,000	2	Oxidation processes
D-Galactose oxidase	Cu	75,000	1	

Uricase	Cu	12,000	1	
Tyrosinase (polyphenyl-oxidase)	Cu	119,000	4	
	Cu	Unknown	Unknown	
	Cu	Unknown	Unknown	
Dopamine-β-hydroxylase	Cu	290,000	2	Synthesis of norepinephrine from dopamine
Laccase	Cu	120,000–141,000	6	Oxidation of diamines and diphenols
Laccase	Cu	~60,000	4	
Cytochrome oxidase	Cu	Unknown	1Cu/heme	Electron Acceptor
MOLYBDENUM				
Nitrate reductase	Mo	Unknown	Unknown	Unknown
CALCIUM				
Amylase	Ca	50,000	1	Polysaccharide clevage
	Ca	50,000	2–3Ca	
	Ca	50,000	1–2Ca	
	Ca	—	4	
	Zn	50,000	0.5Zn	
Pseudomonas protease	Ca	48,000	1–2Ca	Protein hydrolysis
MANGANESE				
Pyruvate carboxylase	Mn	655,000	3–4Mn	Metabolism of carbohydrate

* Vallee has reminded the author that the content of this table is out of date since many additional metalloprotein complexes have been identified in the last six years. However, no literature survey has been carried out in this area recently. Moreover, the reader should come away with an appreciation of the number and type of distinct metalloproteins that occur. This is the major goal of including the table.

[a] Preparation included a Zn heat precipitation of homogenate.

[b] Values given in μg/g.

found associated with them in nature to yield an enzymically active product (Vallee et al., 1958; Folk and Gladner, 1960; Lindskog and Malmström, 1962; Plocke and Vallee, 1962). Similarly, the metal ions of enzymically inactive metalloproteins can be replaced by others. Examples are transferrin, conalbumin, and metallothionein.

There is abundant evidence to support the theory that metal ions bind to the metalloprotein substrates in site-specific modes. Specific ligands of the substrate metalloprotein have been identified with protein–metal ion interactions from X-ray structural investigations. A summary of the specific nature of the metal–ligand complexes in some metalloproteins is presented in Table 12-2. Waara et al. (1972) has pointed out the close proximity in sequence number of at least two protein ligands involved in a complex with a metal ion. Metal ions are bound to at least two protein ligands separated by no more than four residues, except for the zinc ion of insulin. This finding is similar to binding geometries that occur in many chelates. It is becoming possible to identify classes of protein ligand–metal complexes. Calcium ions bind exclusively to ligand oxygen atoms, iron prefers binding to sulfurs from cysteines when not involved with hemes, and zinc generally prefers to complex with the protein through histidyl ligands, although, occasionally, a glutamate side chain may participate. Some metalloproteins contain double metal sites. Concanavalin and thermolysin are examples, and there is evidence that the copper ions in lactase (Malmström and Rydén, 1968) may have double sites.

The most common identification of metal binding sites of proteins has been based largely on the mode of interaction of metals with amino acids, peptides, and their derivatives. Such studies have led to the conclusion that, generally, the amino acid side chain of proteins with dissociable hydrogen ions can serve as ligands for metal interactions, although peptide nitrogens can also participate. There is spectral data (Latt and Vallee, 1971; Coleman and Coleman, 1972) and X-ray information (Quiocho and Lipscomb, 1971; Liljas et al., 1972; Watenpaugh et al., 1971) to support the theory that the ability of certain metal ions to form complexes with irregular geometries may be as important to the understanding of metal–protein interactions from the point of view of metal chemistry as the properties of ionizing amino acid residues of proteins from the standpoint of organic chemistry (Vallee and Williams, 1968*a,b*). The specific metal binding sites of proteins might predispose the donor groups of the ligands to invoke an irregular geometric environment acceptable to only specific metal ions in specific coordination states. Two examples of the irregular geometry of the ion–protein complex have been found in the crystal structures of rubredoxin and ferredoxin. The bond lengths and angles in the FeS_4 complex of rubredoxin, along with standard deviations, are shown in Figure 12-7. These values have been derived from a 2 Å electron density map (Watenpaugh et al., 1971). It is

Table 12-2 Metal Sites in Proteins[a]

Protein	Metal	Ligands[b] 1	2	3	4	5	6	Coordination	Comments	Proposed Role of Metal[c]
Carbonic anhydrase C	Zn	His 93 $N_{\varepsilon 2}$	His 95 $N_{\varepsilon 2}$	His 118 $N_{\delta 1}$	H_2O			tetr	distorted	a, c, e, f
Carboxypeptidase A	Zn	His 69 $N_{\delta 1}$	Glu 72	His 196 $N_{\delta 1}$	H_2O			tetr	distorted	c, e, f
Thermolysin	Zn	His 142 $N_{\varepsilon 2}$	His 146 $N_{\varepsilon 2}$	Glu 166	H_2O			tetr	distorted	a, (e?)
	Ca 1	Asp 138	Glu 177	Asp 185	O'Glu 187	Glu 190	H_2O	oct }	double site, 3.8 Å	a, (b?)
	Ca 2	Glu 177	O'Asn 183	Asp 185	Glu 190	H_2O	H_2O	oct }	apart, distorted	
	Ca 3	Asp 57	Asp 59	O'Gln 61	H_2O	H_2O	H_2O	oct	distorted	
	Ca 4	O'Tyr 193	O_γThr 194	O'Thr 194	O'Ile 197	Asp 200	$2H_2O$			
Insulin	Zn	His B10 $N_{\varepsilon 2}$	His B10 $N_{\varepsilon 2}$	His B10 $N_{\varepsilon 2}$	H_2O	H_2O	H_2O		solvent not well defined	a
Concanavalin A	Mn	Glu 8	Asp 10	Asp 19	His 24 $N_{\varepsilon 2}$	H_2O	H_2O	oct }	double site, 5.3 Å	a, b
	Ca	Asp 10	O'Tyr 12	Asn 14	Asp 19	H_2O	H_2O	oct }	apart	
S. nuclease	Ca	(Asp 19)	Asp 21	Asp 40	Glu 43	H_2O		oct		(e?)
Myogen	Ca 1	Asp 51	Asp 53	Ser 55	O'Phe 57	Glu 59	Glu 62			a
	Ca 2	Asp 90	Asp 92	Asp 94	O'Lys 96	H_2O	Glu 101			a
Rubredoxin	Fe	Cys 6	Cys 9	Cys 38	Cys 41			tetr	distorted, one short Fe–S	c, d
Ferredoxin	Fe_4S_4 1	Cys 8	Cys 11	Cys 14	Cys 45			tetr }	Fe_4S_4 clusters closely	c, d
	Fe_4S_4 2	Cys 18	Cys 35	Cys 38	Cys 41			tetr }	resemble each other,	c, d
HiPIP	Fe_4S_4	Cys 43	Cys 46	Cys 63	Cys 77			tetr }	distorted cube	c, d
Myoglobin	Cu	His A10	Lys A14	Asn GH4				}	Not in native	
	Zn	Lys A14	His GH1	Asn GH4				}	myoglobin	

[a] From Liljas and Rossman (1974).

[b] Protein ligands less than four residues apart are underlined.

[c] a, stabilize the protein structure; b, bring about functionally advantageous conformational changes; c, produce a distorted metal coordination; d, oxidation–reduction or transfer of electrons; e, participate in binding of substrates or cofactors; f, activate the enzyme–substrate complex once formed.

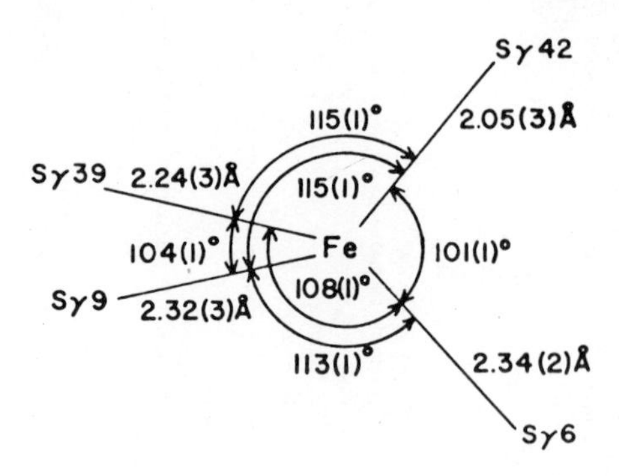

Fig. 12-7 Bond lengths and angles in FeS_4 complex. Jensen (1974).

evident that at least four of the S-Fe-S angles are significantly different from the tetrahedral. More interesting, however, is that one of the Fe-S bonds is nearly 0.3 Å shorter than each of the other three ligand bonds. This short bond length is considered to be "real" and not an artifact of a 2 Å resolution map. The functional significance of this short bond length, if any, remains a mystery. The average geometry of the Fe_4S_4* complex of ferredoxin is shown in Figure 12-8. The synthesis of a structure containing this core was carried out by Herskovitz et al. (1972). In both complexes there is a distortion from tetrahedral symmetry, with the complex actually tetragonal (point group $D_{2d} \approx \bar{4}2$ m).

The Cotten effects observed in the rotatory dispersion curves for metalloproteins provide physical evidence for the asymmetric nature of their metal binding sites in solution. A critical steric arrangement of specific, reactive groups permits the site to serve as a locus of orientation for the binding of metals. The resultant chromophoric and optical rotatory properties of the complexes thus become characteristic and distinct from those of the isolated constituents. The interaction of metal ions at the binding sites of conalbumin has been monitored (Figure 12-9) through the extrinsic Cotton effect. The

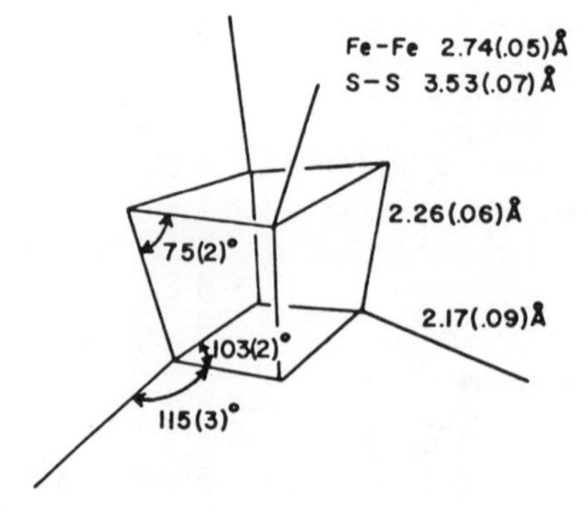

Fig. 12-8 Average dimensions of the ferredoxin complex. Numbers in parentheses are estimated standard deviations in the mean values times 2 to allow for preliminary nature of refinement. Jensen (1974).

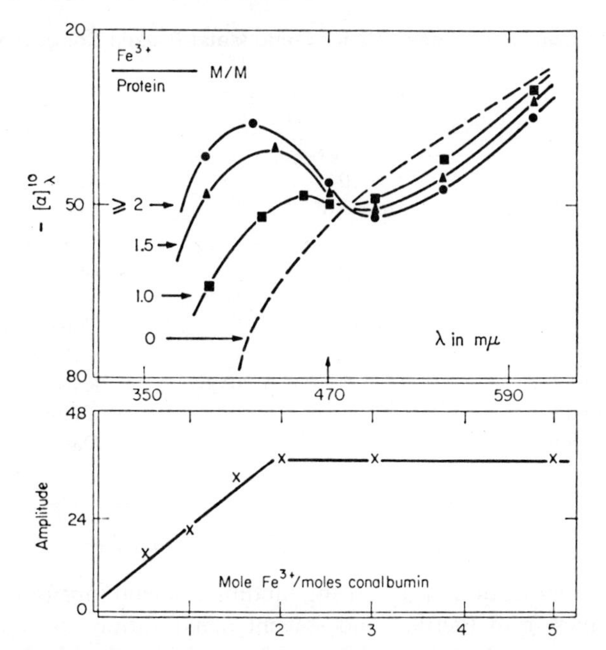

Fig. 12-9 Rotatory dispersion titration of conalbumin with Fe^{3+}. In the upper portion of the figure, specific rotation at 10°, $-[\alpha]^{10}$, is plotted against wavelength. In the presence of Fe^{3+}, a negative Cotton effect appears centered about the absorption maximum at 470 mμ. At a conalbumin concentration of 6.6×10^{-5} *M*, the magnitude of the Cotton effect increases with increasing Fe^{3+} to become maximal at 2 g atoms of iron per mole of conalbumin. In the lower portion, the data are plotted according to the method of molar proportions. The amplitude was calculated from the difference rotation between the metal-free and metal-containing protein at the peak and trough of the Cotton effect (Ulmer and Vallee, 1962).

plain negative rotatory dispersion of metal-free conalbumin between 300 and 675 mμ becomes anomalous in the presence of iron because of the appearance of a negative Cotton effect, and its magnitude is a function of the amount of iron bound to the protein. The inflection point, λ_0, near 470 mμ, corresponds to the absorption maximum of the iron–conalbumin complex.

The stabilities of metal–protein complexes, like those of metals with simpler ligands, may be expected to span a considerable and continuous range, depending on the nature, the number, and the geometry of the donor groups at the binding site. Only in a limited number of cases have the interactions of metals with enzymically active proteins been measured with sufficient precision to yield reasonably reliable estimates of the dissociation constants. Table 12-3 lists some available metal–protein complex dissociation constants.

Table 12-3 Dissociation Constants for Some Metal-Protein Complexes

Metal–Protein Complex	Dissociation Constant M	Reference
Zinc-enolase	2×10^{-5}	Malmström (1961)
Zinc-carboxypeptidase, pH 8	3.2×10^{-10}	Coleman and Vallee (1960)
Zinc-carbonic anhydrase, pH 5.5	3×10^{-10}	Lindskog and Numan (1964)
Zinc-alkaline phosphatase	1×10^{-10}	Cohen and Wilson (1966)
Iron–heme complex copper–proteins metallothione in iron transferrin	1×10^{-20}–1×10^{-30}	Kagi and Vallee (1960) Davis et al. (1962)

An indirect indication of the strong binding of metal ions to the metalloproteins comes from thermal and solvent denaturation and renaturation studies. The remarkable heat stability of thermolysin is highly dependent on the nature of the presence of calcium ions (Feder et al., 1971). Carbonic anhydrase is renatured at higher guanidinium concentrations in the presence of zinc ions (Yazgan and Henkens, 1972). Conformational changes have bèen observed in an X-ray study of binding of calcium and manganese ions to demetallized concanavalin A. The binding of manganese to the innermost part of the double site (Reeke et al., 1974; Hardman and Ainsworth, 1972) must precede the calcium binding, perhaps causing small conformational modifications. The calcium is then able to bind to the outer part of the double site, thereby probably producing a large conformational change in the protein and, as a consequence, an overall organizational modification in the quaternary structure. Hexamers of insulin have also been shown to undergo quaternary structural changes as a function of the amount of bound ion (Dodson, 1966).

V SPECIFIC BINDING OF ANIONS TO PROTEINS AND PROTEIN CRYSTALLIZATION PROCESSES

Just as the metal cations have specific binding sites on proteins, so do a variety of anions. Table 12-4 lists the structural information for protein–anion site-specific interactions.

Table 12-4 Protein Anion Interactions (substrates, coenzymes, or heavy atom modifications are not included)

Protein	Ion in Native Structure	Ligands	Other Ions Binding to this Site	Comment
Carbonic anhydrase	OH^-	Zn^{2+}, Thr 197	I^-, Br^-, Cl^-,	Active site
	?	Lys 22, Arg 243	I^-, Br^-, SCN^-, RSO_2NH^-	
α-Chymotrypsin	SO_4^{2-}	Ser 195, Tyr 146, His 57	SeO_4^{2-}	Active site involved in dimer formation
	SO_4^{2-}	Asn 95, Lys 177	SeO_4^{2-}	
	SO_4^{2-}	Thr 224	SeO_4^{2-}	
	SO_4^{2-}	Arg 154	SeO_4^{2-}	
	SO_4^{2-}	NH_3^+, Ala 149	SeO_4^{2-}	
	SO_4^{2-}	Asn 245	SeO_4^{2-}	
	SO_4^{2-}	Asn 236	SeO_4^{2-}	
Elastase pH 5	SO_4^{2-}	Ser 195, His 57, N′ 193	$2H_2O$ (pH 8.5)	Active site
	SO_4^{2-}	Arg 145, Arg 230	empty (pH 8.5)	
Lactate dehydrogenase	SO_4^{2-}	Arg 171, His 195	citrate	Substrate site
apoenzyme	SO_4^{2-}	Arg 173	citrate	Between subunits
Malate dehydrogenase	SO_4^{2-}			Substrate site
Myoglobin	SO_4^{2-}	His E 7		Absent in deoxy Mb
Ribonuclease	SO_4^{2-}	His 12, His 119, N′ 120, PO_4^{3-}, ASO_4^{3-}, $2H_2O$		Substrate site
	SO_4^{2-}	Lys 7, Lys 41		

Liljas and Rossmann (1974).

Perhaps the most fascinating aspect of the specific ion–protein interactions is the mechanism by which specificity emerges. The action of an ion is less related to direct electrostatic interaction with the protein than to the ability of the ion to modify the structural organization of water about itself and, consequently, the protein when the ion is near the protein. Thus anions and cations can be positioned into a lyotropic series with regard to their ability to modify protein structure. Applying this rule to a set of anions yields the ranking given in Table 12-5.

Table 12-5 Charge Density, Structure-Breaking Ability Ranking in Water, and Estimated Protein Property Modification Ranking of Several Anions

Anion	Charge density e.u./Å^2	Structure-Breaking Potential in Water[a]	Protein Modification Potential
PO_4^{3-}	−0.00777	1	1
Citrate	−0.00572	2	2
SO_4^{2-}	−0.00753	3	3
F^-	−0.04782	4	4
Cl^-	−0.02427	5	5
CH_3COO^-	−0.00445	6	6
Br^-	−0.02044	7	7
NO_3^-	−0.00786	8	8
HCO_3^-	−0.00740	9	9
I^-	−0.01603	10	10
N_3^-	−0.00606	11	11
NCS^-	−0.00426	12	12
NCO^-	−0.00553	13	13

[a] Hofmeister.

The indirect action of ions on protein properties can be realized from an analysis of protein crystals. Most proteins have been crystallized from salt solutions of high molarity. Still, very few specific binding sites for the solvent ions are observed in the electron density maps. Tables 12-2 and 12-4 summarize the structural properties of the limited number of observed solvent ion binding sites in proteins, that is, direct ion–protein interactions. The absence of solvent ions bridging between crystallographically related protein molecules suggests that the total crystallization process is primarily due to the ions structuring water molecules about and between protein molecules, that is, an indirect interaction rather than direct ion-protein encounters. The solvent ions used for crystallization are always structure-making species for water such as sulfate, phosphate, or citrate.

REFERENCES

Baddiel, C. B., M. M. Brever, U. J. Dorian, and I. Rendal (1967). In "Solution Properties of Natural Polymers", Internat. Symp. on Solution Properties of Natural Polymers, Edinburgh, England.

Balasubramanian, D. and R. S. Roche (1970). *Chem. Comm.*, 862.

Balasubramanian, D. and R. Shaikh (1973). *Biopolymers*, **12**, 1639.

Bradbury, J. H. and M. D. Fenn (1968). *J. Mol. Biol.*, **36**, 231.

Brown F. R., III, A. J. Hopfinger, and E. R. Blout (1972). *J. Mol. Biol.*, **63**, 101.

Cohen, S. R. and I. B. Wilson (1966). *Biochemistry*, **5**, 904.

Coleman, J. E. and B. L. Vallee (1960). *J. Biol. Chem.*, **235**, 390.

Coleman, J. E. and R. V. Coleman (1972). *J. Biol. Chem.*, **247**, 4718.

Davis, B., P. Saltman, and S. Benson (1962). *Biochem. Biophys. Res. Commun.*, **8**, 56.

Dodson, E., M. M. Harding, D. C. Hodgkins, and M. G. Rossmann (1966). *J. Mol. Biol.*, **16**, 227.

Dove, W. F. and N. Davidson (1962). *J. Mol. Biol.*, **5**, 467.

Doty, P., A. Wada, J. T. Yang, and E. R. Blout (1957). *J. Polymer Sci.*, **23**, 851.

Drum, D. E., J. H. Harrison, T. K. Li, J. L. Bethune, and B. L. Vallee (1967). *Proc. Natl. Acad. Sci. USA*, **57**, 1434.

Feder, J., L. R. Garrett, and B. S. Wildi (1971). *Biochemistry*, **10**, 4552.

Flory, P. J. (1969). *Statistical Mechanics of Chain Molecules.* Wiley-Interscience, New York.

Folk, J. E. and J. A. Gladner (1960). *J. Biol. Chem.*, **235**, 60.

Frazen, J. S., C. Bobik, and J. B. Harry (1966). *Biopolymers*, **4**, 637.

Friedman, M. (1974). *Protein–Metal Interactions.* Academic Press, New York.

Hanlon, S., S. Brudno, T. T. Wu, and B. Wolf (1975). *Biochemistry*, **14**, 1648.

Hardman, K. D. and C. F. Ainsworth (1972). *Biochemistry*, **11**, 4910.

Harned, H. S. and B. B. Owen (1958). *The Physical Chemistry of Electrolyte Solutions.* Reinhold, New York.

Harrington, W. F. and J. A. Schellman (1956). *C. R. Trav. Lab. Carlsberg, Ser. Chim.*, **30**, 21.

Herskovitz, T. (1972). *Proc. Natl. Acad. Sci. USA*, **69**, 2437.

Hilter, W. A., A. J. Hopfinger, and A. G. Walton (1972). *J. Am. Chem. Soc.*, **94**, 4324.

Homaguchi, J. and E. P. Geiduschek (1962). *J. Am. Chem. Soc.*, **84**, 1329.

Iizuka, E. and J. T. Jaug (1965). *Biochemistry*, **4**, 1249.

Jensen, L. H. (1974). *Annu. Rev. Biochem.*, **43**, 461.

Joubert, F. J., N. Lotan, and H. A. Scheraga (1970). *Biochemistry*, **9**, 2197.

Kagi, J. J. R. and B. L. Vallee (1960). *J. Biol. Chem.*, **235**, 3460.

Kotin, L. (1963). *J. Mol. Biol.*, **7**, 309.

Krimm, S. and J. E. Mark (1968). *Proc. Natl. Acad. Sci. USA*, **60**, 1122.

Latt, S. A. and B. L. Vallee (1971). *Biochemistry*, **10**, 4263.

Liljas, A., K. K. Kannan, P. C. Bergsten, I. Waara, K. Fridborg, B. Standberry, U. Carlbom, L. Jarlip, S. Lougren, and M. Petef (1972). *Nature* (*New Biol.*), **235**, 131.

Liljas A. and M. G. Rossmann (1974). In *Annual Review of Biochemistry*, **43**, 475.
Review of Biochemistry, **43**, 475.

Lindskog, S. and B. G. Malmström (1962). *J. Biol. Chem.*, **237**, 1129.

Lindskog, S. and P. O. Nyman (1964). *Biochim. Biophys. Acta*, **85**, 462.

Lotan, N. (1973). *J. Phys. Chem.*, **77**, 242.

Lotan, N. (1974). In *Peptides, Polypeptides and Proteins* (E. R. Blout, F. A. Bovey, M. Goodman and N. Lotan, Eds.) Wiley-Interscience, New York, p. 157.

Lotan, N., M. Bixon, and A. Berger (1967). *Biopolymers*, **5**, 69.

Lotan, N., A. Yaron, and A. Berger (1966). *Biopolymers*, **4**, 365.

Malmström, B. G. (1961). In *The Enzymes*, 2nd ed. (P. D. Boyer, H. Lardy, and K. Myrback, Eds.). Academic Press, New York, Vol. 5, p. 471.

Malmström, B. G. and L. Rydén (1968). In *Biological Oxidations* (T. P. Singer, Ed.). Wiley-Interscience, New York, p. 415.

Mandelkern, L. and D. E. Roberts (1961). *J. Am. Chem. Soc.*, **83**, 4292.

Manning, G. S. (1972). *Biopolymers*, **11**, 951.

Marmur, J. and P. Doty (1962). *J. Mol. Biol.*, **5**, 109.

Pauling, L. (1960). *The Nature of the Chemical Bond*. Cornell University Press, Ithaca, N. Y.

Peggion, E., L. Strassorier, and A. Cosani (1970). *J. Am. Chem. Soc.*, **92**, 381.

Plocke, D. J. and B. L. Vallee (1962). *Biochemistry*, **1**, 1039.

Quiocho, F. A. and W. N. Lipscomb (1971). *Adv. Protein Chem.*, **25**, 1.

Record, M. T., Jr. (1975). *Biopolymers*, **14**, 2137.

Reeke, G. N. (1974). *Ann. N.Y. Acad. Sci.* (in press).

Rippon, W. B., J. L. Koenig, and A. G. Walton (1970). *J. Am. Chem. Soc.*, **92**, 7455.

Rippon, W. B. and A. G. Walton (1971). *Biopolymers*, **10**, 1207.

Rippon, W. B. and A. G. Walton (1972). *J. Am. Chem. Soc.*, **94**, 4319.

Robinson, R. A. and R. H. Stokes (1959). *Electrolyte Solutions*, 2nd ed. Buttersworths, London, p. 476.

Ross, P. D. and R. L. Scruggs (1964*a*). *Biopolymers*, **2**, 79.

Ross, P. D. and R. L. Scruggs (1964*b*). *Biopolymers*, **2**, 231.

Schildkraut, C. and S. Lifson (1965). *Biopolymers*, **3**, 195.

Shiraki, M. and K. Imahori (1966). *Sci. Pap. Coll. Gen. Educ., Univ. Tokyo*, **16**, 215.

Steigman, J., E. Peggion, and A. Cosani (1969). *J. Am. Chem. Soc.*, **91**, 1822.

Strauss, B. P., C. Helfgott, and H. Pink (1967). *J. Phys. Chem.*, **71**, 2550.

Tiffany, M. L. and S. Krimm (1968). *Biopolymers*, **6**, 1967.

Tiffany, M. L. and S. Krimm (1969). *Biopolymers*, **8**, 347.

Traub, W. and K. A. Piez (1971). In *Adv. Protein Chem.* (C. B. Anfensen, J. T. Edsall and F. M. Richards, Eds.). Academic Press, New York, Vol. 25, p. 243.

Ulmer, D. D. and B. L. Vallee (1962). *Biochem. Biophys. Res. Commun.*, **8**, 331.

Urry, D. W., J. R. Krivacic, and J. Haider (1971). *Biochem. Biophys. Res. Commun.*, **43**, 6.

Vallee, B. L., J. A. Rupley, T. L. Coombs, and H. Neurath (1958). *J. Am. Chem. Soc.*, **80**, 4750.

Vallee, B. L. and W. E. C. Wacker (1970). *The Proteins*, Vol. V, *Metalloproteins* (H. Neurath, Ed.). Academic Press, New York.

Vallee, B. L. and R. J. P. Williams (1968*a*). *Proc. Natl. Acad. Sci. USA*, **59**, 498.

Vallee, B. L. and R. J. P. Williams (1968*b*). *Chem. Brit.*, **4**, 397.

von Hippel, P. H. (1967). In *Treatise on Collagen*, (G. N. Ramachandran, Ed.). Academic Press, New York, Vol. 1, p. 253.

von Hippel, P. H. and T. Schiech (1969). In *Structure and Stability of Biological Macromolecules* (S. N. Timasheff and G. D. Fasman, Eds.). Marcel Dekker, New York, Vol. 2, p. 417.

von Hippel, P. H. and K.-Y. Wong (1963*a*). *Biochemistry*, **2**, 1387.

von Hippel, P. H. and K.-Y. Wong (1964). *Science*, **145**, 577.

von Hippel, P. H. and K.-Y. Wong (1965). *J. Biol. Chem.*, **240**, 3909.

Waara, I., S. Lövgren, A. Liljas, K. K. Kannan, and P. C. Bergsten (1972). *Adv. Exp. Med. Biol.*, **28**, 169.

Watenpaugh, K. D., L. C. Sieker, J. R. Herriott, and L. H. Jensen (1971). *Cold Spring Harbor Symp. Quant. Biol.*, **36**, 359.

Wyman, J. (1964). *Adv. Protein Chem.*, **19**, 223.

Yazgan, A. and R. W. Henkens (1972). *Biochemistry*, **11**, 1314.

CHAPTER 13

Biomolecular Aggregation and Association

Intermolecular organization requires initial aggregation or at least association processes. This chapter begins with a discussion of probably the most studied and best understood biomolecular aggregates—lipid-based membranes. The intermolecular association is governed to a larger degree by entropy than are individual intramolecular or crystalline processes. The aggregation phase, irrespective of biologic implication, has not yet been extensively studied and characterized. Foster (1975) has recently edited the first in what will hopefully be a series of volumes dealing with molecular association. Consequently, a relatively limited number of biomolecular associated systems are discussed in regard to their specific properties rather than in general terms. Still, some general remarks have been attempted for polyelectroylite aggregates. We recognize that a fine line separates the classification of some of the systems reported in this chapter from those presented in Chapter 15, which deals with ultrastructural organization.

I AGGREGATION OF LIPIDS IN AQUEOUS MEDIA (MEMBRANE MODELS)

A Structural Features

Biologic membranes are composed primarily of lipids and proteins. Accurate knowledge of the composition of membranes is not yet available. This is primarily because of the difficulty in isolating membranes free of cytoplasm and other components of the organism. Some analytical data on the overall compositions of typical membranes are given in Table 13-1. In this section, we consider the molecular thermodynamics of the structure and organization of the lipid component in aqueous solution. This is probably the most extensively studied class of biomolecular aggregates.

Table 13-1 Analytical Protein and Lipid Content of Several Membranes[a]

	Percent of Dry Weight	
Membrane Sample	Protein	Lipid
Myelin	18	79
Human erythrocyte	49	43
Bovine retinal rod	51	49
Mitochondria (outer membrane)	52	48
Mycoplasma laidlawii	58	37
Sarcoplasmic reticulum	67	33
Gram-positive bacteria	75	25
Mitochondria (inner membrane)	76	24

[a] Based on data from Guidotti (1972). Where the figures given do not add up to 100%, the remainder is listed as "carbohydrate." Such carbohydrate would not normally be in free form, but a constitutive part of glycoproteins or glycolipids.

Biologic lipids are amphiphile ions (see Tanford, 1973, for a discussion of this term) that have the capacity to interact and to form a wide variety of intermolecular structural organizations in aqueous media when the lipid concentration is sufficiently high so as to self-associate. A listing of the chemical structures of some known biologic lipids is presented in Figure 13-1. A common structural feature of all these molecules is two relatively long, linear hydrocarbon chains per molecule. All lipids containing two hydrocarbon chains per molecule aggregate into a bilayer structural arrangement in aqueous solution. A schematic illustration of a bilayer structure is shown in Figure 13-2. It is probable that the bilayer molecular organization is relatively uniform, at least locally, for a given lipid concentration and temperature. The polar-ionic head groups point out into the aqueous media to minimize their mutual repulsive electrostatic interactions by taking advantage of the high dielectric nature of the aqueous solution. It is also reasonable to suggest that strongly favorable interactions between the ionic head groups and individual water molecules may also contribute to the stabilization of the bilayer. The lipid hydrocarbon chains aggregate side by side and roughly tail to tail in order to maximize bilayer hydrophobic interactions.

The overall size and shape of a lipid bilayer structure depends on the type of lipid, concentration, temperature, and method of preparation (Saunders et al., 1962; Dervichian, 1964; Bangham et al., 1965; Huang, 1969; Lewis and Gottlieb, 1971; Miyamoto and Stoeckenius, 1971; Sheetz and Chan, 1972; Smith and Tanford, 1972). The bilayer shapes predominantly found are (1) extended planar bilayers, (2) multiwalled vesicles, and (3) single-walled vesicles.

A relatively large number of X-ray diffraction studies of lipid bilayers have been attempted. Bear et al. (1941) and Palmer and Schmitt (1941) were the first to study lipids from natural sources using X-ray techniques. These workers were able to demonstrate the existence of the bilayer structure and to assign reasonable atomic dimensions for it. They were also the first to observe the 4.2 Å spacing characteristic of hexagonally packed hydrocarbon chains perpendicular to the bilayer plane in, essentially, dry samples that broadened to an average value of 4.6 Å when water was added to the sample. They correctly deduced that the interior of the bilayer is liquidlike, as postulated by Schmitt (1939). Most subsequent X-ray studies are summarized in a review by Luzzati (1968).

The most precise investigations of lipid bilayer structure are probably those Levine et al. (1968) and Levine and Wilkens (1971), who were able to prepare highly ordered stacks of bilayers separated by water layers. Figure 13-3, taken from Levine et al. (1968), shows the electron density profile of one repeat of an oriented stack of diplamitoyl phosphatidylcholine. The peaks of high electron density represent the location of phospholipid head groups. The two peaks separated by about 43 Å in the dry state represent the location of the two head groups on either side of a single bilayer. The increase in this separation distance upon wetting the sample very likely represents the projection of head groups of the lipids out into the aqueous medium to undergo favorable interactions with water molecules. The other pair of adjacent high-electron-density peaks that are separated by about 16 Å in dry samples and by 25 Å

Fig. 13-1 Some lipid primary structures.

sphingolipids

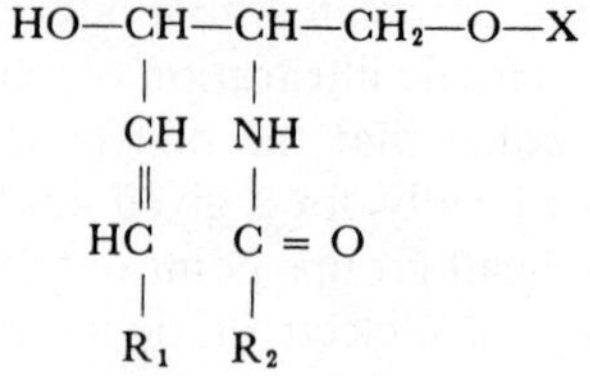

```
HO—CH—CH—CH₂—O—X
   |   |
   CH  NH
   ||  |
   HC  C = O
   |   |
   R₁  R₂
```

The carbohydrate moiety X may be a single neutral hexose sugar (as in cerebrosides), a sulfated sugar (cerebosides), a sulfated sugar (cerebroside sulfate or sulfatide), or a more complex oligosaccharide. The head group of sulfatides bears a negative charge, as does the head group of ganglisosides, in which X is a complex oligosaccharide containing one or more moles of sialic acid.

Mitochondria contain considerable quantities of cardiolipin, in which four hydrocarbon chains are attached to a single head group. This lipid is essentially a dimeric form of a phosphoglyceride. Two molecules of the diacyl glycerophosphate moiety are linked by one molecule of glycerol (i.e.,—X is replaced by —$CH_2CHOHCH_2$—). In the most common forms of the phosphoglycerides,

phosphoglycerides

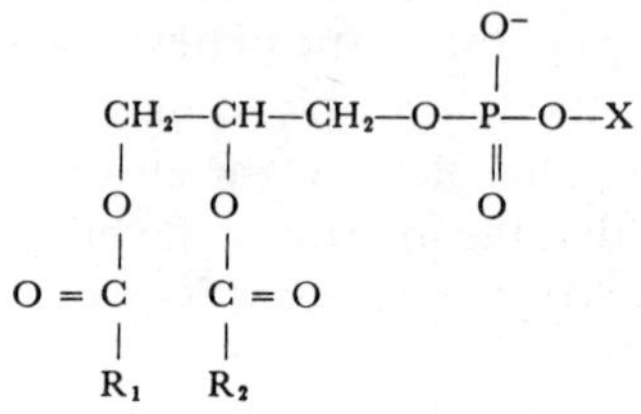

```
                          O⁻
                          |
    CH₂—CH—CH₂—O—P—O—X
    |    |                ||
    O    O                O
    |    |
O = C    C = O
    |    |
    R₁   R₂
```

R_1 and R_2 are hydrocarbon chains. The head group bears a net negative charge at neutral pH if the X group in this formula is neutral or zwitterionic, as follows:

Phosphatidic acid	—X = —H
Phosphatidylserine	—X = —$CH_2CH(NH_3^+)COO^-$
Phosphatidylinositol	—X = —$C_6H_6(OH)_5$
Phosphatidylglycerol	—X = —$CH_2CHOHCH_2OH$

Fig. 13-1 (*Continued*)

Alternatively, the head group may be zwitterionic at neutral pH, if the X group bears a positive charge as in

Phosphatidylethanolamine $-X = -CH_2CH_2NH_3^+$

Phosphatidylcholine $-X = -CH_2CH_2N(CH_3)_3^+$

Aminoacyl derivatives of phosphatidyl glycerol occur in some bacteria (Houtsmuller and Ban Deenen, 1965; Lennarz, 1972). Included in this category is the lysyl ester of phosphatidylglycerol, which bears two positive charges, conferring a net positive charge on the head group as a whole. Apart from this instance, phosphoglycerides with positively charged head groups are rare or nonexistent.

sphingomyelin

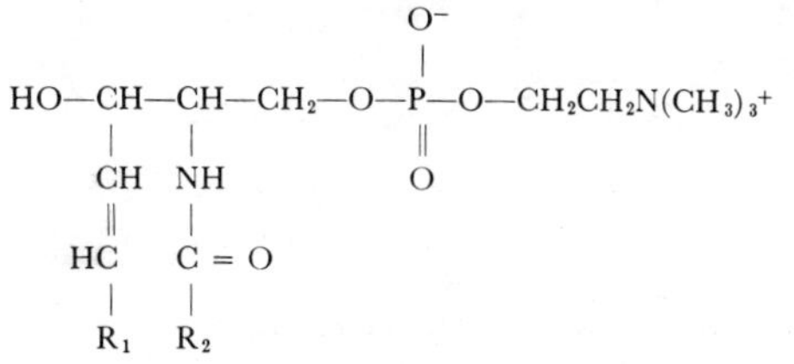

```
                              O⁻
                              |
HO—CH—CH—CH₂—O—P—O—CH₂CH₂N(CH₃)₃⁺
   |   |                      ||
   CH  NH                     O
   ||  |
  HC   C = O
   |   |
   R₁  R₂
```

cholesterol

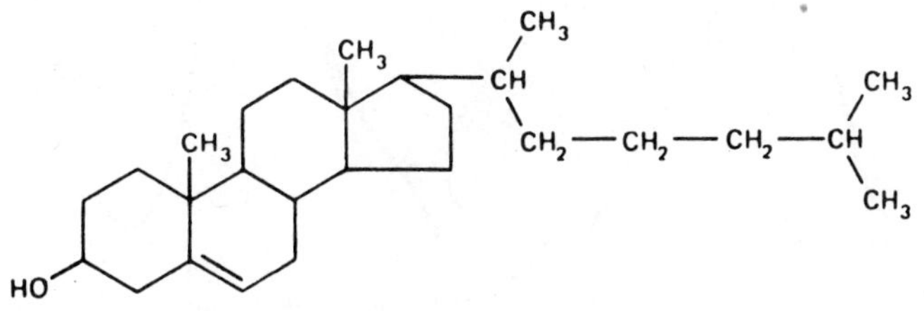

CH3 CH3 CH CH3 CH3 CH2—CH2—CH2—CH CH3 HO

This figure was generated from data given in Tanford (1973).

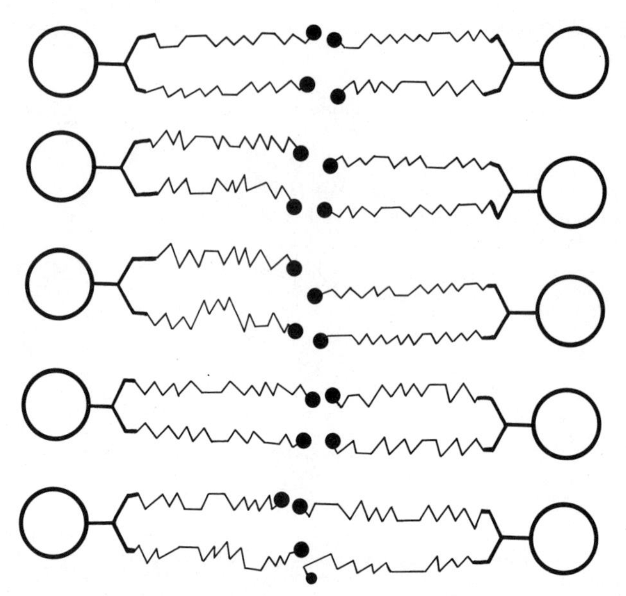

Fig. 13-2 Bilayer structure. Circles represent the polar head groups. The distorted zig-zags represent the hydrocarbon chains that are essentially in an extended conformation. The distortions are indicative of deviations from the extended conformation. The dark circles at the ends of the hydrocarbon chains indicate terminal methyl groups.

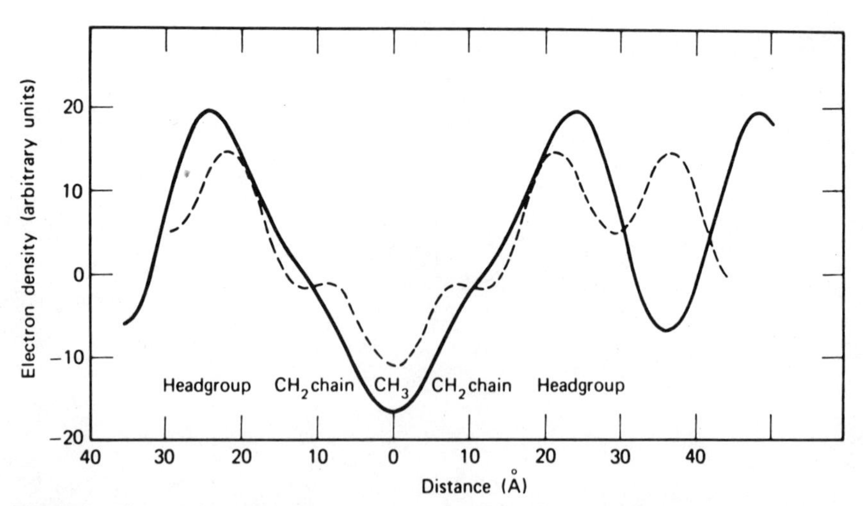

Fig. 13-3 Electron-density profile of one repeat of an oriented stack of bilayers of dipalmitoyl phosphatidylcholine (taken from Levine et al., 1968). Electron densities are given relative to liquid water as zero. The dashed line represents lipid in the dry state; the solid line represents data obtained when the system is under water.

in wet samples corresponds to the location of head groups in adjacent bilayers. The large dip in electron density mid-way between the intrabilayer headgroup electron density maxima has been thought to correspond to the location of the methyl groups at the ends of the hydrocarbon chains.

In the direction parallel to the bilayer surface, only one reflection, the 4.2 Å spacing corresponding to the separation between closely packed extended hydrocarbon chains was observed. This spacing did not change as a function of water content, and it was concluded that the bilayer has a core consisting of ordered hydrocarbon chains that are fully extended in a direction perpendicular to the bilayer surface.

The preceding discussion presents the salient features of the structural organization of biologic lipids in aqueous solution. The remaining portion of this section is concerned with a molecular thermodynamic explanation for the realization of such structures. Singer (1972) and Tanford (1973) have written reviews dealing with the structural and thermodynamic properties of lipids.

B Thermodynamic Properties

When lipid molecules are dissolved in water, they can achieve segregation of their hydrophobic portions from the solvent by self-aggregation. The aggregated products are known as micelles. The hydrocarbon chains form a hydrocarbon core for the micelle, with the polar heads on the surface of the micelle serving to maintain solubility in water. The hydrocarbon chains in such micelles are thermodynamically disordered; thus the hydrophobic core is, effectively, a small volume of liquid hydrocarbon. The simplest evidence for the liquid-like nature of the micelle interior comes from the ability to dissolve hydrocarbons and other hydrophobic substances within them (Wishnia, 1963). Table 13-2 contains the free energies, enthalpies, and entropies to transfer simple hydrocarbons from water to the interior of a dodecyl sulfate

Table 13-2 Thermodynamics of Transfer from Water to the Interior of a Dodecyl Sulfate Micelle[a]

Hydrocarbon	$\mu^\circ_{\text{mlc}} - \mu^\circ_W$ (cal/mole)	$\bar{H}^\circ_{\text{mlc}} + \bar{H}^\circ_W$ (cal/mole)	$\bar{S}^\circ_{\text{mlc}} - \bar{S}^\circ_W$ (cal/mole)
Ethane	−3450	+2000	+18.3
Propane	−4230	+1000	+17.5
Butane	−5130	0	+17.2
Pentane	−5720	−1100	+15.6

[a] From Wishnia (1963). Results are for 25°C.

micelle. Spectroscopic relaxation methods have proved very useful for estimation of fluidity in micelles in terms of the rate of motion of molecules within the systems. Fluorescence depolarization (Stryer, 1968) spectroscopy using hydrophobic fluorescent probes (Shinitzky et al., 1971) and electron-spin resonance spectroscopy (Hamilton and McConnell, 1968; Waggoner et al., 1967; Griffith and Waggoner, 1969) have most often been employed.

The size and shape of the micelles depend mainly on temperature and concentration and can be predicted reasonably well by solution thermodynamics. The theory of the solution thermodynamics of micelle formation has been neatly packaged in what Tanford (1973) terms the principle of opposing forces. Micelle formation clearly requires the existence of two opposing forces: an attractive force favoring aggregation and a repulsive force that prevents growth of the aggregates to large size. The attractive force arises from the hydrophobic effect acting on the hydrocarbon tails of the lipids. An important feature is that a minimum number of hydrocarbon chains must associate with one another before an effective reduction of the hydrocarbon–water interface can be achieved. Consequently, micelle formation is necessarily a cooperative process requiring joint participation by many lipids.

The repulsive force in micelle formation comes from electrostatic interactions between like charges on the head groups and a preference for hydration over self-association by the head groups. The thermodynamic parameter that determines the equilibrium between micelles and free lipid in solution, as well as the distribution between micelles of different size, is the difference between the standard unitary free energy of the lipid in a micellar aggregate and the corresponding quantity for the free lipid in aqueous solution.

In so far as micelle formation can be regarded as a phase separation process, it is convenient to talk of a critical micelle concentration (CMC) that corresponds to that concentration of lipid for which micelles just begin to form. Micelle formation is not equivalent to phase separation, but only approaches a true phase separation in that the micellar state occurs over a very narrow, critical range of concentration. Tanford (1973) discusses this behavior in detail. Figure 13-4 contains plots of the ln(CMC) versus the size of the hydrocarbon tails for several lipid-like molecules. With the exception of ionic micelles, curve F of Figure 13-4, the various lipid-like molecules show about the same CMC dependence on hydrocarbon chain size as reflected by similar slopes of the lines. Table 13-3 contains calculated and observed critical micelle concentrations of two biologic lipids in aqueous solution at 25°C. Standard free energies of the micelles are also given.

For additional reviews on the thermodynamics of micelle formation see Shinoda et al. (1963), Hill (1964), Becker (1967), Hall and Pethica (1967), Mukerjee (1967), Anacker (1970), and Tanford (1973).

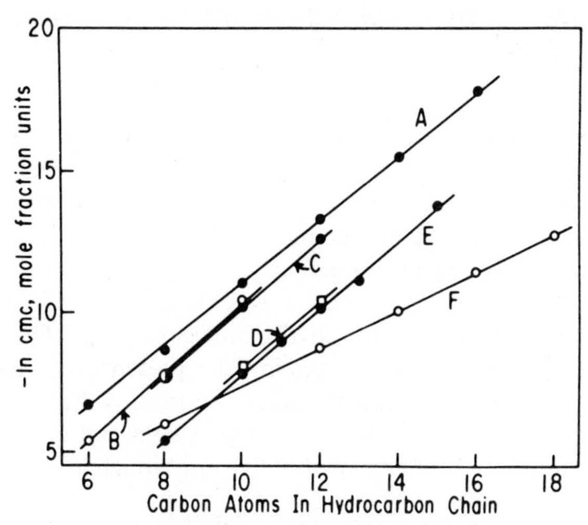

Fig. 13-4 Plots of ln(cmc) versus hydrocarbon chain length at 25°C, unless otherwise stated. A, Alkyl hexaoxyethylene glycol monoethers from Becher (1967). B, Alkyl sulfinyl alcohols (Corkill et al., 1966). C, Alkyl glucosides from Becher (1967). D, Alkyl trimethylammonium bromides in 0.5 *M* NaBr (Emerson and Holtzer, 1967*a*; Geer et al., 1971). E, N-alkyl betaines (Swarbrick and Daruwala, 1969). F, Alkyl sulfates in the absence of added salt, at 40°C (Evans, 1956). From Tanford (1973).

Table 13-3 Critical Micelle Concentrations of Biologic Lipids in Aqueous Solution at 25°C[a]

Lipid	cmc (moles/liter)	cmc (mole fraction)	$\mu^\circ_{\text{mlc}} - \mu^\circ_W$ (kcal/mole)
Palmityol[b] *Lysophosphatidylcholine*			
Observed[c]	$<12 \times 10^{-5}$	$<2.1 \times 10^{-6}$	<-7.7
Calculated[d]	6×10^{-5}	1.0×10^{-6}	—
Dipalmitoyl[b] *Phosphatidylcholine*			
Observed[e]	4.7×10^{-10}	8.4×10^{-12}	-15.1
Calculated[d]	11×10^{-10}	21×10^{-12}	—

[a] From Tanford (1973).
[b] The palmitoyl moiety contains a C_{15} hydrocarbon chain.
[c] Lewis and Gottlieb (1971). Robinson and Saunders (1958) observed a cmc in the range of 2 to 20 × 10^{-5} *M* for a lysophosphatidylcholine with mixed hydrocarbon chains.
[d] Based on extrapolation of data using the betaine head group as representative of the phosphatidylcholine head group.
[e] Smith and Tanford (1972).

II THE POLY(L-LYSINE)-CHONDROITIN 6-SULFATE AGGREGATE

Circular dichroism studies discussed in Chapter 11 have shown that an interaction occurs between anionic glycosaminoglycans and cationic homopolypeptides when mixed in dilute aqueous solution such that a conformational change occurs for the polypeptide.

Unfortunately, the CD technique only reveals conformational changes that may occur for the polypeptide, and reveals no other information about the mechanism of the interaction at the molecular level. The solutions in some cases are visibly turbid, suggesting the presence of multimolecular aggregates in solution, and the CD spectra often show light-scattering distortions, that is, red shifts are observed for the troughs and crossover together with a dampening in ellipticity of the 209 nm trough relative to the 222 trough.

These observations led Schodt et al. (1967*a*) to use quasielastic laser light scattering to determine the sizes of the particles in solution under conditions identical to those used for the CD measurements. This technique analyzes the spectrum of light scattered by microscopic fluctuations in concentration (or refractive index) caused by particles undergoing Brownian motion. The power spectrum of the scattered light is given by Dubin (1972):

$$S_i(\nu) \;\alpha\; \frac{2\Gamma/2}{\nu^2 + (2\Gamma/2\pi)^2} \tag{13-1}$$

where ν is the frequency of scattered light and Γ is the rate of decay of the fluctuation that produced the scattering. This is a Lorentzian curve of half-width Γ/π. Γ in turn can be related to the translational diffusion coefficient, D, by the scattering factor, K:

$$\Gamma = DK^2 = D\left[\frac{4\pi n}{\lambda_0}\sin\theta/2\right]^2 \tag{13-2}$$

where θ is the scattering angle, n is the index of refraction of the scattering medium, and λ_0 is the wavelength of incident light. For very dilute solutions, provided the particles are optically isotropic and assuming they are spherical in shape, the hydrodynamic size of the particles may be calculated using the Stokes-Einstein equation (Tanford, 1961):

$$R = \frac{kT}{6\pi\eta D} \tag{13-3}$$

where R is the radius, k is the Boltzman constant, and η is the viscosity of the solvent.

Initial work by Schodt et al. (1976*a*) has been done on maximum interacting mixtures of poly(L-lysine) and chondroitin 6-sulfate. The CD spectrum in Figure 13-5 indicates the presence of at least 80% α-helical content for poly-(L-lysine) in a 1:1 mixture with chondroitin 6-sulfate. Figure 13-6 shows the best single Lorentzian fit to the Rayleigh spectrum of the same mixture. Analysis of the half-width obtained from the scattered spectrum indicates the presence of fairly large aggregates in solution. From a series of measurements for these specimens of poly(L-lysine) and chondroitin 6-sulfate under the conditions used, the aggregates have radii in the range of 800–1100 Å. Maximum induced α-helical content for this system occurs only after the mixture is cooled below 10°C, or when the components are mixed at 10°C. If the components are mixed at room temperature, there is no conformational change induced and the CD spectrum resembles that of curve B in Figure 13-5. However, the half-width of the Rayleigh spectrum for the solution prior to the conformational change reveals the presence of particles of the same size ($R \simeq 1100$ Å). Cooling to 10°C induces the α-helical conformation, which is then stable at room temperature.

In addition, the aggregates are stable with increasing temperature, as determined by equation (13-3) with a correction for the change in solvent viscosity. The radii remain constant at least up to 60°C, whereas the CD shows a melting transition whereby the α helix breaks down at 47°C. As the ionic strength is increased, the aggregate shows an increase in radius. At

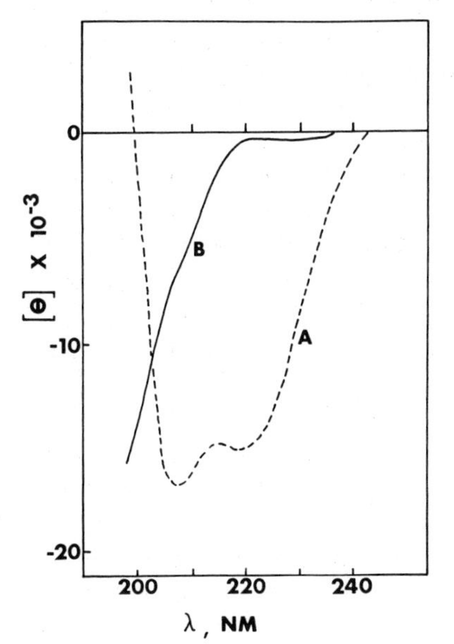

Fig. 13-5 Circular dichroism spectra of 1/1 mixtures of poly(L-lysine) and chondroitin 6-sulfate.
a. Experimentally observed spectrum after formation of α helix.
b. Calculated spectrum of simple addition of spectra of the two components.

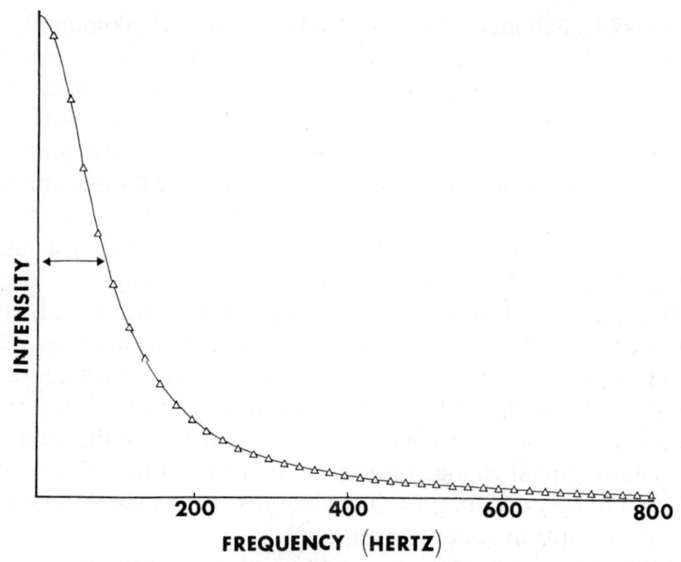

Fig. 13-6 Power spectrum of scattered light from a 1/1 mixture of poly(L-lysine) and chondroitin 6-sulfate at 40° scattering angle: half width = 90 Hertz; hydrodynamic radius = 1140 Å. Schodt et al. (1976).

1.0 *N* NaCl, the aggregates are on the order of 2500 Å in radii. In contrast, the CD shows a disruption of the α-helical directing effect at 0.27 *M* NaCl.

Correlation of the CD studies and laser light-scattering results leads to a clearer understanding of the poly(L-lysine)-chondroitin 6-sulfate interaction at the molecular level. The first step in the mechanism is the formation of a fairly large-sized aggregate, as opposed to the existence of isolated chains in solution. At low temperature, after the aggregate is formed, the polypeptide undergoes a slow (probably nucleation-controlled) conformational rearrangement from a disordered or possibly charged coil form to an α-helical form that does not disturb the macroscopic structure of the aggregate. Furthermore, the temperature and ionic strength melting effects relate to a conformational change of the individual polypeptide molecules in the aggregate and do not involve disruption of the aggregate.

Very recently, Schodt et al. (1976*b*) have reported the sizes of additional polypeptide-glycosaminoglycan aggregates, all of which appear to be spherical in shape. The results are presented in Table 13-4. Although it is far too early to propose any interrelationships, there do appear to be very definite changes in the radii of the aggregates as a function of both choices of polypeptide and glycosaminoglycan as well as the stoichiometric ratio of peptides

Table 13-4 Aggregate radii (Å) for Four Different Homopolypeptide-Glycosaminoglycan Mixtures

System	(Peptide) to (Disaccharide)	Radius (Å)
Poly(L-lysine)[a]-chondroitin 6-sulfate[b]	(1) to (1)	1200
Poly(L-lysine)-chondroitin 4-sulfate[c]	(1) to (1)	1756
Poly(L-arginine)[d]-chondroitin 6-sulfate	(1) to (1)	1904
	(2) to (1)	1865
Poly(L-lysine)-heparin[e]	(2.3) to (1)	2909

[a] Mol. wt. 100,000
[b] Mol. wt. 60,000
[c] Mol. wt. 38,000
[d] Mol. wt. 58,000
[e] Mol. wt. 11,000
From yet unpublished results of Schodt et al. (1976*b*).

to disaccharides. The effect of molecular weight cannot be ascertained to even qualitative cause-effect status.

III NUCLEOSIDE BASE AGGREGATION IN AQUEOUS SOLUTION

Studies by Ts'o and others (Ts'o, 1970; Ts'o et al., 1963; Ts'o and Chan, 1964) on the monomeric units of nucleic acids in aqueous solution indicate that mononucleosides associate in "stacks" made up of the essentially planar base moiety. The stacks are held together primarily by so-called "hydrophobic" interactions as opposed to hydrogen bonding. The stacking process has been described by a model that assumes that the free-energy change as well as the enthalpy change for the addition of a single base molecule to a stack is independent of the size of the stack (Ts'o, 1970). For some nucleoside bases, however, particularly those which associate to a higher degree, it appears that this model overestimates the degree of association at higher base concentrations. Attempts to correct for this overestimation include placing an artificial limit on the size of the stack (Ts'o et al., 1963; Ts'o and Chan, 1964), and introducing activity coefficients for the aggregates, which results in a sequence of decreasing association equilibrium constants (Hill and Chen, 1973). In the case of N^6, N^9-dimethyladenine, the dimerization equilibrium constant, K_2, was allowed to vary independently of the equal sequential constant, K, with the result that $K_2 > K$ (Porschke and Eggers, 1972).

Kinetic investigations of the aggregation of nucleoside bases are quite rare, consisting essentially of ultrasonic absorption studies of N^6, N^9-dimethyl-adenine (Porschke and Eggers, 1972) and 6-methyl-purine (Garland and Patel, 1974). The relaxation curves observed are somewhat broader than would be predicted on the basis of a single relaxation time, indicating the presence of more than one equilibrium reaction step. This has led to the postulation of a two-state model to describe the sound spectra (Porsche and Eggers, 1972).

There is a growing body of evidence (Ts'o, 1970, Porschke and Eggers, 1972) that indicates that the stacking process is somewhat different from the "hydrophobic bonding" interaction in which aggregate formation is characterized by positive changes in enthalpy, entropy, and volume (Kauzmann, 1959). Porschke and Eggers (1972), for example, report, for the stacking of N^6, N^9-dimethyladenine, $\Delta V = -6.5$ ml/mole. Stacking interactions between the base molecules might have been expected to lead to a positive volume change because of the effect of removal of oriented water molecules from contact with the hydrophobic faces of the purine molecules. However, this effect seems to be outweighed by the relatively strong binding between the π-electronic systems of the bases. Studies of the denaturation of ribonuclease seem to indicate that negative volume changes accompanying hydrophobic exposure of hydrocarbon residues (in proteins) are less than had been predicted from studies of smaller molecular systems (Brandts et al., 1970). Hopfinger (1975), in a review of biopolymer solvation properties, presents relevant data supporting both sides of the issue as to whether "classic hydrophobic interactions" drive the base-stacking process. No conclusion could be made concerning this phenomenon.

IV AGGREGATION OF POLYPEPTIDES

Since the first observations by Doty et al. (1956) and Tinoco (1957), it is now well known that poly-γ-benzyl-L-glutamate (PBLG) molecules form aggregates when dissolved in some of their helicogenic solvents such as benzene, dioxane, ethylene dichloride, and chloroform. These aggregates can be gradually destroyed by the addition of small amounts of dichloroacetic acid (DCA) or dimethyl formamide (DMF). Dielectric dispersion and electric birefringence measurements have often been employed to study these aggregates. Dielectric dispersion was used by Wada (1960) to explain the structure of PBLG aggregates in dioxane: DMF mixtures. Applications of the electric birefringence technique to this problem have been made by Watanabe (1965) and Powers and Peticolas (1970), who studied the aggregation of PBLG in several solvent systems. Most investigators agree that PBLG in helicogenic solvent can form aggregates that are one or a combination of the four types of structures:

1. linear head-to-tail type (type A);
2. linear head-to-head type (type B);
3. side-by-side type with parallel orientation (type C);
4. side-by-side type with antiparallel orientation (type D).

As shown in Table 13-4 for the case of aggregation of two identical helical molecules, distinction of type A from type C or of type B from type D is impossible on the basis of the mean square dipole moment per unit solute mass, μ_a^2/M_a, values alone. A knowledge of critical frequency, f_c, enables this distinction between linear (types A and B) and side-by-side (types C and D) aggregation, because of the difference in molecular dimensions. For linear aggregation, f_c for an aggregated molecule will be same as that of a helical molecule of the same molecular weight as that of the aggregate, whereas, for side-by-side aggregation, f_c for the aggregates will not be much different from that of the unaggregated molecules. These criteria of distinction are summarized in Table 13-4. For the case of more than two molecules, one can give variation rules as a function of the degree of aggregation, which are also summarized in the last two columns of Table 13-4.

Watanabe (1965) has suggested that the aggregation of PBLG in ethylene dichloride (EDC) is of the head-to-tail type and consists of two or three solute molecules. Gupta et al. (1974) have carried out dielectric dispersion measurements on PBLG in dioxane and dioxane–dichloroacetic acid (DCA) mixtures in order to study the structure of molecular aggregates. They conclude that the structure of aggregates is explained on the basis of the variation of dipole moment and relaxation time with degree of aggregation. PBLG was found to form linear head-to-tail type aggregates in dioxane. These aggregates gradually reduce in size without losing their α-helical structure during the process of disaggregation obtained by either adding DCA to the solution in dioxane or by heating. Addition of 30 wt % DCA completely destroys the aggregation of PBLG in dioxane at 30°C. Thermal disaggregation, however, is not complete even at a temperature approaching the boiling point of the solvent. Gupta et al. (1974) have proposed a reaction scheme for aggregation from which equilibrium constants have been calculated at various stages of aggregation. The enthalpy of aggregate formation is found to be -3 kcal/mole in their model. Figure 13-7 shows the dependence of the reduced specific viscosity of PBLG in EDC–DMF mixtures at 25°C

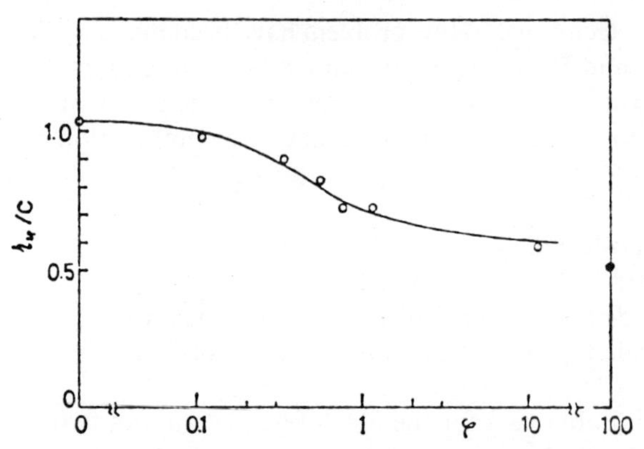

Fig. 13-7 Reduced viscosity of PBLG in EDC with a small fraction of DMF as a function of φ (vol % of DMF to EDC): The solute concentration, 6.9×10^{-3} g/ml; ○, data in EDC-DMF mixtures; temp, 25°C; ●, data in DMF; temp, 25°C. Kihara (1975).

on the volume percent, ϕ, of DMF. According to Boeckel et al. (1962), this result may be taken to suggest that aggregates present in pure EDC become molecularly dispersed by the addition of DMF up to 1%. An extensive study of PBLG in EDC–DMF mixtures has led Kihara (1975) to conclude that the change in rotational relaxation times, τ, with ϕ for the system PBLG–EDC–DMF reflects the change in the stage of aggregation, with the dipole moment kept constant at the value for dispersed solutes. Kihara (1975) concludes that types C and D aggregation may be rejected in the PBLG–EDC–DMF system because it does not exhibit a faster relaxation than unaggregated molecules, as would be expected in these types of aggregates. From structural considerations, PBLG is not likely to aggregate in head-to-head fashion, according to Kihara (1975). Therefore, type A aggregation, as suggested by Watanabe (1965), is thought to be most probable. The data in Figure 13-8 show that τ in the aggregated state is larger by a factor of about 2 than in the dispersed state. Yu and Stockmayer (1967) calculated τ of a hinged polymer composed of two equal rods connected by a flexible joint, and found that it is four times as large as that of a straight rod with half its total length. This has led Kihara (1975) to the supposition that the number of polymer molecules per aggregate is two or three and is in equilibrium with singly dispersed molecules, again in support of Watanabe's model (1965). Recently Gupta et al. (1976) have analyzed the effect of concentration on PBLG in dioxane. A critical concentration ($\simeq 10^{-3}$ g/g for the range of molecular weights studied) was determined below which the aggregates were found to have linear head-to-tail type structure. Above the critical concentration, a different mode of aggregation is observed, although it could not

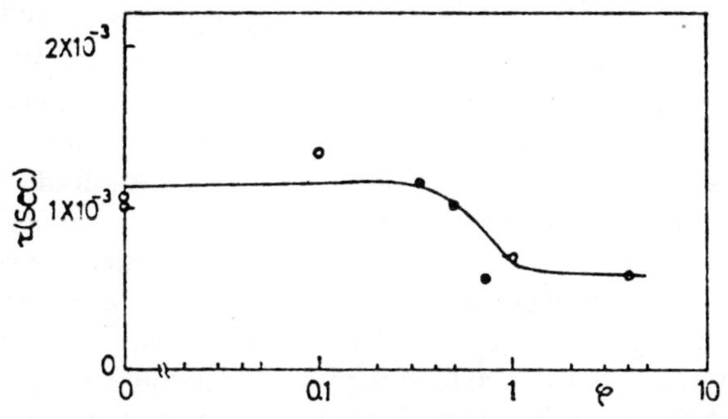

Fig. 13-8 Rotational relaxation times, τ, at 25°C of PBLG in EDC with a small fraction of DMF as a function of φ. They are reduced under the condition of 1 poise. The symbols are same as those in Figure 13-7. Kihara (1975).

be fully analyzed. Gupta et al. (1976) suggest that the formation of "ordered domains" that might lead to a nematic liquid crystalline phase might be present above the critical concentration range.

The dielectric absorption of poly-DL-phenylalanine and poly-γ-benzyl-L-aspartate (PBLA) has been measured in very dilute solutions by Marchal (1974) to determine the type of molecular association and to locate the helix-coil transition. Both polypeptides are present as associated helices in chloroform. The mode of aggregation, which was determined by measuring the dipole moment and the critical frequency, does not depend on the polarity of the side chain but rather on that of the solvent. In both polymers, the dissociation of the aggregates in chloroform was observed on addition of small amounts of dichloroacetic acid; further addition of the acid led to the helix-coil transition.

V PEPTIDE AGGREGATES

A Gramicidin A

Gramicidin A facilitates the passive diffusion of the alkali cations and hydrogen ions through natural (Harold and Baarda, 1967; Chappell and Crofts, 1965; Harris and Pressman, 1967; Bamberg et al., 1975) and artificial lipid bilayer membranes (Mueller and Rudin, 1967; Myers and Haydon, 1972). There is considerable evidence that gramicidin A forms a channel spanning the membrane hydrocarbon (Hladky and Haydon, 1972; Krasne et al., 1971), and some evidence that the channel requires two molecules of gramicidin A (Tosteson et al., 1968; Bamberg and Lauger, 1973). Gramicidin A is a linear peptide with the sequence (Sarges and Witkop, 1965): formyl-L-Val-Gly-L-Ala-D-Leu-L-Ala-D-Val-L-Val-D-Val-(L-Try-D-Leu)$_3$-L-Trp-ethanolamine. If glycine is considered a pseudo D residue, the sequence is alternating LDLD. All side chains are relatively hydrophobic.

Several groups of workers (Glickson et al., 1972; Urry et al., 1972; Isbell et al., 1972; Rothschild and Stanley, 1974) have examined the conformation and aggregation of gramicidin A in solution in order to assess models for the conformation of gramicidin A involved in the transmembrane channel. In particular, Veatch et al. (1974) have identified four conformational species of gramicidin A from a single nonpolar solvent system. Three distinct conformations were found: two are probably helices of opposite handedness with largely parallel-β hydrogen bonding; another has largely antiparallel-β hydrogen bonding. All four of the conformational species appear to be dimers. To account for some of these and other observations, a new family of parallel-β and antiparallel-β double helices have been postulated by Veatch et al. (1974). Table 13-5 summarizes the parameters of the lower diameter parallel- and antiparallel-β double-helical dimers and compares them with those of the head-to-head π (LD) helical dimers, which are the same for either handedness. It should be noted that all of the structures in Table 13-5 have NH-C_αH dihedral angles near 180°. Also, all have approximately the same number of hydrogen bonds and would be stabilized by nonpolar solvents (Veatch and Blout, 1974). Figure 13-9 shows schematic illustrations of the proposed antiparallel (a) and parallel (b) β helices of gramicidin A aggregates. These species, isolated from a single nonpolar solvent system, have been denoted 1, 2, 3, and 4 in order of increasing mobility in thin-layer chromatography. [Veatch et al. (1974) conclude that species 1, 2, and 4 are likely the helical structures with largely parallel-β hydrogen bonding; species 1 and 2 have a handedness opposite that of species 4. Species 3 probably possesses largely antiparallel-β hydrogen bonding.]

The aggregation of gramicidin in nonpolar solvents is probably a dimerization process. It is likely that all of the isolated gramicidin species are dimers. Even species 1 and 2, which have very similar conformations, may have the same molecularity. Veatch et al. (1974) find that the dimerization constant is much higher in relatively nonpolar solvents, such as ethyl acetate and dioxane, than in more polar solvents, such as dimethyl sulfoxide. This suggests that the dimers are stabilized by a substantial number of hydrogen bonds. The very slow aggregation rates for the isolated species in nonpolar solvents suggest structures containing many intermolecular hydrogen bonds, which is again indicative of the formation of parallel- and antiparallel-β double helices. It should be noted that the gramicidin analog, *N*-acetyl-desformylgramicidin, has almost identical dimerization constant and species

Table 13-5 Parameters of the Lower Diameter Parallel- and Antiparallel-β Double-Helical Dimers

	Summary of Double-Helix Parameters[a]				Summary of Head-to-Head π(LD) Helical-Dimer Parameters		
Parameter	Parallel-β		Antiparallel-β				
Residues/turn	6	7	6	7	4.4[b]	6.3[b]	8.4[b]
Symmetry (orientation of rotation	C_2 or C_1	C_1	C_2	C_2	C_2	C_2	C_2
axis relative to helix axis)	$\parallel$		$\perp$	$\perp$	$\perp$	$\perp$	$\perp$
Length, Å	32	25	32	28	37[b]	28[b]	21[b]
Rise/turn, Å	11	10	12	11	5.5	5.0	5.0
Inside diameter, Å	~3	~5	~3	~4	1.4[b]	4[b]	6[b]
Outside diameter, Å							
C-terminal	16–18	17–20	16	18–22			
N-terminal	10–13	14–16	Uniform	Uniform			
Hydrogen bonds (including hydroxyl)	30	28	30	28	Inter 4	6	8
					Intra 26	22	18
					Total 30	28	26

[a] Dimensions were measured from CPK models.

[b] From Urry et al., 1971.

Veatch et al. (1974).

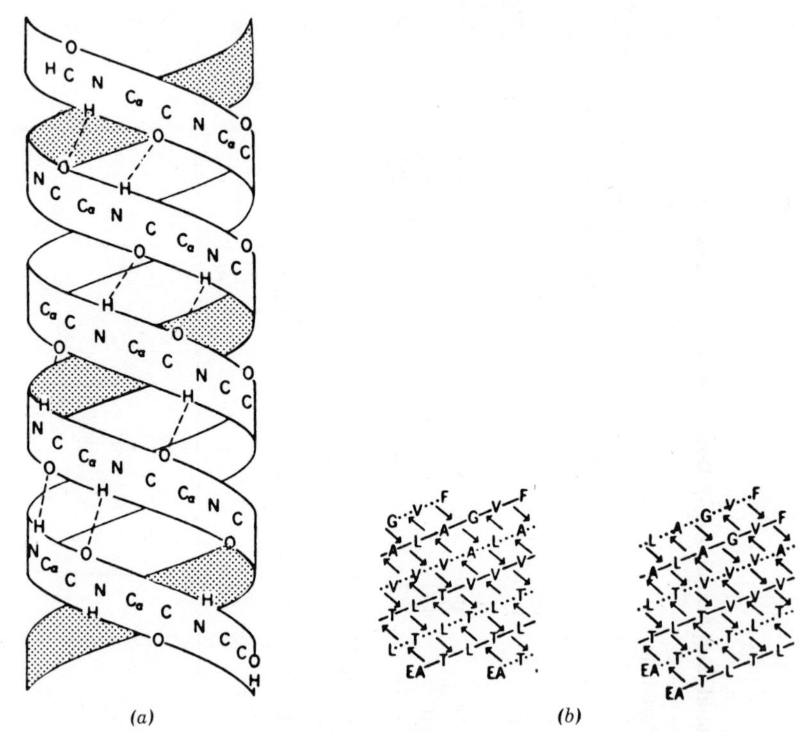

Fig. 13-9 (*a*) Schematic diagram of antiparallel-β double-helical gramicidin dimer with seven residues per turn and "even ends." The helix diagrammed is left-handed; however, it is not now possible to specify the handedness of the species. One chain is shaded, and the other is not. The dotted lines denote hydrogen bonds. The formyl group of the unshaded chain is at the upper left, and the C-terminal hydroxyl of the unshaded chain is at the lower right. Bonds between atoms have been omitted because of the geometry of the main-chain atoms is distorted for clarity. (*b*) Schematic diagram of right-handed parallel-β double-helical gramicidin dimers with six residues per turn. These diagrams look at helices slit along the back and laid out flat. The letters denote the amino acid residues, and the two chains are shown as a solid line and dashed line, respectively. The arrows denote hydrogen bonds and go from CO to NH. In (*a*) the ends are "even," and the structure has a twofold rotation axes along the helix axis. In (*b*) the ends are not "even" and lack the formal twofold axis, but the total number of hydrogen bonds is the same as in (*a*). (Veatch et al., 1974).

mole fractions as gramicidin in ethanol (Veatch et al., 1974). These findings imply that none of the isolated species are energetically destabilized by the substitution of a methyl group for the less bulky proton.

Further studies by Blout and co-workers (Fossel et al., 1974), using ^{13}C NMR, indicate from the spin-lattice relaxation time, T_1, measurements that, in dimethyl-d_6 sulfoxide, the gramicidin A peptide backbone undergoes slowest motion near the center of the molecule ($T_R = 5.0$ nsec), whereas the N-terminal residue moves more rapidly ($T_R = 0.7$ nsec). In methanol-d_4, a solvent in which it has been shown that gramicidin A exists predominantly as a helical dimer (Veatch et al., 1974; Veatch and Blout, 1974), the T_1 measurements show that motion in the backbone of the molecule has been greatly reduced ($T_R = 30$ nsec). This extreme decrease in motion in the peptide backbone going from random coil to dimer form is again consistent with the proposed (Veatch et al., 1974) double-helical dimer model.

These studies suggest that the transmembrane channel is an ion-conducting dimer in rather slow equilibrium with nonconducting monomer (Bamberg and Lauger, 1973). Some, if not all, of the gramicidin isolated species are dimers in slow equilibrium with monomer. The channel probably contains some water (Myers and Haydon, 1972) and is less than 30 Å in length (Hladky and Haydon, 1972). The only direct estimate of the dimensions of any of the isolated species is for species 3; two of three possible dimensions suggested by the crystallographic data are cylinders 27–32 Å in length (Veatch et al., 1974). There is no direct evidence that any of the gramicidin isolated species have internal cavities. However, species 1, 2, and 4 are helical and consequently must have at least a small hole down their middles. The data for species 3, which support an antiparallel-β double helix, also imply a substantial cylindrical hole, since this model cannot be made without one (Veatch et al., 1974). Although it has been postulated that the transmembrane channel dimer involves a left-handed helix and is symmetric (Urry, 1971; Urry et al., 1971), no direct membrane conformational data support these hypotheses. The gramicidin isolated species 1, 2, and 4 demonstrate the potential of gramicidin to form helical aggregates of either handedness.

B Some Homo-oligopeptides

A study of the dependence of the NMR chemical shifts on concentration make it possible to distinguish NH protons involved in intermolecular from those involved in intramolecular hydrogen bonds. Similarly, it is also possible to distinguish the above two types of NH protons from those fully solvated. Wittstruck and Cronan (1968) have investigated the concentration dependence of the NMR spectra of several ketoalcohol derivatives in order to determine the ratio of intermolecular to intramolecular hydrogen bonds. As the concentration is varied, the peak assigned to the proton of the ketoalcohols that is involved in intramolecular hydrogen bonding shifts appreciably less than that of the proton of the ketoalcohols participating in intermolecular hydrogen bonds. Ovchinnikov et al. (1970) have compared the chemical shift of the NH proton of alanine with those of the alanine dimer. They conclude that the chemical shift of the NH proton that participates in an intramolecular hydrogen bond in the dimer is different from that of the free solvated NH proton. Shields and co-workers (Shields and McDowell, 1967; Shields et al., 1968) carried out a detailed infrared analysis that shows that oligomers as short as the tetramer can form intramolecular hydrogen bonds in deuterochloroform. These authors speculated that a possible stable structure for Boc-L-Val-L-Val-L-Ala-Gly-OEt in this solvent would involve a β turn.

Goodman et al. (1975*a*) have carried out NMR studies of the isoleucine oligopeptides that give clear evidence for the onset of aggregation in going from the dimer to the tetramer. The chemical shifts of the NH protons in the dimer are relatively insensitive to concentration. Although minor changes are observed for the NH protons of Boc(Ile)$_3$OMe, they are restricted to the concentration region below 0.015 *M*. Goodman et al. (1975*a*) conclude that the three NH protons of the tripeptide are completely solvated in this concentration range. In contrast, the large concentration dependence of these protons in the tetrapeptide suggests that they are involved in intermolecular hydrogen bonding at a concentration of 0.02 *M*. The findings are in accord with the IR investigation of Boc(Ile)$_4$OMe by Shields et al. (1968), which show that no intramolecular hydrogen bonds are present in this peptide. Goodman et al. (1975*b*) postulate that, at concentrations of 0.02 *M* (1.1% w/v) and above, simple molecular aggregation of the isoleucine tetramer occurs. The analogous dimer and trimer are thought to be fully solvated. They find no evidence that any of the isoleucine oligomers assume a preferred conformation involving an intramolecular hydrogen bond. The chemical shift dependence of the three NH protons of the alanine trimer are quite similar to those of Boc(Ile)$_2$OMe and to those of Boc(Ile)$_3$OMe above 0.01 *M*. The lack of a concentration dependence for a chemical shift of an NH proton could indicate that it is fully solvated or involved in an intramolecular hydrogen bond. Goodman et al. (1975*b*) believe that MeO(C_2H_4O)$_2$Ac(Ala)$_3$OEt is fully solvated. This is supported by their IR studies, which show that no intramolecular hydrogen bonds are present in this oligopeptide in deuterochloroform.

NMR studies of the alanine tetramer give definite indications that there are at least two different types of NH protons (Goodman et al., 1975*b*). The chemical shifts of two of the NH protons in this oligomer exhibit sharp changes with increasing concentration at 23°C or decreasing temperature at 0.022 *M*. The remaining NH protons are relatively independent of concentration. The temperature dependence of these protons changes dramatically when the oligomer concentration is decreased to 0.006 *M*. These findings indicate that two of the protons in the tetramer are affected by intermolecular hydrogen bonding for concentrations greater than 0.012 *M*. The remaining two NH protons must then be either fully solvated or involved in intramolecular hydrogen bonds. The results of IR studies show that the latter is probably the case (Goodman et al., 1975*b*). These results agree with the conclusion of Shields et al. (1968) that intramolecular hydrogen bonds exist in Boc(Ala)$_4$OEt in deuterochloroform. Goodman et al. (1975*b*) suggest that the alanine tetramer exists in two forms in chloroform. At high concentration, aggregation occurs leading to the observed intermolecular hydrogen bonds, whereas at lower concentrations a folded form, containing intramolecular hydrogen bonds, persists. It is possible that this folded form is similar to a β-type bend. Although the coupling constants found by Goodman et al. (1975*b*) for the tetramer are slightly smaller than those calculated by Nemethy and Printz (1972) for the β-I turn, they are consistent with a β turn in rapid equilibrium with an extended form. Goodman et al. (1975*b*) have proposed a tentative model to explain the physicochemical behavior of the tetrapeptide. At high concentration (0.022–0.04 M; 1.1–2.0%) the molecules exist essentially as side-by-side dimers of folded species as shown in Figure 13-10. As the concentration is lowered, the dimers break up to give intramolecularly hydrogen-bonded species that are in equilibrium with the open-chain molecule. Goodman et al. (1975*b*) are not fully satisfied with all of the features of this model. Although it accurately predicts the NMR double peaks at high concentration and the intramolecular hydrogen bond at low concentration for the tetramer, it is not immediately apparent why the protons resonating at low field should move to higher field on dilution.

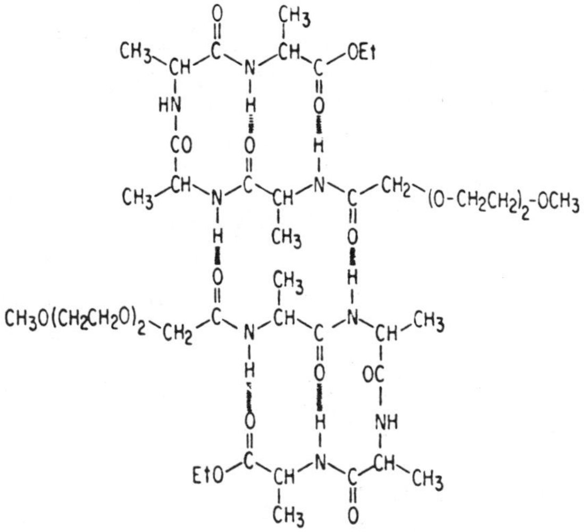

Fig. 13-10 Schematic representation of the side-by-side model of MeO$(C_2H_4O)_2$Ac-(Ala)$_4$OEt in which the N-terminal regions interact with each other. Goodman et al. (1975*b*).

VI NUCLEOHISTONE AGGREGATION AND CHROMATIN STRUCTURE

Evidence is accumulating that indicates that chromatin structure is particulate in nature, consisting of periodic units containing DNA and sets of histone clusters. These particles most likely contain fewer than 200 base pairs of DNA and may alternate with other regions of DNA that are relatively free of histone. The data for this emerging model come from studies of electron microscopy (Olins and Olins, 1974), nuclease digestion (Clark and Felsenfeld, 1971; Hewish and Burgoyne, 1973; Sahasrabuddhe and Van Holde, 1974), and from considerations of stoichiometry (Kornberg, 1974). Li et al. (1975) proposed a model in which the chromatin molecule is composed of condensed and extended regions; the condensed regions correspond to DNA segments bound by the subunits of whole histones (minus histone I) and have more α-helical structure with respect to histones and more base tilting with respect to DNA. The more extended regions correspond to free DNA segments or DNA segments bound by histone I or nonhistone proteins and have less α-helical structure with respect to proteins and less base tilting with respect to DNA. This model could explain both the particulate model of Olins and Olins (1974) and the supercoil model of Pardon and Richards (1967) shown in Figure 13-11. In addition, it is suggested that histone I with little α-helical structure might not protect DNA from nuclease digestion as well as other histones.

Clusters (probably tetramers) of the arginine-rich f2a1 and f3 histones have been found to exist in chromatin (Clark and Felsenfeld, 1972; Varshavsky and Georgiev, 1972) and in histone preparations extracted from chromatin by gentle methods (Kornberg and Thomas, 1974; Roark et al., 1974). It is possible that the interhistone forces that hold together these clusters are similar to those interactions between portions of f2a1 molecules that are responsible for the specific, CD-distorting, cooperative complex formation between DNA and f2a1 in model systems (Adler et al., 1975). The role that f2a1 plays in associated clusters of both homo and hetero complexes of histones may be viewed in terms of its conformational potential. Upon addition of salt, the molecule assumes partial α-helical structure and retains a large component of β sheet. The sites for aggregation might well be the β regions of the polypeptide. Although it has been suggested that the hydrophobic surfaces of α helices may be the sites of histone–histone interaction (Boublik et al., 1970*a*), a survey of the literature shows many more examples of β sheet–β sheet interactions, for example, crystalline insulin hexamers (Blundell et al., 1972). Thus it should be noted that β-sheet interactions probably form more stable assemblies than associated α helices. Furthermore, kinetic studies on f2a1 show that aggregation accompanies β-sheet formation

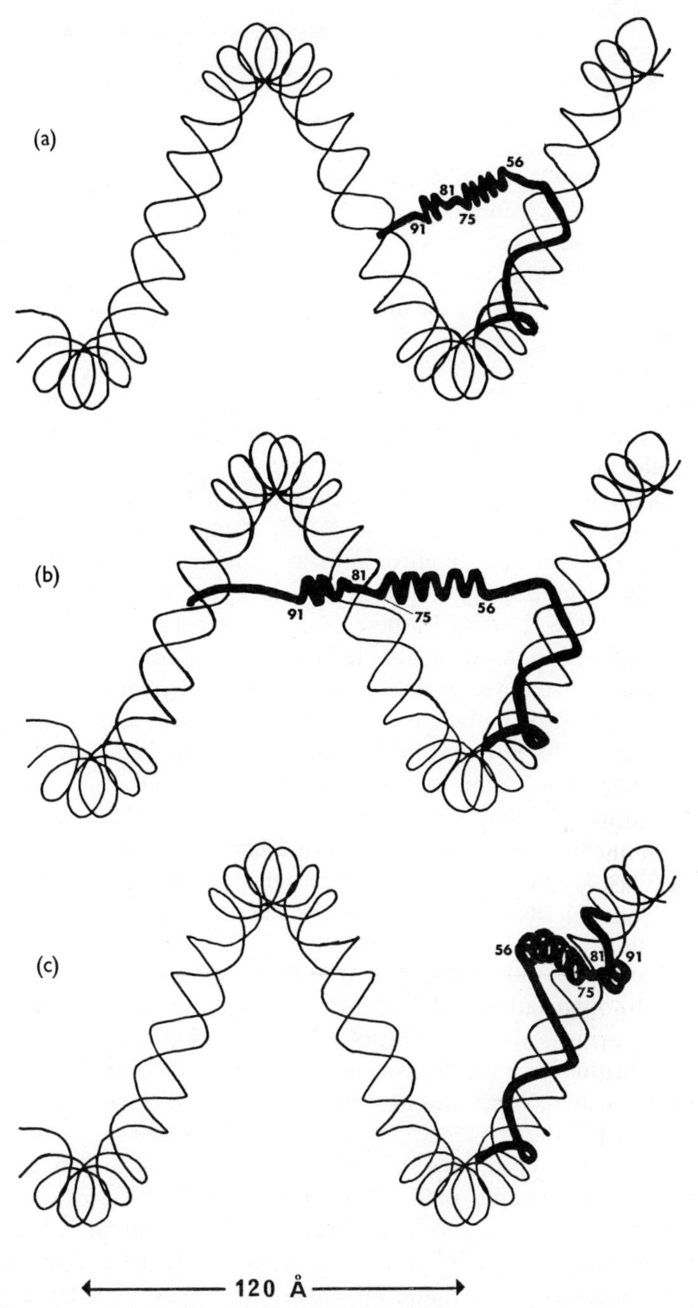

Fig. 13-11 Alternative arrangements for the attachment of histone type f2al to supercoiled DNA in which the C-terminal half of the protein forms an intramolecular crosslink. The histone is represented by the thick line, and the α-helical regions are included. Histone f2al is unlikely to form links across a complete gyre of the supercoil either diametrically (*a*) or parallel to the supercoil axis (*b*). The most likely arrangement, in which the α-helix forms small loops away from the DNA, is shown in (*c*). Pardon and Richards (1967).

but not α-helix formation (Wickett et al., 1972; Smerdon and Isenberg, 1974). This hypothesis for histone interactions is also supported by the data of Ziccardi and Schumaker (1972, 1973), and Bradbury and co-workers (Bradbury and Rattle, 1972; Boublik et al., 1970*a*). In addition to the arginine-rich histones, f2a1 and f3, found associated together in small oligomers when histones are extracted gently from chromatin (Kornberg and Thomas, 1974; Roark et al., 1974), there is also some evidence that f2a1 and f3 are clustered together on sites in native chromatin and can be separated from other histones (Clark and Felsenfeld, 1972; Varshavsky and Georgiev, 1972). Thus any conclusions for the interactions of f2a1 may have to be interpreted along with similar findings for histone f3.

The treatment of chromatin with varying concentrations of urea (Bartley and Chalkley, 1973; Ilyin et al., 1971) appears to disrupt the compactness of the chromatin and change the circular dichroism (CD) in the 270-nm region to nearly that of free DNA. These results suggest that protein–protein interactions are important to chromatin structure and are in general agreement with the histone self-aggregation models of Bradbury and Rattle (1972) and Hayashi and Iwai (1971). In these models, parts of histones are bound to DNA and other portions are available for histone–histone interactions.

It has been known for several years that histones aggregate upon addition of salts or at extremes of pH (Cruft et al., 1958; Edwards and Shooter, 1969; Boublik et al., 1970*b*; Barclay and Eason, 1972; Diggle and Peacocke, 1971; Li et al., 1972). Also, some workers have alluded to possible interactions between histones of differing primary structure (Cruft et al., 1958; Laurence, 1966; Shih and Bonner, 1970; Edwards and Shooter, 1970). However, only within the last couple of years has the existence of specific cross complexes been verified and their characterizations begun (Skandrani et al., 1972; D'Anna and Isenberg, 1973, 1974; Kelley, 1973).

Work by Hjelm and Huang (1974) indicates that the response of each chromatin to dissociation is different as to the amount and type of histone removed at various salt concentrations. The exception to this is the unique removal of histone I from all the chromatins with 0.6 ionic strength solutions. Correlation of these observations with the circular dichroism (CD) above 260 nm of each of the depleted nucleoproteins shows that no change occurs in the CD of the chromatins with removal of the lysine-rich histone I. Furthermore, the changes that occur with removal of the other (slightly lysine- and arginine-rich) histones are found to be approximately linear with the amount, not the type, of total histone in the depleted nucleoprotein sample. On this basis, Hjelm and Huang (1974) suggest that the secondary structure of chromatin DNA is determined equally and independently by each of the slightly lysine-rich and arginine-rich histone fractions. This point is also supported by an inverse linear correlation observed between the magnitude of the CD above 260 nm of depleted nucleoproteins and the amount of slightly lysine- and arginine-rich histones bound to a DNA as estimated by analysis of thermal denaturation data. In contrast, the data show that the nonhistone proteins do not effect the CD of DNA in chromatin, although they do bind the DNA (Hjelm and Huang, 1974).

Skandrani et al. (1972) reported that histones LAK (IIb1; f2a2) and KAS (IIb2; f2b) interact during guanidine hydrochloride gradient chromatography on amberlite resin. From column work and amino acid analyses of the LAK-KAS band, Skandrani et al. (1972) have concluded that the histones form an equimolar complex. Kelley (1973) has also reported complex formation based on studies of chromatographic fractions of mixed histones. D'Anna and Isenberg (1973) have reported complex formation between histone KAS and GRK. CD and fluorescence continuous variation curves imply an equimolar complex. On the basis of a dimer complex, the interaction is quite strong ($K_A = 10^6\ M^{-1}$), and there is an increase of eight residues of α helix in the complex as compared to the individual histone. Measurements of fluorescence anisotropy, relative fluorescence intensity, and circular dichroism (CD) also indicate that histones LAK and KAS form a 1:1 complex in solutions of sodium phosphate or sodium chloride, pH 7.0. The order of addition of histone LAK, histone KAS, or salt is not important. The complex is strong and has an association constant of about $10^6\ M^{-1}$. Upon complexing, the number of α-helical residues increases by about 15. The addition of urea reduces complexing.

The LAK–KAS complex has many characteristics in common with that of KAS–GRK. Both are formed only in the presence of salt, they are strong ($K_A = 10^6\ M^{-1}$), and they are characterized by increases in CD, fluorescence intensity, and anisotropy as compared to noninteracting solutions. Upon complex formation, there are increases in the α-helical content. These results are in general agreement with the equilibrium scheme proposed for histones KAS and GRK (D'Anna and Isenberg, 1973).

On the other hand, the LAK–GRK complex is about two orders of magnitude weaker than LAK–KAS or KAS–GRK. Nevertheless, even this weak interaction interferes with the slow changes of histone GRK. D'Anna and Isenberg (1974) have also seen that the fluorescence continuous variation curves of LAK and GRK do not deviate from ideality, although the CD and kinetic data indicate complexing. This implies that the rotational relaxation and the fluorescence intensity of the tyrosine residues in the KAS–GRK complex are, essentially, the same as those in isolated KAS or GRK. The three histones, GRK, KAS, and LAK, all show about the same number of residues in the α-helical state and, in each histone, the α-helical regions are in about equivalent locations along the protein primary structure (D'Anna and Isenberg, 1974).

It seems that histone cross complexing may be important in determining the structure of chromatin. Recent studies suggest that histones bind with some degree of base specificity (Clark and Felsenfeld, 1972; Combard and Vendrely, 1970; Sponar and Sormova, 1972). Varshavsky and Georgiev (1972) have reported that histones ARE and GRK bind to DNA in blocks. While cross-complexing experiments are incomplete, the trends prompt one to suggest the following scheme: histones bind to specific regions of DNA. These regions may be determined by the relative affinity of the histones for the base sequence or perhaps by histone modification, as suggested by Louie et al. (1972), or perhaps by a specific interaction between histones and non-histone proteins that, in turn, interact with specific DNA sequences. These regions then bind specifically to other histone-bound regions via histone cross complexing. Such interactions could effect much of the ordering and packaging of chromatin and changes in morphology during the cell cycle.

Combining the information reported in this section with that given in Chapter 9 on DNA–histone interactions leads to the condensed phase chromatin model proposed in Figure 13-12. This model is in agreement with the salient features of the Pardon & Richards model (1967) and the Li et al., (1975) organizational picture of chromatin. It does, however, infer more detailed molecular information. Specifically, it suggests (1) that the histones bind to the DNA chains in an α-helical conformation, with ionic interactions forming

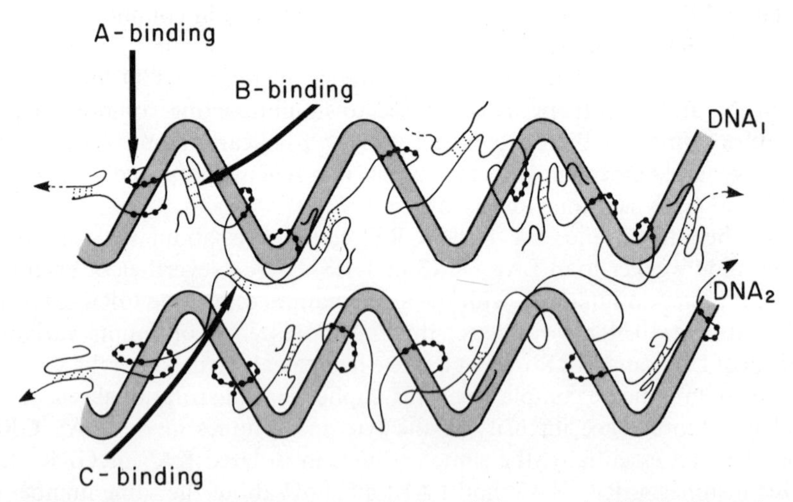

Fig. 13-12 A proposed model for the organization of histone and DNA in the condensed regions of chromatin. It is important to point out that this is only a possible topological distribution of interhistone organization, and the two suggested DNA binding sites on the histone molecule could also be inter-DNA specific.

the cross-molecular complex stability, (2) that there are two binding regions per histone to the DNA, (3) that the ends of the histone chains associate through interchain β-like structures (no head-to-head or tail-to-tail preference is required), and (4) that interhistone interactions for histones complexed to different DNAs occur in the central regions of the histones between the α-helical units. One cannot rule out a mixture of end-to-middle β-like associations both of intra- and inter-DNA histone interactions, although the proposed organization in Figure 13-12 is more appealing from a symmetry point of view.

A number of different histone nomenclatures are used in this section and throughout Chapter 9. It is unfortunate, but the author thought it better to adopt the rule of using the nomenclature of each original paper rather than generate errors by attempting the transformation to an acceptable common nomenclature. It is suggested that the reader consult the original references in regard to cross referencing histone nomenclatures.

VII POLYELECTROLYTE COMPLEX AGGREGATES

When a polyelectrolyte reacts with an oppositely charged polyelectrolyte in an aqueous solution, a polyelectrolyte complex is formed. The complex formation with polyelectrolytes may differ from salt formation with microelectrolytes from the viewpoint of molecular conformations, that is, the formation of a polyelectrolyte complex may not be stoichiometric, if, for example, the chains are rigid. The reaction to form a polyelectrolyte complex is influenced by the degree of polymerization, the ionic strength of the medium, the reaction temperature, etc., and depends on pH if at least either the polyanion or the polycation is a weak polyelectrolyte. The polyelectrolyte complex is obtained in the state of a precipitate, a gel, or a sol, depending on conditions (for a general discussion see Sélégry et al., 1974).

In most studies of complex formation, stoichiometric symmetry is in effect. The colloid titration (Terayama, 1948, 1952) is performed under such a condition. Maeda (1975) suggests that the effect of aggregation on the titration properties is composed of essentially two terms, representing the effects on the chemical potentials of polyion and solvent. The nonelectric free energy change associated with aggregation is unaffected even if the intrinsic dissociation constant of dissociable sites changes as a result of aggregation. For certain types of phase separation, Maeda (1975) claims that the nonelectric free energy change can be obtained from the titration data in the same manner as the case without phase separation. Michaels (1965) has pointed out that the reaction of two strong polyelectrolytes with opposite charges, sodium poly(styrene sulfonate) and poly(vinyl-benzyl trimethyl ammonium chloride), is stoichiometric. Hosono (1972) has examined the complex formation from partially sulfated poly(vinyl alcohol) and poly(vinyl alcohol) partially acetallyzed with diethoxy ethyl trimethyl ammonium salt, both of which are strong polyelectrolytes, and has pointed out that the reaction is stoichiometric and does not depend on the pH of the medium.

In regard to complexes made by reaction of a strong polyelectrolyte with a weak polyelectrolyte, Fuoss et al., (1949) showed that poly(4-vinyl-n-N-butyl pyridonium bromide) (strong polybase) reacts with sodium polyacrylate (weak polyacid) in nearly stoichiometrical equivalence, and Hosono et al., (1970) have pointed out that partially sulfated poly(vinyl alcohol) reacts with partially aminoacetalyzed poly(vinyl alcohol) (weak polybase) stoichiometrically as a function of the degree of dissociation of the weak polyelectrolyte.

Sato and Nakajima (1975) have studied the effect of chain flexibility on the stoichiometry of interactions, that is, the equivalence of ammonium (N^+) in polyethylenimine (PEI) and carboxylate (COO^-) in carboxymethyl cellulose (CMC), as a function of pH. Both of these are weak polyelectrolytes. The PEI molecule is a branched and flexible chain, and its density of ionizable groups is high, whereas the CMC molecule has a rather rigid backbone chain. Experimental data on turbidity and conductometric and potentiometric titrations led to the conclusion that the formation of their polyelectrolyte complex does not obey stoichiometry. Sato and Nakajima (1975) suggest that such behavior for this polyelectrolyte complex may be attributed to the less flexible nature of CMC chains, high density of ionizable groups in PEI, and/or branching in PEI. The fully extended conformations of CMC and PEI are shown in Figure 13-13. The distance between neighboring ionizable groups of PEI is considerably shorter than that of CMC, which should leave free positive charges in the complex. Consequently, the remarkable deviation from stoichiometric ideality might be explained by the following two reasons: first, release or incorporation of protons by a polyelectrolyte is induced by the partner polyelectrolyte having the opposite charge; second, geometrical hindrance in ionic linkages may leave free charges of PEI, particularly in the region of the acidic side. Wajnerman et al., (1972) have reported that the complex formation between arginic acid, a weak and rigid polyanion, and gelatin, a weak polycation, depends on pH and is stoichiometric. This observation suggests that conformational effects are not significant to complex formation. These findings are at partial odds to the CMC–PEI system, in which complex formation is apparently not stoichiometric because of conformational effects of both components and of the induced ionization of undissociated residues by oppositely charged residues.

The phenomenon of phase separation in aqueous solutions of two oppositely charged polyions into two liquid phases has been known for many years. Bungenberg de Jong (1952), who studied the gelatin–gum arabic system

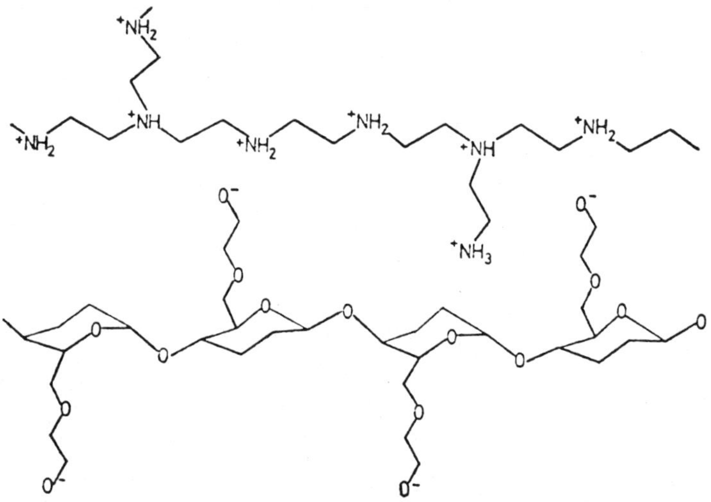

Fig. 13-13 Extended conformations of PEI and CMC. Sato and Nakijima (1975).

extensively, suggested the name "complex coacervation" for this phenomenon. Although many systems of oppositely charged polyions (Morawetz and Gobran, 1954; Feitelson and Joseph, 1963; Rogacheva and Zezin, 1969, 1972; Rogacheva et al., 1970*a*,*b*; Lutsenko et al., 1971; Zezin, 1972; Nakajima and Sato, 1972), and especially biopolymers, (Latt and Sober, 1967; Perlmann and Grizzuti, 1971; Caroll, 1972; Bettelheim, 1970), have been studied since then, little is known about the mechanism of complex coacervation, or about the interaction between oppositely charged polyions.

A theoretical interpretation of the experimental results of Bungenberg de Jong (1952) has been suggested by Voorn and co-workers (Voorn, 1956, 1957; Michaeli et al., 1957). Veis and co-workers (Veis and Aranyi, 1960; Veis, 1961, 1970; Veis et al., 1967) have attempted to verify the Voorn theory with experiments on two gelatins with different isoelectric points. A disagreement was found. Based on their experiments, Veis and co-workers proposed a different mechanism for complex coacervation. Veis assumed interaction between the polyions before phase separation occurs. Such interactions would result in the formation of aggregates of the two polyions. Polderman (1975) has studied a system consisting of polyethylenimine $(EtNH)_n$ and an acrylic acid-acrylamide copolymer $(AcrOH,AcrNH)_n$ with 25% acrylic acid monomer units. Changing the pH in a mixture of $(EtNH)_n$ and $(AcrOH,AcrNH)_n$, leads to complex coacervation in a certain pH region. The dependence of the pH region of phase separation on the polymer composition is given in Figure 13-14. The results of the potentiometric titrations are consistent with the interaction mechanism of the electrostatic cooperative

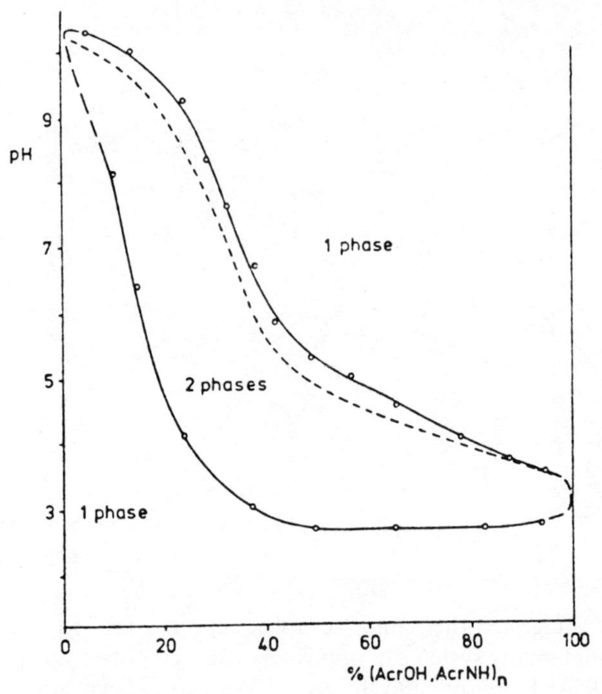

Fig. 13-14 pH of initial phase separation as a function of polymer composition. (experimentally uncertain for %-$(AcrOH,AcrNH)_n$ >95% and <5%); (——)-pH of initial phase separation, (– – – –)—pH of zero net charge. Polderman (1975).

physical association. The behavior of the system of poly(sodium styrenesulfonate) $(StySO\text{-}_3N_a)_n$ and poly(vinylbenzyltrimethylammonium chloride) $(CH_2CHBzlN^+Me_3Cl^-)_n$, studied extensively by Michaels and co-workers (Michaels and Miekka, 1961; Michaels et al., 1965*a,b*), is quite different. Michaels et al. found a stoichiometric precipitate in all mixtures with a $(StySO_3Na)_n/(CH_2CHB\ lN^+Me_3Cl^-)_n$ ratio from 9 to 0.1. The excess of one polymer remains in solution, practically without the other polymer. In this system one can describe the interaction as an electrostatic cooperative chemical association. Polderman (1975) has suggested two reasons why chemical association is found in the Michael's system and physical association in his system:

1. The monomer structure and the distance between two charges on a chain are almost equal for $(StySO_3Na)_n$ and $(CH_2CHBalN^+Me_3Cl^-)_n$ but quite different for $(EtNH)_n$ and $(AcrOH,AcrNH)_n$.

2. $(StySO_3Na)_n$ and $(CH_2CHBzlN^+Me_3Cl^-)_n$ are polymers with a low degree of branching and a relatively stiff chain. The same holds for (AcrOH, $AcrNH)_n$, but $(EtNH)_n$ is a compact, highly branched molecule.

In the system $(StySO_3Na)_n/(CH_3CHBzlN^+Me_3Cl^-)_n$ a "ladder-like" interaction may be favorable, whereas such a cooperative ion binding is not possible in the system $(EtNH)_n/(AcrOH,AcrNH)_n$. The charge density of both polymers may also play a role in the molecular organization. If the charge densities are high and equal to each other, a "ladder-like" interaction may be favorable. The system $(EtNH)_n/(AcrOH,AcrNH)_n$ might be considered as an ideal asymmetric system: the steric properties as well as the maximum charge density of $(EtNH)_n$ and $(AcrOH,AcrNH)_n$ are quite different. Conversely, Michaels' system is an ideal symmetric system. The asymmetry of Polderman's system results in cooperative physical association, whereas the symmetry of Michaels' system results in a cooperative ion binding. From this discussion it is possible to postulate why precipitation is much more common than coacervation in systems with polymers of opposite charge.

REFERENCES

Adler, A. J., A. W. Fulmer, and G. D. Fasman (1975). *Biochemistry*, **14**, 1446.

Anacker, E. W. (1970). In *Cationic Surfactants* (E. Jungermann, Ed.). Marcel Dekker, New York.

Bamberg, E., H.-A. Kolb, and P. Laüger (1975). In *The Structural Basis of Membrane Function* (Y. Hatefi and L. Djavadi-Ohaniance, Eds.). Academic Press, New York, p. 186.

Bamberg, E. and P. Laüger (1973). *J. Membrane Biol.*, **11**, 177.

Bangham, A. D., M. M. Standish, and J. C. Watkins (1965). *J. Mol. Biol.*, **13**, 238.

Barclay, A. B. and R. Eason (1972). *Biochim. Biophys. Acta*, **269**, 37.

Bartley, J. and R. Chalkley (1973). *Biochemistry*, **12**, 468.

Bear, R. S., K. J. Palmer, and F. O. Schmitt (1941). *J. Cell. Comp. Physiol.*, **17**, 355.

Becher, P. (1967). In (M. J. Schick, Ed.). Marcel Dekker, New York.

Bettelheim, F. A. (1970). *Biological Polyelectrolytes* (Veis, A., Ed.). Academic Press, New York.

Blundell, T., G. Dodson, D. Hodgkin, and D. Mercola (1972). *Adv. Protein Chem.*, **26**, 279.

Boeckel, G., J. C. Genzling, G. Weill, H. Benoit (1966). *J. Chem. Phys.*, **59**, 999.

Boublik, M., E. M. Bradbury, and C. Crane-Robinson (1970*a*). *Eur. J. Biochem.*, **14**, 486.

Boublik, M., E. M. Bradbury, C. Crane-Robinson, and E. W. Johns (1970*b*). *Eur. J. Biochem.*, **17**, 151.

Brandts, J. F., R. J. Oliveria, and C. Westwort (1970). *Biochemistry*, **9**, 1038.

Bradbury, E. M. and H. W. E. Rattle (1972). *Eur. J. Biochem.*, **27**, 270.

Bungenberg de Jong, H. L. and H. H. Druyt (1952). *Colloid Science* Springer-Verlag Amsterdam, Vol. 2.

Caroll, D. (1972). *Biochemistry*, **11**, 421; 426.

Chappell, J. B. and A. R. Crofts (1965). *Biochem. J.*, **95**, 393.

Clark, R. J. and G. Felsenfeld (1972). *Nature (New Biol.)*, **240**, 226.

Combard, A. and R. Vendrely (1970). *Biochem. J.*, **118**, 875.

Corkill, J. M., J. F. Goodman, R. Robson, and J. R. Tate (1966). *Trans. Faraday Soc.*, **62**, 987.

Cruft, H. J., G. M. Mauritzen, and E. Stedman (1958). *Proc. Roy. Soc., Ser. B*, **149**, 21.

D'Anna, J. A., Jr. and I. Isenberg (1973). *Biochemistry*, **12**, 1035.

D'Anna, J. A., Jr. and I. Isenberg (1974). *Biochemistry*, **13**, 2093.

Dervichian, D. G. (1964). *Prog. Biophys. Mol. Biol.*, **14**, 263.

Diggle, J. H. and A. R. Peacocke (1971). *FEBS Lett.*, **18**, 138.

Dobin, S. B. (1972). *Methods Enzymol.* **26**, 119.

Doty, P., J. H. Bradbury, and A. M. Holtzer (1956). *J. Am. Chem. Soc.*, **78**, 947.

Edwards, P. A. and K. V. Shooter (1969). *Biochem. J.*, **114**, 227.

Emerson, M. F. and A. Holtzer (1967). *J. Phys. Chem.*, **71**, 1898.

Evans, H. C. (1956). *J. Chem. Soc.*, 579.

Feitelson, J. and R. Joseph (1963). *Biopolymers*, **1**, 331.

Fossel, E. T., W. R. Veatch, Y. A. Ovchinnikov, and E. R. Blout (1974). *Biochemistry*, **13**, 5264.

Foster, R. (1975). *Molecular Association*. Academic Press, New York, Vol. 1.

Garland, F. and R. C. Patel (1974). *J. Phys. Chem.*, **78**, 848.

Geer, R. D., E. H. Eylar, and E. W. Anacker (1971). *J. Phys. Chem.*, **75**, 369.

Glickson, J. D., D. F. Mayers, J. M. Settine, and D. W. Urry (1972). *Biochemistry*, **11**, 477.

Goodman, M., N. Ueyama, and F. Naider (1975*a*). *Biopolymers*, **14**, 901.

Goodman, M., N. Ueyama, F. Naider, and C. Gilon (1975*b*). *Biopolymers*, **14**, 915.

Griffith, O. H. and A. S. Waggoner (1969). *Acc. Chem. Res.*, **1**, 17.

Guidotti, G. (1972). *Annu. Rev. Biochem.*, **41**, 731.

Gupta, A. K., C. Dufour, and E. Marchal (1974). *Biopolymers*, **13**, 1293.

Gupta, A. K., C. Strazielle, and E. Marchal (1976). *Biopolymers*, submitted for publication.

Hall, D. G. and B. A. Pethica (1967). In *Nonionic Surfactants* (M. J. Schick, Ed.). Marcel Dekker, New York.

Hamilton, C. L. and H. M. McConnell (1968). In *Structural Chemistry and Molecular Biology* (A. Rich and N. Davidson, Eds.). W. H. Freeman, San Francisco, p. 115.

Harold, F. M. and J. R. Baarda (1967). *J. Bacteriol.*, **94**, 53.

Harris, E. J. and B. C. Pressman (1967). *Nature (Lond.)*, **216**, 918.

Hayashi, H. and K. Iwai (1971). *J. Biochem.*, **70**, 543.

Hewish, D. R. and L. A. Burgoyne (1973). *Biochem. Biophys. Res. Commun.*, **52**, 504.

Hill, T. L. (1964). *Thermodynamics of Small Systems*, W. A. Benjamin, New York, Vol. 2.

Hill, T. L. and Y. D. Chen (1973). *Biopolymers*, **12**, 1285.

Hjelm, R. P. and R. C. C. Huang (1974). *Biochemistry*, **13**, 5275.

Hladky, S. B. and D. A. Haydon (1972). *Biochim. Biophys. Acta*, **274**, 294.

Hopfinger, A. J. (1975). *Intern. J. Polymeric Mater.*, **4**, 79.

Hosono, M., S. Sugii, O. Kusudo, and W. Tsuji (1972). *Report of the Poval Committee, Kyoto*, **61**, 79.

Husono, M., O. Kusudo, and W. Tsuji (1970). *Report of the Poval Committee, Kyoto*, **57**, 99.

Houtsmuller, U. M. T. and L. L. M. Van Deenen (1965). *Biochim. Biophys. Acta*, **106**, 564.

Huang, C. (1969). *Biochemistry*, **8**, 344.

Huberman, J. A. (1973). *Annu. Rev. Biochem.*, **42**, 355.

Ilyin, Y. V., A. Y. Varshavsky, U. N. Michelsaar, and G. P. Georgiev (1971). *Eur. J. Biochem.*, **22**, 235.

Isbell, B. E., C. Rice-Evans, and G. H. Beaven (1972), *FEBS Lett.*, **25**, 192.

Kauzmann, W. (1959). *Adv. Protein Chem.*, **14**, 1.

Kelley, R. I. (1973). *Biochem. Biophys. Res. Commun.*, **54**, 1589.

Kihara, H. (1975). *Polymer J.*, **7**, 407.

Kornberg, R. D. (1974). *Science*, **184**, 868.

Kornberg, R. D. and J. O. Thomas (1974), *Science*, **184**, 865.

Krasne, S., G. Eisenman, and G. Szabo (1971). *Science*, **174**, 412.

Latt, S. A. and H. A. Sober (1967). *Biochemistry*, **6**, 3293; 3307.

Laurence, D. J. R. (1966). *Biochem. J.*, **99**, 419.

Lennarz, W. J. (1972). *Acc. Chem. Res.*, **5**, 361.

Levine, Y. K., A. I. Bailey, and M. H. F. Wilkins (1968). *Nature*, **220**, 577.

Levine, Y. K. and M. H. F. Wilkins (1971). *Nature* (*New Biol.*), **230**, 69.

Lewis, M. S. and M. H. Gottlieb (1971). *Fed. Proc.*, **30**, 1303 abs.

Li, H. J., C. Chang, Z. Evangelinou, and M. Weiskopf (1975). *Biopolymers*, **14**, 211.

Li, H. J., R. R. Wickett, A. M. Craig, and I. Isenberg (1972). *Biopolymers*, **11**, 375.

Louie, A., P. Candido, and G. H. Dixon (1972). *Fed. Proc.*, **31**, 1121.

Lutsenko, V. V., A. B. Zezin, and A. R. Rudman (1971). *Vysolomolekul. Soedin.*, **B13**, 396.

Luzzati, V. (1968). In *Biological Membranes* (D. Chapman, Ed.). Academic Press, New York, Chapt. 3.

Marchal, E. (1974). *Biopolymers*, **13**, 1309.

Michaels, A. S., G. L. Falkenstein, and N. S. Schneider (1965*b*). *J. Phys. Chem.*, **69**, 1456.

Michaels, A. S. and R. S. Miekka (1961). *J. Phys. Chem.*, **65**, 1765.

Michaels, A. S., L. Mir, and N. S. Schneider (1965*a*). *J. Phys. Chem.*, **69**, 1447.

Michaeli, I., J. Th. G. Overbeek, and M. J. Voorn (1957). *J. Polymer Sci.*, **23**, 443.

Miyamoto, V. K. and W. Stoeckenius (1971). *J. Membrane Biol.*, **4**, 252.

Morawetz, H. and R. H. Gobran (1954). *J. Polymer Sci.*, **12**, 133.

Mueller, P. and D. O. Rudin (1967). *Biochem. Biophys. Res. Commun.*, **26**, 398.

Mukerjee, P. (1967). *Adv. Colloid and Interface Sci.*, **1**, 241.

Myers, V. B. and D. A. Haydon (1972). *Biochim. Biophys. Acta*, **274**, 313.

Nakajima, A. and H. Sato (1972). *Biopolymers*, **11**, 1345.

Nemethy, G. and M. P. Printz (1972). *Macromolecules*, **5**, 755.

Olins, A. L. and D. E. Olins (1974). *Science*, **183**, 330.

Ovchinnikov, Yu. A., V. T. Ivanov, V. F. Bystrov, A. I. Miroshikov, E. N. Shepel, N. D. Abdullaev, E. S. Efremov, and L. B. Senyavina (1970). *Biochem. Biophys. Res. Commun.*, **39**, 217.

Palmer, K. J. and F. O. Schmitt (1941). *J. Cell. Comp. Physiol.*, **17**, 385.

Pardon, J. F. and B. M. Richards, (1973). In *Subunits in Biological Systems* Part B (G. D. Fasman and S. N. Timasheff, Eds.). Marcel Dekker, New York, p. 1.

Perlmann, G. E. and K. Grizzuti (1971). *Biochemistry*, **10**, 4169.

Polderman, A. (1975). *Biopolymers*, **14**, 2181.

Porschke, D. and F. Eggers (1972). *Eur. J. Biochem.*, **26**, 490.

Powers, J. C. and W. L. Peticolas (1970). *Biopolymers*, **9**, 195.

Roark, D. E., T. E. Geoghegan, and G. H. Keller (1974). *Biochem. Biophys. Res. Commun.*, **59**, 542.

Robinson, N. and L. Saunders (1958). *J. Pharm. Pharmacol.*, **10**, 755.

Rogacheva, V. B., S. Ya, and V. A. Kargin (1970*a*). *Vysokomolekul. Soedin.*, **B12**, 340.

Rogacheva, V. B. and A. B. Zezin (1969). *Vysokomolekul. Soedin.*, **B11**, 327.

Rogacheva, V. B., A. B. Zezin, and V. A. Kargin (1970*b*). *Vysolomolekul. Soedin.* **B12**, 826.

Rotschild, K. J. and H. E. Stanley (1974). *Science*, **135**, 616.

Sahasrabuddhe, C. G. and K. E. Van Holde (1974). *J. Biol. Chem.*, **249**, 152.

Sarges, R. and B. Witkop (1965). *J. Am. Chem. Soc.*, **87**, 2011.

Sato, H. and A. Nakajima (1975). *Polymer J.*, **7**, 242.

Saunders, L., J. Perrin, and D. Gammack (1962). *J. Pharm. Pharmacol.*, **14**, 567.

Schmitt, F. O. (1939). *Physiol. Rev.*, **19**, 270.

Schodt, K., M. McDonnell, A. Jamieson, and J. Blackwell (1976*a*). *Macromolecules* (in press).

Schodt, K., M. McDonnell, A. Jamieson, and J. Blackwell (1976*b*). Unpublished results.

Sélégny, E., M. Mandel, and U. P. Strauss (1974). *Charged and Reactive Polymers*, Vol. 1: *Polyelectrolytes*. Reidel-Dordrecht, Holland.

Sheetz, M. P. and S. I. Chan (1972). *Biochemistry*, **11**, 4573.

Shields, J. E. and S. T. McDowell (1967). *J. Am. Chem. Soc.*, **89**, 2499.

Shields, J. E., S. T. McDowell, J. Pavlos, and G. R. Gray (1968). *J. Am. Chem. Soc.*, **90**, 3549.

Shih, T. Y. and J. Bonner (1970). *J. Mol. Biol.*, **47**, 469.

Shinitzky, M., A. C. Dianoux, C. Gitler, and G. Weber (1971). *Biochemistry*, **10**, 2106.

Shinoda, K., T. Makagawa, B. Tamamushi, and T. Isemura (1963). *Colloidal Surfactant* Academic Press, New York.

Singer, S. J. (1971). In *Structure and Function of Biological Membranes* (L. I. Rothfield, Ed.). Academic Press, New York, p. 146,

Skandrani, E., J. Mizon, P. Sautiere, and G. Biserte (1972). *Biochemie*, **54**, 1267.

Smerdon, M. J. and I. Isenberg (1974). *Biochemistry*, **13**, 4046.

Smith, R. and C. Tanford (1972). *J. Mol. Biol.*, **67**, 75.

Sponar, J. and Z. Sormova (1972). *Eur. J. Biochem.*, **29**, 99.

Stryer, L. (1968). *Science*, **162**, 526.

Swarbrick, J. and J. Daruwala (1969). *J. Phys. Chem.*, **73**, 2627.

Tanford, C. (1961). *Physical Chemistry of Macromolecules*. Wiley-Interscience, New York.

Tanford, C. (1973). *The Hydrophobic Effect*. Wiley-Interscience, New York.

Terayama, H. (1948). *Kagaku no Kenkyu*, **1**, 75.

Terayama, H. (1952). *J. Polymer Sci.*, **8**, 243.

Tinoco, I., Jr. (1957). *J. Am. Chem. Soc.*, **79**, 4336.

Tosteson, D. C., T. E. Andreoli, M. Tieffenberg, and P. Cook (1968). *J. Gen. Physiol.*, **51**, 373S.

Ts'o, P. O. P. (1970). In *Fine Structure of Proteins and Nucleic Acids* (G. D. Fasman and S. M. Timesheff, Eds.). Marcell Dekker, New York, f. 49.

Ts'o, P. O. P. and S. I. Chan (1964). *J. Am. Chem. Soc.*, **86**, 4176.

Ts'o, P. O. P., I. S. Melvin, and A. C. Olson (1963). *J. Am. Chem. Soc.*, **85**, 1289.

Urry, D. W. (1971). *Proc. Natl. Acad. Sci. USA*, **68**, 672.

Urry, D. W., J. D. Glickson, D. F. Mayers, and J. Haider (1972). *Biochemistry*, **11**, 487.

Urry, D. W., M. C. Goodall, J. D. Glickson, and D. F. Mayers (1971). *Proc. Natl. Acad. Sci. USA*, **68**, 1907.

Varshavsky, A. and G. P. Georgiev (1972). *Biochim. Biophys. Acta*, **281**, 669.

Veatch, W. R. and E. R. Blout (1974). *Biochemistry*, **13**, 5257.

Veatch, W. R., E. T. Fossel, and E. R. Blout (1974). *Biochemistry*, **13**, 5249.

Veis, A. and C. Aranyi (1960). *J. Phys. Chem.*, **64**, 1203.

Veis, A. (1970). *Biological Polyelectrolytes.* (Veis, A., Ed.). Academic Press, New York.

Veis, A., E. Bodor, and S. Mussell (1967). *Biopolymers*, **5**, 37.

Voorn, M. J. (1956). *Rev. Trav. Chim.*, **75**, 317; 405; 925; 1021.

Voorn, M. J. (1957). *Fortschr. Hochpolym. Forsch.*, **1**, 192.

Wada, A. (1960). *J. Polymer. Sci.*, **45**, 145.

Waggoner, A. S., O. H. Griffith, and C. R. Christensen (1967). *Proc. Natl. Acad. Sci. USA*, **57**, 1198.

Wajnerman, E. S., W. Ja. Grinberg, and W. B. Tolstogusow (1972). *Kolloid-Z. Z. Polymer*, **250**, 945.

Watanabe, H. (1965). *Nippon Kagaku Zasshi*, **86**, 179.

Wickett, R. R., H. J. Li, and I. Isenberg (1972). *Biochemistry*, **11**, 2952.

Wishnia, A. (1963). *J. Phys. Chem.*, **67**, 2079.

Wittstruck, T. A. and J. F. Cronan (1968). *J. Phys. Chem.*, **72**, 4243.

Yu, H. and W. H. Stockmeyer (1967). *J. Chem. Phys.*, **47**, 1369.

Zezin, A. B. (1972). *Vysokomolekul. Soedin.*, **A14**, 772.

Ziccardi, R. and V. Schumaker (1972). *Biopolymers*, **11**, 1701.

Ziccardi, R. and V. Schumaker (1973). *Biopolymers*, **12**, 3231.

CHAPTER 14

Theories of Intermolecular Interactions

The theoretical treatment of intermolecular interactions can be divided into two major areas: conformational and configurational energy calculations, and statistical mechanical models that are almost always grounded in an Ising framework. In this chapter, we emphasize the the conformational and configurational energy calculations done to model solute–solvent and counterion interactions. Particular attention should be given to the estimated energetics of the intermolecular interactions. A survey of the statistical mechanical theories of ligand binding and aggregration processes is presented for completeness.

I INTRODUCTION TO MOLECULAR ENERGY CALCULATIONS

The earliest attempts to take into account solute–solvent interactions focused on varying the dielectric constant in electrostatic energy terms used to describe pairwise interactions between species in the solute molecule [Brant and Flory, 1965; Ooi et al., 1967; Gibson and Scheraga, 1967*a*; Lipkind et al., 1970; Hopfinger, 1973 (a review)]. Workers have qualitatively realized that, in addition to modifying the self-electrostatic potential of the solute molecule, solvent molecules could also be expected to bind to the solute molecule at certain solute sites and dislike interacting with the solute species at other sites. The favorable solute–solvent interaction sites have come to be termed "preferred hydration sites" without any restriction of only water as a solvent implied by this term. Moreover, the two types of solvent-originating phenomena, dielectric modification and hydration binding, are not independent, but strongly coupled.

The resulting time average picture of the solute moleculue in dilute solution is that of a supramolecular species—the solute molecule with ligand-like solvent molecules bound at certain sites immersed in a homogeneous dielectric medium (the bulk solvent). However, because of the intrinsic dielectric properties of the solute and the voids produced by unfavorable solute group/solvent interactions the "local dielectric" behavior about the solute is heterogeneous and can be substantially different in magnitude from that of bulk solvent. The total dielectric anisotropy can be represented by encapsulating the supramolecular species in an artificial ellipsoid. Within the ellipsoid, the dielectric behavior is variable, whereas outside, it is that of homogeneous bulk solvent. Figure 14-1 shows schematic illustrations of various conceptual levels of solute–solvent organization in dilute solutions. Note that these ideas of solute–solvent interactions are independent of time, or at least time averaged. One would intrinsically expect time-dependent fluctuations in both dielectric and solvent-binding properties. In addition, the logical construct of an encapsulating ellipsoid skirts around the very difficult problem of conceptualizing how solvent organization is redistributed as one goes out from the first "layer" of solvent molecules surrounding the solute.

Subsequent to the early dielectric constant modification studies to include solvent effects (Figure 14-1*b*), the theoretical study of solute–solvent interactions has diverged into two paths: refined dielectric modeling and hydration site/supramolecular studies. Each of these areas can, at least to some degree, be partitioned with respect to the method of energy determination—quantum mechanical or pairwise empirical.

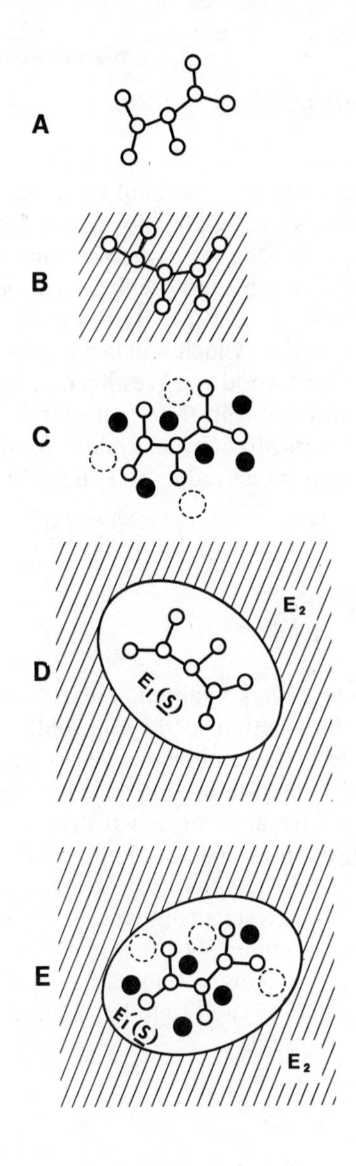

Fig. 14-1 Solute-Solvent Organization Models.

a. Solute in free space.

b. Solute in some homogeneous dielectric medium having dielectric constant ε.

c. Solute-solvent supramolecular complex in a free-space dielectric medium.

● solid solvent molecule,

◌ open solvent voids.

d. Solute in a bulk solvent of homogeneous dielectric constant ε_2. The $\varepsilon_1(\mathbf{s})$ indicates a local dielectric heterogenity.

e. Solute-solvent supramolecular complex in bulk solvent ε_2. The $\varepsilon_1'(\mathbf{s})$ indicates that the local heterogeneous dielectric medium is modified by solvent molecules from that of $\varepsilon_1(\mathbf{s})$.

II DIELECTRIC STUDIES

To begin, no one has attempted to develop the model in Figure 14-1*d* for the case in which $\varepsilon_1(\mathbf{s})$ varies with spatial position **s**. Only $\varepsilon_1(\mathbf{s})$ as a constant value over space has been considered. Moreover, the published results have only considered the case in which the ellipsoid degenerates to a sphere (Sinanoglu, 1974; Beveridge et al., 1974; Hylton et al., 1974*a,b*). Figure 14-2 shows the geometry for this case, which is in fact a classic problem in electrostatics (Kirkwood, 1934; Kirkwood and Westheimer, 1938). The total electrostatic potential $V(\mathbf{r})$ at any point in the sphere for a collection of charges, $\{q\}$, immersed in the spherical cavity of radius, a, and homogeneous dielectric ε_1, which, in turn, is immersed in bulk dielectric ε_2 is

$$V(\mathbf{r}) = \sum_{k=1}^{M} \frac{q_k}{\varepsilon_1|\mathbf{r} - \mathbf{r}_k|} + \sum_{k=1}^{M} \left(\frac{q_k}{\varepsilon_1 a}\right) \sum_{l=0}^{\infty} \left[\frac{(l+1)(1-\varepsilon_2)}{\varepsilon_2(l+1)+l}\right] \cdot \left(\frac{|\mathbf{r} - \mathbf{r}_k|}{a^2}\right)^l P_l(\cos\theta_k) \tag{14-1}$$

where $P_l(\cos\theta_k)$ is the Legendre polynominal of order l. Recently, Beveridge (1975) has determined the solution to the ellipsoidal geometry and is employing it in his work. Beveridge and Schnuelle (1975) have also determined the free energy of a charge distribution in concentric dielectric continua.

Beveridge et al. (1974) have used molecular orbital techniques in a study of the conformational properties of acetylcholine and some of its congeners in dilute aqueous solution. The spherical dielectric cavity model (Figure 14-1*d*) was employed and the results compared to those achieved using the free space approximation. (Figure 14-1*a*). Figure 14-3 shows the conformational energy maps of τ_1 versus τ_2 (1) in a free space medium and (2) in aqueous solution (the spherical cavity model). Figure 14-4 shows the geo-

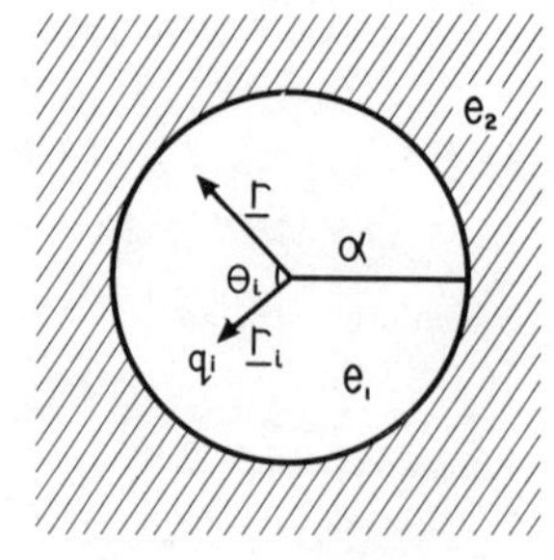

Fig. 14-2 Geometry of the charge-cavity dielectric model. Hylton et al. (1974).

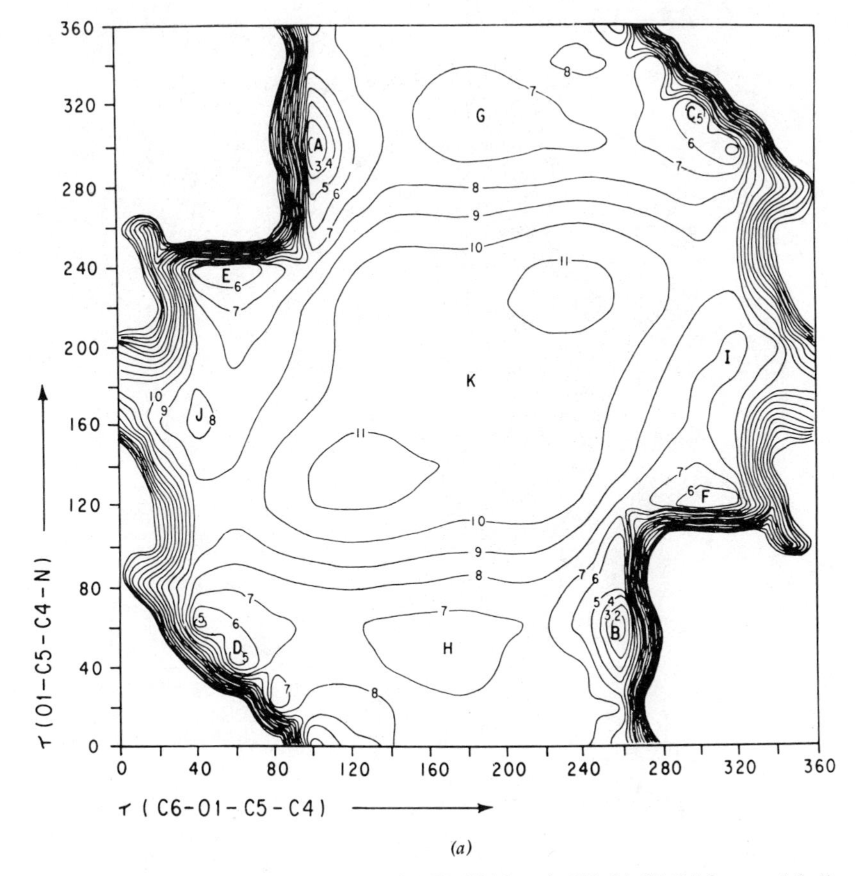

Fig. 14-3(*a*) INDO calculated energy vs τ(O1-C5-C4-N) and τ(C6-O1-C5-C4) for acetylcholine in the free-space approximation.

metry of the acetylcholine molecule for reference. The spherical cavity model indicates that solute–bulk solvent dielectric-dependent interactions promote the compact gauche conformation over a more extended, near trans, conformation. This is to be expected, since the spherical cavity model requires that the solvation energy varies directly with the dipole moment of the molecule, inversely with respect to molecular volume in the electrostatic term (negative), and directly with respect to the molecular volume in the cavity term (positive). Consequently, to the extent that the spherical cavity model is valid, molecular conformations yielding low solvation energies possess relatively large dipoles balanced against relatively small molecular volumes. For acetylcholine, the dipole is largest for the fully trans conformation and

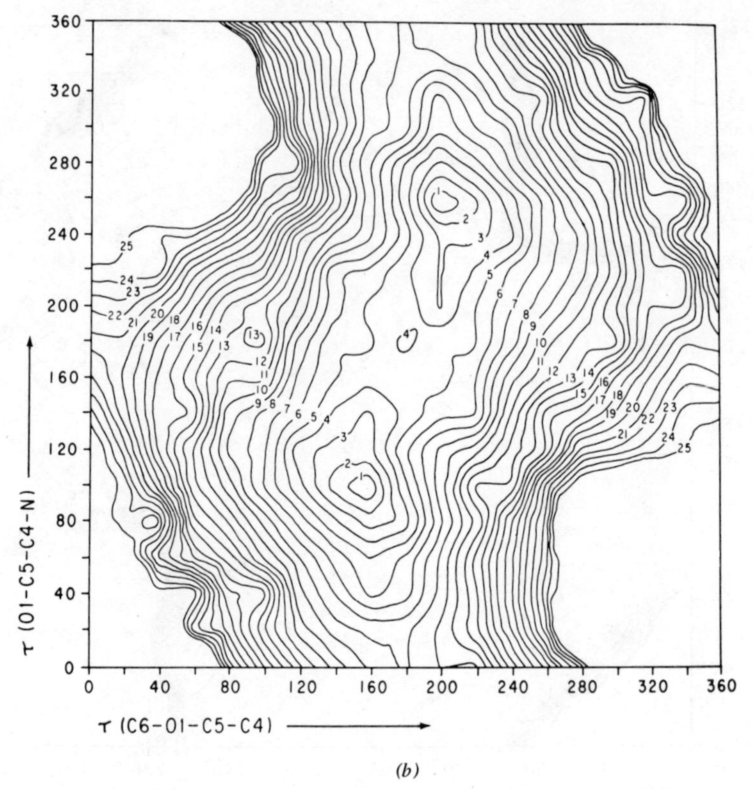

(b)

Fig. 14-3(*b*) Calculated energy of acetylcholine vs τ(O1-C5-C4-N) and τ(C6-O1-C5-C4) in water. Beveridge et al. (1974).

Fig. 14-4 Acetylcholine.

least for the cis conformation (with respect to τ_2). The molecular volume behaves in a directly opposite fashion. Consequently, the intermediate gauche conformation represents the compromise conformation leading to the lowest solvation energy. This finding is consistent with the observation that the constant dielectric model, or even the free space model (Bergmann and Pullman, 1973; Pullman and Courrière, 1972), predict the gauche conformation as the most stable for acetylcholine. However, it is wrong to conclude that the solvent contribution is unimportant. The statistical rotamer populations of acetylcholine are very much altered by the solvent-term, as is clear from the energy maps in Figure 14-3. Christoffersen and co-workers (McCreery et al., 1976*a,b*) have made progress in consolidating a dielectric cavity model with an ab initio molecular fragment, quantum mechanical model in order to account for solute–solvent interactions.

Some studies of a spatially variable dielectric medium and its consequence on solute conformation using pairwise empirical energy functions have been attempted (Hopfinger, 1973, 1974; Weintraub and Hopfinger, 1976). Qualitatively, the dielectric medium between two solute species should have a shielding value (dielective constant) of approximately unity when the interaction distance is less than or near to the diameter of a solute molecule. At long interactive separations, those larger than two–three molecular diameters of the solvent molecule, the shielding value becomes that of the bulk-solvent dielectric constant. But what happens at intermediate distances? The shielding value should be a function of both interactive distance and angular variation with respect to some molecular reference frame. Hopfinger and Weintraub (1976) have chosen different distance-dependent dielectric functions to compute the conformational properties of acetylcholine. A linear function is the simplest distance/dielectric relationship to employ. Figure 14-5 shows this relationship and the corresponding pairwise electrostatic potential

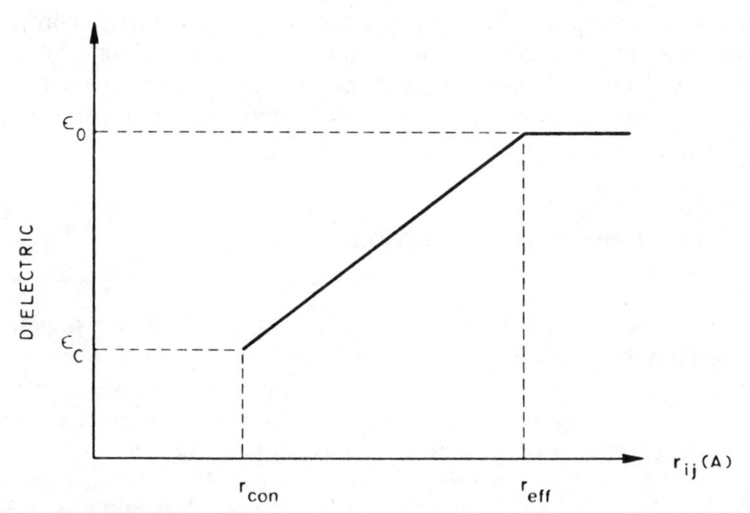

Fig. 14-5 The dependence of the dielectric $\varepsilon(r_{ij})$ on the interaction distance, r_{ij}, for a linear dielectric model. The ε_c is the value of the dielectric at the contact distance, r_{con}, and ε_0 is the effective bulk value of the dielectric at some effective interaction distance, r_{eff}. Hopfinger (1973).

energy is given as

$$V(r_{ij}) = \frac{kq_iq_j}{b}\left\{\frac{1}{r_{\text{eff}}} - \frac{1}{r_{ij}}\right\} + kq_iq_j\frac{a}{b^2}\left\{\ln\left[\frac{ar_{\text{eff}} + b}{r_{\text{eff}}}\right] - \ln\left[\frac{ar_{ij} + b}{r_{ij}}\right]\right\} + \frac{kq_iq_j}{\varepsilon_0 r_{\text{eff}}} \tag{14-2}$$

where $a = \dfrac{\varepsilon_0 - \varepsilon_c}{r_{\text{eff}} - r_{\text{con}}}$ and $b = \varepsilon_c - ar_{\text{con}}$

The determination of the effective distance, r_{eff}, necessary to achieve the bulk dielectric constant shielding value is not obvious, and reasonable estimates must suffice.

The values of the individual pairwise electrostatic energies for short-range interactions differ markedly for the linear versus constant dielectric model (Hopfinger, 1973; Weintraub and Hopfinger, 1976). However, in an unionized solute molecule, the total electrostatic potential energy is the result of the sum total of equal numbers of positive and negative interactions of which there are approximately equal positive-negative pair contributions. Hence the total electrostatic potential energy is relatively insensitive to the choice of the dielectric function, since large differences in electrostatic pair potential energy interactions tend to cancel one another. This should not be the case for ionized solute species where the balance of positive and negative pair-potential energy interactions no longer exists because of the excess charge build-up. To the extent that pairwise electrostatic potentials represent a realistic description of the molecular electrostatic potential, one concludes that local dielectric inhomogenity will have minor effects on the conformational behavior of neutral solute species, but potentially major contributions to the spatial behavior of charged solute molecules.

III HYDRATION SITE STUDIES

Gibson and Scheraga (1967*b*) were the first to quantitatively attempt to describe both the molecular geometry and energetics of solvent organization about a solute species. Earlier work on hydrophobicity and water structure by Nemethy and Scheraga (1962) produced a combination of experimental and theoretical estimates of the free energy of aqueous solvation of a variety of solvated chemical groups. Gibson and Scheraga (1967*b*) then decomposed peptides into appropriate solvating groups. Using space-filling molecular models, they estimated the number of water molecules "bound" directly to these solvation groups along with the corresponding interaction geometrics.

The bound water molecules were assigned to the first hydration shell about a solute group. Intersection of hydration shells as a function of solute conformation was then considered as an energy-dependent process, since water molecules would have to be ejected from, or added to, the constituent hydration shells. Through the use of a step function to relate solvation free energy to hydration shell excluded volumes, Gibson and Scheraga (1967*a,b*) were able to estimate the solvation energy of a supramolecular, solute–solvent complex. Table 14-1 lists the results (Gibson and Scheraga, 1967*a*) when this model is applied to the conformational analysis of oxytocin and vasopressin. The choice of the dielectric constant, *D*, was also varied in the solvent-independent calculations on oxytocin. Neither solvent or dielectric have appreciable effects on the preferred local chain conformation except at the ends of the chain. The starting conformation in an energy minimization is the crucial factor in dictating conformation according to the results of Table 14-1.

Hopfinger and co-workers (Hopfinger, 1971, 1973; Forsythe and Hopfinger, 1973; Weintraub and Hopfinger, 1973, 1974) used the concept of a first hydration shell and both extended and revised the original solvent-binding model of Gibson and Scheraga (1967*a,b*). Configurational energy calculations were first carried out for water molecules interacting with solute groups (Hopfinger, 1971). The good agreement between the theoretically computed solvation free energies from these simulation studies to the experimental values (Nemethy and Scheraga, 1962; Gibson and Scheraga, 1967*a,b*) provided impetus to extend the revised first hydration shell model to solvents other than water. At the time of this writing, the hydration shell parameters for water, methanol, ethanol, formic acid, acetic acid, *n*-butanol, and 1-octanol have been computed. All parameters, except for 1-octanol (Hopfinger and Battershell, 1976), have been published elsewhere (Hopfinger, 1973). Table 14-2 contains the latest set of water and 1-octanol hydration shell parameters.

In Chapter 3, it was pointed out that Hansch and co-workers were often able to correlate drug activity with the log of the 1-octanol/water partition coefficient. To do this, the partition coefficient of a variety compounds had to be known in order to generate the necessary set of substituent constants (π constants). Hopfinger and Battershell (1976) have made use of the available octanol/water partition coefficient data to test the first hydration shell model by predicting partition coefficients. Under equilibrium conditions, $\log (P_{wo})$ the log of the octanol/water partition coefficient is

$$\log (P_{\mathrm{wo}}) = \frac{-1}{RT}(F_{\mathrm{w}} - F_{\mathrm{o}}) \tag{14-3}$$

where F_{w} and F_{o} are the minimum conformational free energies of the solute

Table 14-1 Minima of Oxytocin and Vasopressin[a]

Peptide Final Structure;[b] Residue	Vasopressin ϕ	Vasopressin ψ	Oxytocin ϕ	Oxytocin ψ	Oxytocin (no solvent; $D = 3.0$) ϕ	Oxytocin (no solvent; $D = 3.0$) ψ	Oxytocin (no solvent; $D = 1.0$) ϕ	Oxytocin (no solvent; $D = 1.0$) ψ
1. $\frac{1}{2}$-Cystine	—	327.2	—	315.4	—	306.9	—	288.8
2. Tyrosine	110.0	7.5	120.8	334.4	121.9	339.4	122.6	352.5
3. Isoleucine[c]	106.6	266.1	132.7	276.8	135.2	276.9	132.7	279.0
4. Glutamine	31.8	6.6	22.1	7.5	25.5	6.0	21.3	355.8
5. Asparagine	86.1	239.5	86.0	238.5	84.5	234.4	91.8	266.2
6. $\frac{1}{2}$-Cystine	91.6	325.1	95.2	324.9	93.5	324.8	37.2	324.6
7. Proline	120.1	289.2	120.3	298.9	297.9	120.5	120.5	299.6
8. Leucine[d]	76.9	258.8	87.8	326.4	90.5	321.5	77.9	328.8
9. Glycine[e]	105.0	292.9	119.8	295.6	90.2	327.6	124.6	295.3
Final energy (kcal/mole)	−6.17		10.86		−45.32		−106.55	
1. $\frac{1}{2}$-Cystine	—	214.2	—	286.9	—	237.2	—	259.6
2. Tyrosine	85.8	141.4	43.1	121.6	93.3	168.1	44.2	112.7
3. Isoleucine[c]	137.6	109.5	95.2	104.9	97.5	101.2	90.6	145.6
4. Glutamine	351.1	243.7	18.8	242.0	6.4	231.7	56.0	23.1
5. Asparagine	62.7	112.5	83.8	144.2	82.7	124.3	248.6	97.8
6. $\frac{1}{2}$-Cystine	111.2	126.9	111.3	126.3	110.6	125.8	122.9	120.1
7. Proline	120.4	137.5	120.3	146.2	119.9	151.8	119.3	115.4
8. Leucine[d]	127.2	118.4	258.0	344.8	254.5	96.8	229.1	32.4
9. Glycine[e]	109.1	84.0	125.3	142.0	90.0	70.8	244.2	52.5
Final energy (kcal/mole)	30.96		66.14		−16.35		−150.61	

[a] Starting conformation, upper group: $\phi = 60°, \psi = 300°$, except for proline, for which $\phi = 120°, \psi = 300°$; lower group, $\phi = 120°, \psi = 130°$.
[b] For conventions defining dihedral angles, see chapters 4 and 5.
[c] Phenylalanine in vasopressin.
[d] Lysine in vasopressin.
[e] C-terminal amide in both peptides.
Gibson and Scheraga, 1967*a*.

Table 14-2 The aqueous and 1-octanol Hydration Shell Parameters

Solute Group		Contact Radius	Solvent							
			Water				1-Octanol			
Name	Symbol	r Å	n	Δf^a kcal/mole	R_v Å	V_f Å^3	n	Δf^a kcal/mole	R_v Å	V_f Å^3
(Amide	>N—	1.35	2	0.63	4.3	35.8	2	0.18	6.4	53.6
(Amide)	>C=	1.50	2	0.63	3.9	14.3	2	0.15	5.9	60.6
(Ester)	>C=	1.50	2	0.46	3.9	38.3	2	0.16	5.9	60.5
Carbonyl	=O	1.35	2	1.88	3.9	67.8	2	0.90	6.3	52.4
(Ester)	—O—	1.35	1	0.21	3.9	59.6	1	0.12	6.2	88.9
(Hydroxyl)	—O—	1.35	2	1.58	3.9	55.2	1	0.52	5.6	56.6
(Carboxyl)	—O—	1.35	2	4.20	4.1	64.1	1	1.10	6.1	46.8
(Carboxylate anion)	—O	1.35	4	4.20	4.1	42.5	2	2.45	6.5	60.9
(Amide)	—H	1.20	2	0.31	3.5	31.3	1	0.28	5.7	51.7
(Hydroxyl)	—H	1.20	2	0.31	3.5	54.7	1	0.52	4.3	53.8
(Carboxyl)	—H	1.20	2	0.31	3.5	54.7	1	0.88	6.1	51.6
(Bromo)	—Br	2.00	6	−0.12	6.6	93.5	3	−0.04	7.7	138.6
(Cloro)	—Cl	1.80	6	0.12	4.9	53.2	3	0.28	6.7	70.8
(Fluoro)	—F	1.35	3	0.21	4.3	69.7	2	0.18	5.5	44.5
(Sulfide)	—S—	1.85	8	−0.05	6.2	79.3	3	−0.06	7.8	99.2

Table 14-2 *(continued)*

Solute Group Name	Symbol	Contact Radius r Å	Water n	Water Δf^a kcal/mole	Water R_v Å	Water V_f Å^3	1-Octanol n	1-Octanol Δf^a kcal/mole	1-Octanol R_v Å	1-Octanol V_f Å^3
(Sulfoxide)	—S(=O)—	2.20	4	0.71	5.8	58.8	1	0.88	7.0	82.3
(Sulfone)	O←S→O	2.35	6	0.95	6.1	46.6	2	0.89	7.1	69.9
(Sulfonate)	—S(O)(O)→O	2.50	6	1.38	6.3	38.9	3	0.72	7.4	59.6
		2.50	6	1.38	6.3	38.9	3	0.72	7.4	59.6
(Sulfate)	O—S(O)(O)—O	2.60	8	1.45	6.4	26.5	4	0.70	7.6	37.9
(Nitro)	>N—	1.35	2	0.56	4.3	35.8	2	0.32	6.4	53.6
(Nitro)	—O	1.35	2	2.70	4.1	42.5	2	2.55	6.5	70.1
(Cyano)	—C≡N	2.10	6	0.71	5.7	96.8	3	0.83	7.2	89.6
(Acetylinc)	—C≡C— or —C≡C—H	2.15	2	0.53	5.5	123.6	2	0.46	7.0	133.7

(Amnonium ion)	N^+H_3	1.80	3	15.40	4.3	22.1	[b]	—	—	—
(Tri-methyl amnonium ion)	$N^+(CH_3)_3$	2.10	9	2.08	7.3	54.0	—	—	—	—
(Phosphate ester oxygen)	—O—P—	1.75	2	2.68	3.8	47.8	—	—	—	—
(Phosphate ester PO_2^-)	O=P—O^-	2.05	4	2.82	4.9	73.5	—	—	—	—
(Aromatic)	>C—H	1.65	3	0.11	3.9	3.3	2	0.40	6.5	48.6
(Aromatic)	>C—X ≠ H	1.50	2	0.06	3.9	43.6	2	0.36	6.5	56.7
(*t*-butyl-carbon)	—C—	1.60	—	0.00	—	—	—	0.00	6.5	—
(Methine)	—C—H	1.75	2	0.13	5.5	104.8	2	0.36	7.1	56.7
(Methylene)	H—C—H	1.85	4	—0.10	5.5	60.8	3	0.39	7.4	36.8
(Methyl)	H—C(H)—H	2.05	8	−0.13	5.5	41.8	4	0.41	7.1	31.5
(Vinyl)	>C=C<	2.35	2	0.26	5.5	89.6	2	0.51	6.7	103.6
(Methoxy)	—O—	1.35	2	1.18	4.2	68.5	1	0.20	—	72.3

[a] The convention used to report Δf is stating the amount of free energy required to *remove* the solute group from the solvent medium. [b] — indicates the calculation has not been made. Hopfinger and Battershell, 1976.

molecule in octanol and water, respectively. Table 14-3 lists the predicted and observed log (P_{wo})'s for 20 different compounds. Also given in Table 14-3 are the predicted log (P_{wo})'s using Hansch analysis.

The most significant observation made from an inspection of Table 14-3 is the good agreement between the observed log (P_{wo}) values and those computed using the hydration shell model. The maximum difference for benzyl alcohol is -24.5%, whereas the average difference is approximately 9%. An inspection of Table 14-3 suggests that hydration shell conformational analysis predicts log (P_{wo}) values with about the same reliability as that obtained using π constants. Both appear adequate for quantitative usage. Most molecules studied are relatively rigid and nearly independent of conformational freedom. Those molecules which possess potential conformational flexibility, for example, pentane, are still relatively rigid because of intramolecular interactions. Consequently, it is not possible to discern from the work reported here how much better, if at all, the hydration shell theory will predict log (P_{wo}) than Hansch analysis in flexible molecules where the additive principle (as assumed by Hansch for molecular groups) is least reliable, and conformational effects most pronounced.

The hydration shell model appears to perform most poorly for molecules containing large aliphatic moieties. The source of error has been assigned to an inadequate description of the solvent organization about the aliphatic solute groups. Interhydration shell solvent organization is neglected, although it is likely to be extensive in aliphatic molecules. Hansch analysis also seems to be least reliable for molecules possessing large aliphatic groups. These observations may be indirect evidence to support the existence of significant solvent structuring about certain sites on the solute molecule.

Brown et al. (1972) used a very simple hydration shell-like model to rather successfully predict melting temperatures of synthetic collagen models. This model assigns exposure coefficients to certain key polar groups on the biopolymer in order to estimate solute–solvent interactions. In addition, quantitative application of the model depends on an experimental calibration that assigns the characteristic polymer–solvent enthalpy term descriptive of the solvent and the set of congeneric polymers.

The hydration shell model is able to estimate the free energy of interaction of solvent with the entire solute molecule by sacrificing a great deal of detailed information concerning the spatial organization of solvent molecules about the solute. The solvent geometry reduces to a set of spherically symmetric, free energy densities of specific sizes centered about each of the solute groups. If solvent molecules organize themselves between solute groups as bridges, the hydration shell concept will not correctly account for the conformational stabilization, but will, in some cases, treat the interaction as destabilizing. For example, Avignon et al. (1973) have shown that the C_7 ring conformation

Table 14-3 Values of log (P_{wo}) for 20 Compounds Determined Experimentally, by Hansch Analysis, and by the Hydration Shell Model. Also included are values for the F_o and F_w for the Hydration Shell Calculations as well as the relative errors in the log (P_{wo})s based upon [% Rel. Error] = [(Obser. − Calc.)/Obser.] × 100

				Hydration Shell Calculations			
		log (P_{wo}) Hansch Analysis[b]		log (P_{wo}) Calc.			
Compound	log (P_{wo}) Obser.[a]	Value	Percent Rel. Error[c]	Value	Percent Rel. Error[d]	F_o kcal/mole	F_w kcal/mole
Benzene	2.13	2.13	0.0	2.23	−4.7	4.65	1.62
Aniline	0.90	0.90	0.0	0.92	−2.2	5.44	4.19
Propylbenzene	3.68	3.63	+1.4	3.52	+4.4	5.55	0.76
2-butanone	0.29	0.29	0.0	0.24	+17.2	3.80	3.47
Cyclohexanol	1.23	1.07	+13.0	1.22	+0.8	4.21	2.55
2,2 dimethyl propanol	1.36	0.94	+30.9	1.43	−5.1	3.98	2.03
2-butanol	0.61	0.61	0.0	0.50	+18.0	3.05	2.37
Ethyl acetate	0.73	0.73	0.0	0.59	−19.2	3.96	3.16
Chloroform	1.97	1.67	+15.2	2.11	−7.1	4.65	1.78
Chlorobenzene	2.84	2.84	0.0	2.82	+0.7	6.01	2.17
2-methyl-2-butanol	0.89	0.91	−2.2	0.75	+15.7	3.08	2.06
Propionitrile	0.16	0.16	0.0	0.19	−18.8	3.71	3.44
1-pentyne	1.98	1.98	0.0	1.96	+1.0	2.80	0.13
Benzyl alcohol	1.10	1.47	−33.6	1.37	−24.5	7.01	5.05
Chlorobutane	2.39	2.39	0.0	2.17	+9.2	3.82	0.87
Toluene	2.69	2.63	+2.2	2.62	+2.6	4.64	1.08
Ethylbenzene	3.15	3.21	−1.9	3.01	+4.4	5.04	0.96
Flourobenzene	2.27	2.27	0.0	2.28	−0.4	5.23	2.13
Nitrobenzene	1.85	1.85	0.0	2.08	−12.4	15.38	12.66
Pentane	2.50	2.50	0.0	2.17	+13.2	4.20	1.25

[a] As reported by Hansch et al., (1968).
[b] From π values given by Tute (1971). In some instances, the π values used were taken from the log (P_{wo}) *Obser.* experiments. Hence agreement is forced and artificial.
[c] Average absolute error = 5.0%.
[d] Average absolute error = 9.0%.
Hopfinger and Battershell, 1976.

of dipeptides is stabilized in aqueous solution by a water molecule bridge as shown in Figure 14-6. The hydration shell model, in contradiction (Forsythe and Hopfinger, 1973) predicts that diglycine and dialanine are destabilized in water as compared to free space.

Consequently, some workers have chosen to study the detailed geometry of interaction of one, or at the very most, two, solvent molecules with the solute molecule as a simultaneous function of solute conformation and solute–solvent geometry. Venkatachalam and Krimm (Krimm and Venkatachalam, 1971; Venkatachalam and Krimm, 1973) have used empirical energy functions to describe the interaction of water and a few other solvent molecules with certain polar binding sites (carbonyl oxygens and amide hydrogens) on peptides. The changes in the conformational properties of the solute molecule as a function of the selective solvation processes have been monitored. Rein and co-workers (Rein et al., 1976; Rein and Renugopalakrishman, 1976) have carried out similar selected solute binding-site studies involving peptides but employing molecular orbital techniques.

The majority of investigations employing the assumption of selective hydration sites that lead to specific representations of the supramolecular

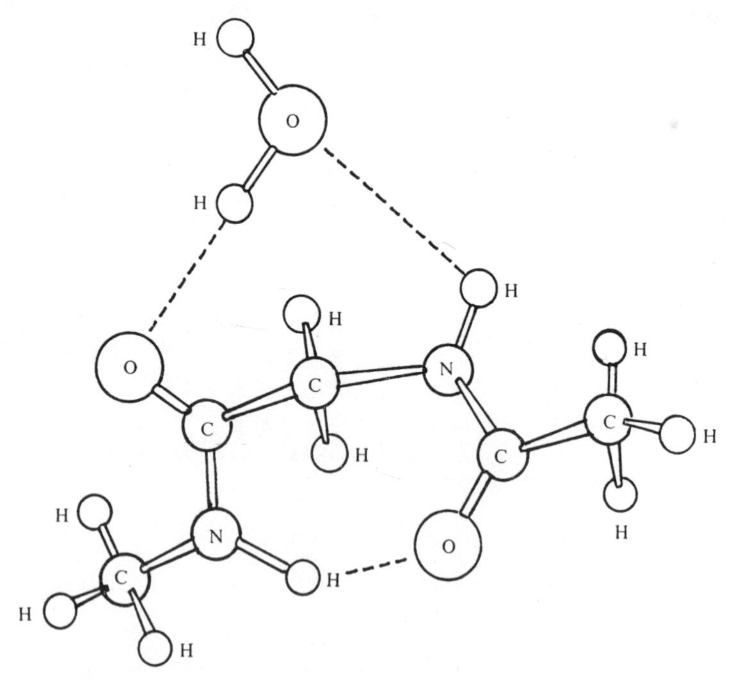

Fig. 14-6 Stabilization of the C_7 form of dipeptide by water. Avignon et al. (1974).

solute–solvent complex (Figure 14-1c) have been carried out by the Pullmans (Pullman and Pullman, 1975). A convenient semiquantitative means of representing the geometry and energy of solute–solvent interactions is shown in Figure 14-7(I–XII) for several solute species interacting with a single water molecule. The relative orientation of a water molecule to a solute species, as shown in the figure, along with the displayed characteristic interaction energy, was found by initially placing the water molecule at a "likely" hydration site and subsequently scanning the local configurational water molecule space to locate an energy minimum. For each solute molecule, this process was repeated for each prechosen "likely" hydration site. The physical significance of the results of these limited supramolecular models can be considered valid only in so far as simultaneous hydrated water–water, water–solute, and bulk water-hydrated water interactions do not modify the single water molecule–solute molecule configurational geometries and energies. The authors also caution that the particular basis sets used in the molecular orbital calculations may be sufficiently incomplete so as to overestimate hydration energies. In any event, it is of interest to note the magnitudes of the solute–water molecule

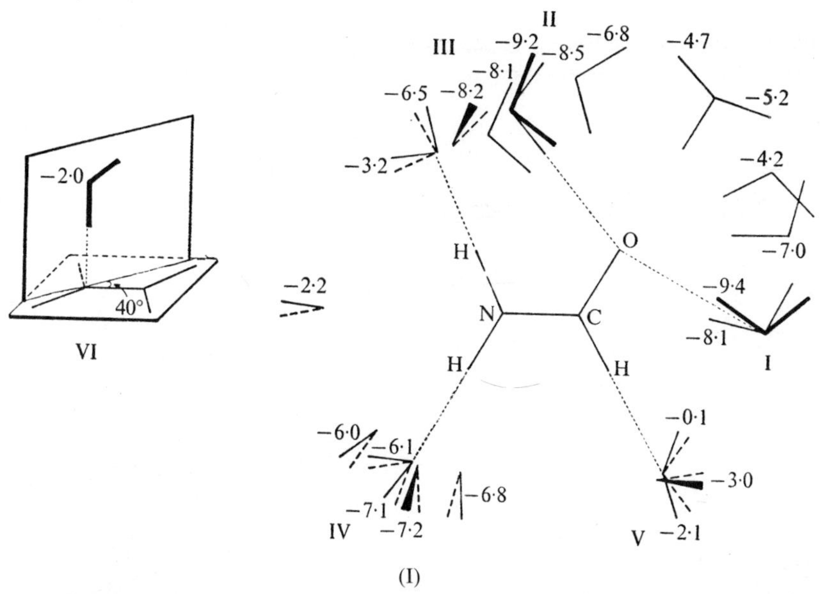

(I)

Fig. 14-7I. Part A Hydration of formamide. Water molecules in heavy lines: most favorable associations in the region considered (see Alagona et al., 1973). Water molecules with one full and one dotted line indicate that the water plane is perpendicular to the formamide plane. Alagona et al. (1973).

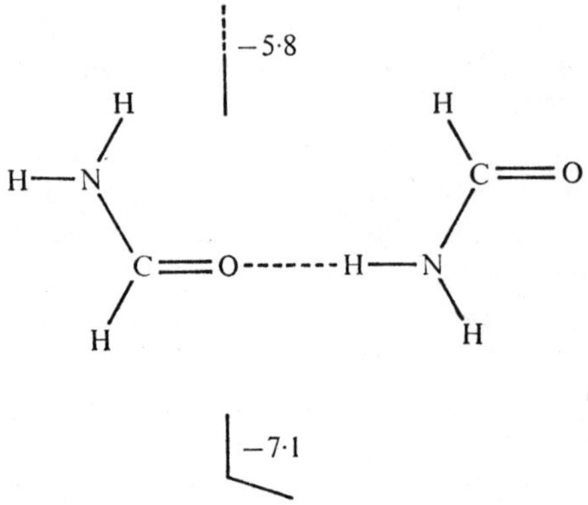

−5·8

H H

H—N C═O

C═O-------H—N

H H

−7·1

(II)

Fig. 14–7 (*continued*)

Fig. 14-7I. Part B Geometry and energy characteristics of the stable hydrates of *trans*-N-methylacetamide based on the type of analysis given in Part A. Pullman et al. (1974). **II.** The linear antiparallel dimer of formamide showing the most favorable hydration sites with their calculated interaction energies (kcal/mole). Port and Pullman (1974). **III.** (*a*) Hydration sites in ethanol with their calculated energies of interaction (kcal/mole). (*b*) Hydration sites in *p*-cresol with their calculated energies of interaction (kcal/mole). Pullman and Pullman (1975). **IV**. (*a*) Hydration sites in formic acid with their calculated energies of interaction (kcal/mole). (*b*) Hydration sites in the formate ion with their calculated energies of interaction (kcal/mole). Pullman and Pullman (1975).

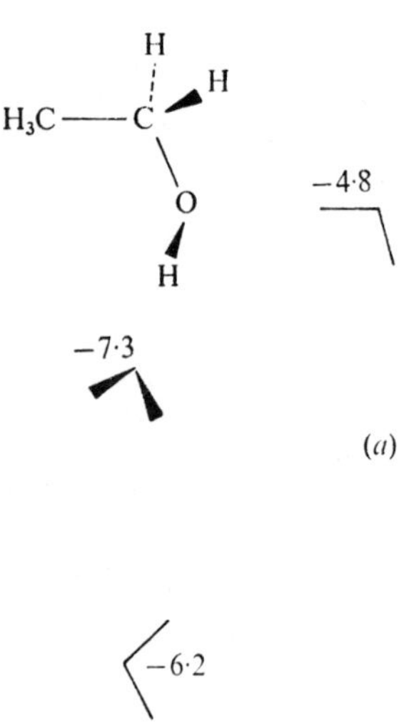

H

H

H_3C—C

O −4·8

H

−7·3

(*a*)

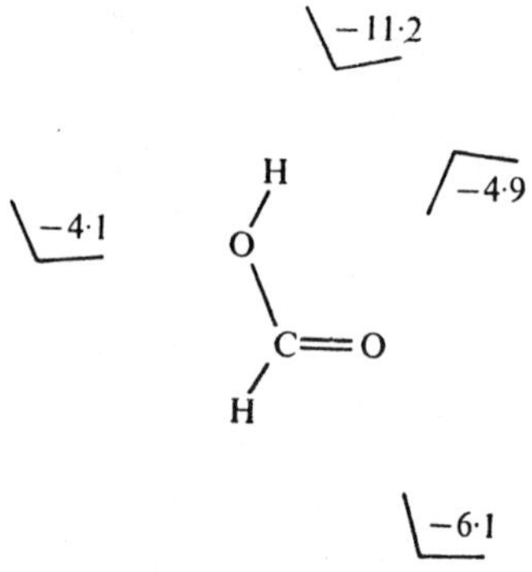

−6·2

O—H

−10·0

CH_3

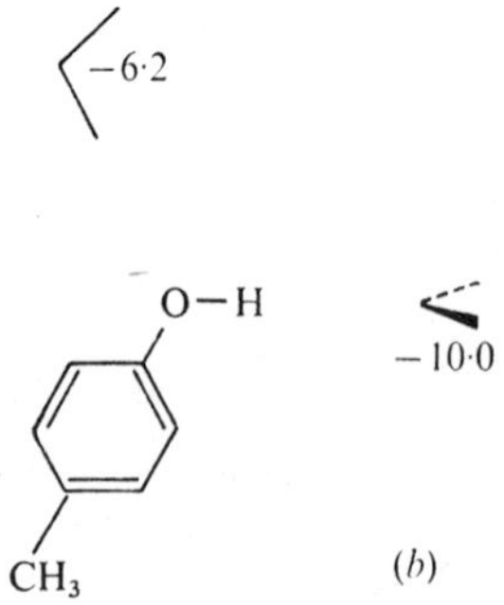

(*b*)

(III)

−11·2

H

−4·1 O −4·9

C═O

H

−6·1

(*a*)

−25

−22 O

⊖

C====O

H

−22

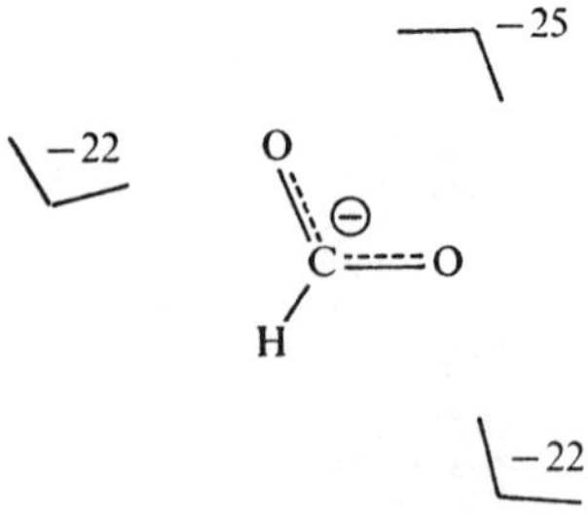

(*b*)

(IV)

Fig. 14–7 (*continued*)

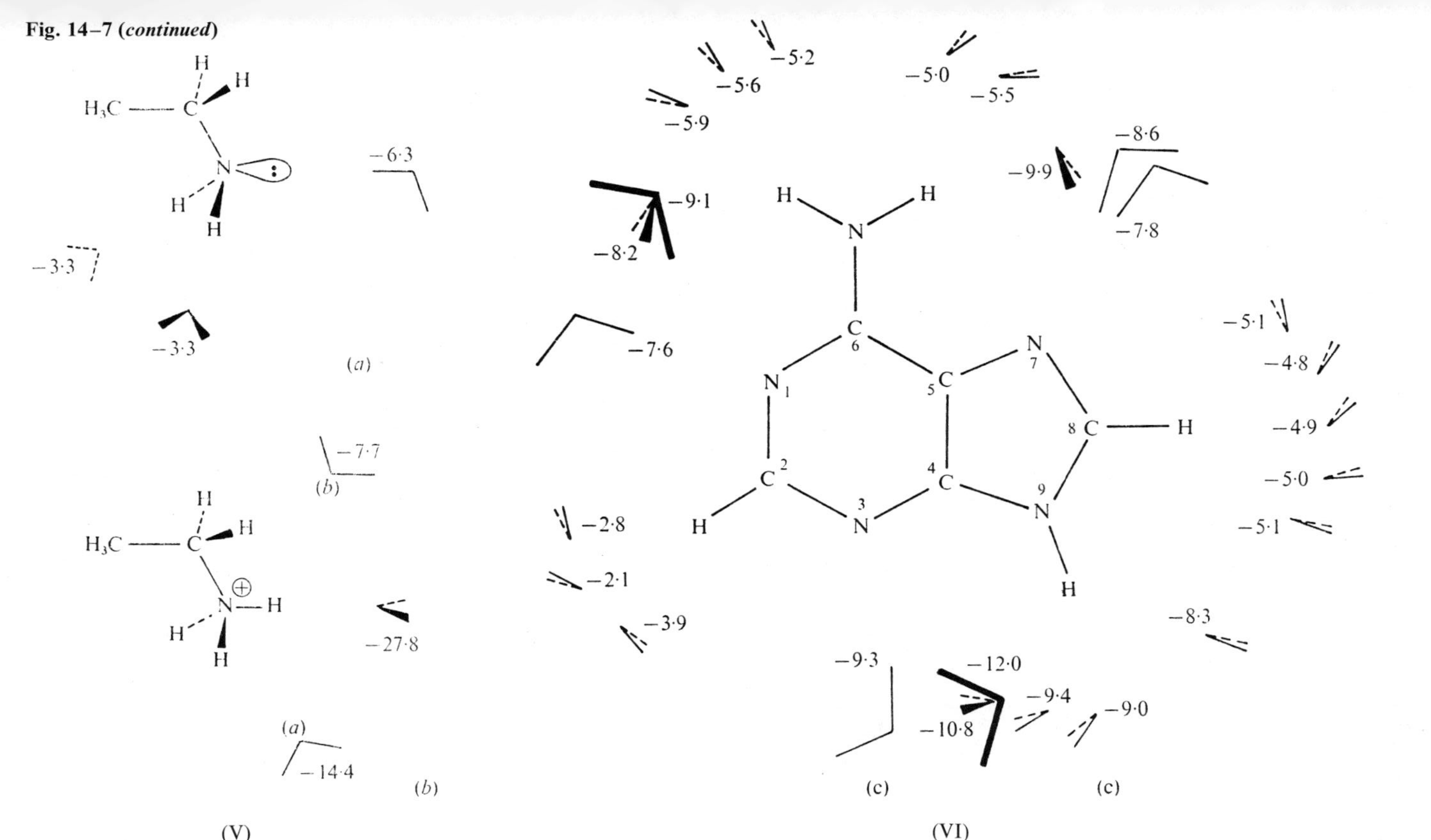

V. (*a*) Hydration sites in ethylamine with their calculated energies of interaction (kcal/mole). (*b*) Hydration sites in ethylammonium ion with their calculated energies of interaction (kcal/mole). Pullman and Pullman (1975). **VI.** Hydration sites in adenine. Energies in kcal/mole. Heavy lines: preferred hydration sites. Full lines: coplanar arrangement of water and base. Half-dashed: perpendicular arrangement of water with respect to the plane of the base. (*c*) refers to configuration akin to conventional hydrogen bonding. Port and Pullman (1973*a*).

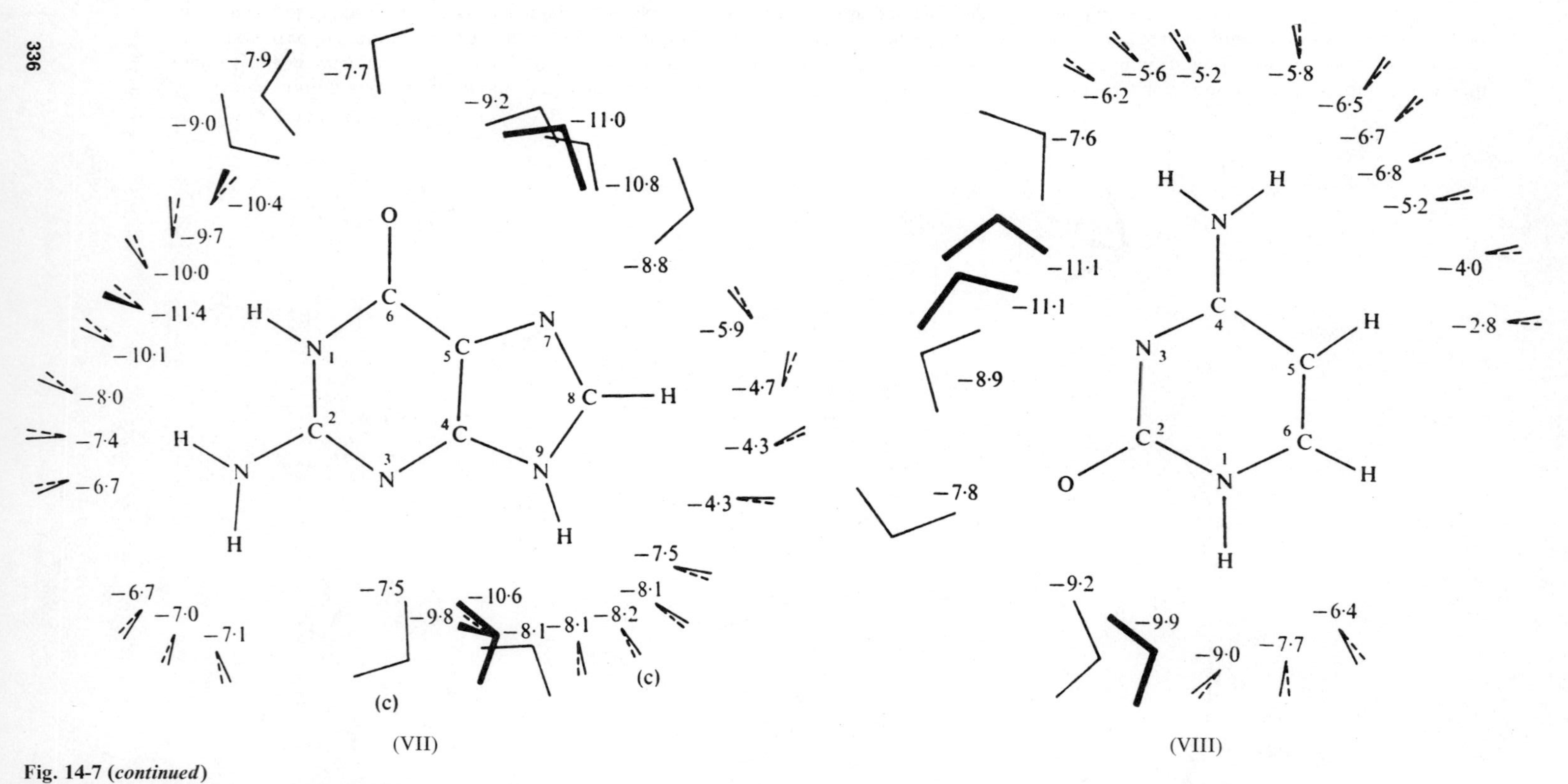

Fig. 14-7 (*continued*)
VII. Hydration sites in guanine. Energies in kcal/mole. Heavy lines: preferred hydration sites. Full lines: coplanar arrangement of water and base. Half-dashed: perpendicular arrangement of water with respect to the plane of the base. (*c*) refers to configurations akin to conventional hydrogen bonding. Port and Pullman (1973*a*). **VIII.** Hydration sites in cytosine. Energies in kcal/mole. Heavy lines: preferred hydration sites. Full lines: coplanar arrangement of water with respect to the plane of the base. Port and Pullman (1973*a*).

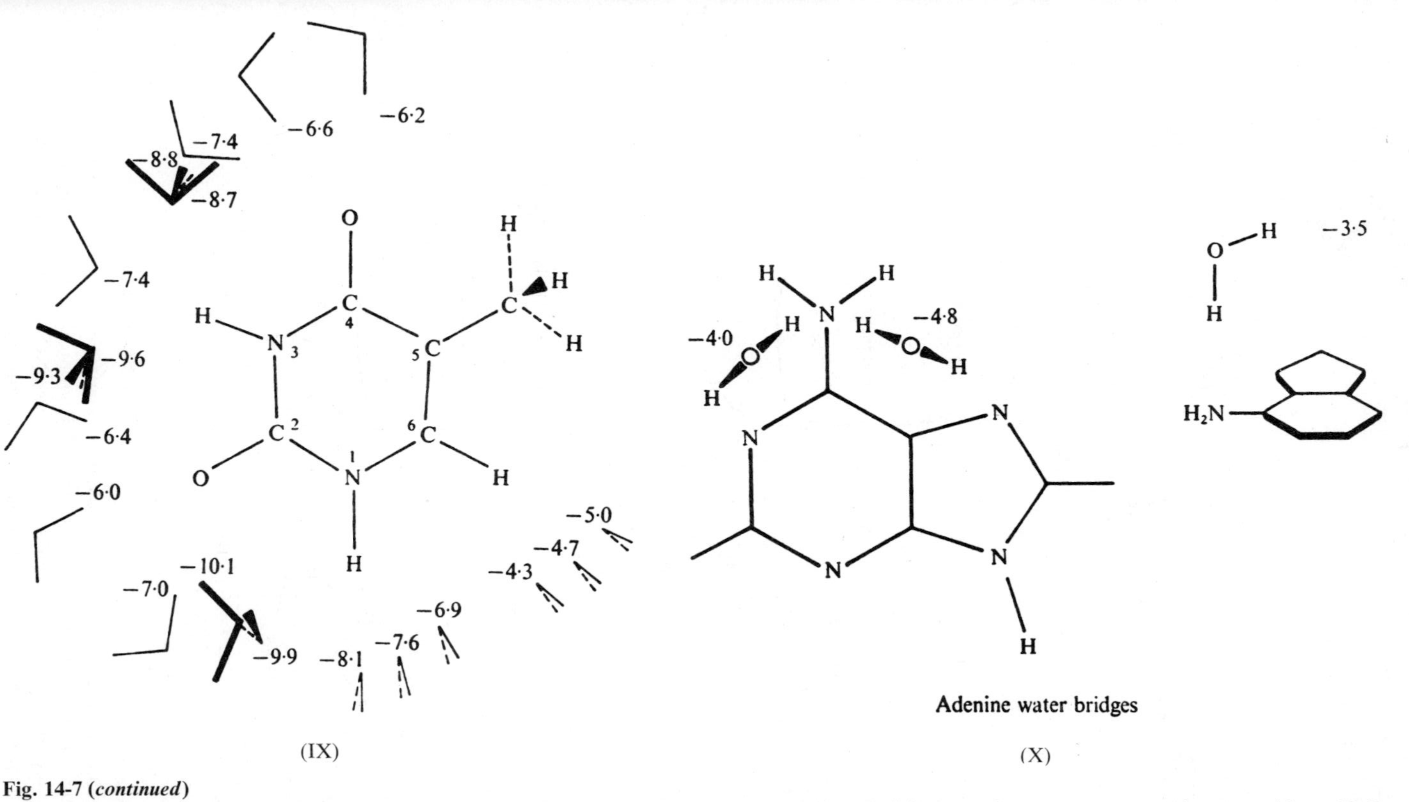

Fig. 14-7 (*continued*)
IX. Hydration sites in thymine. Energies in kcal/mole. Heavy lines: preferred hydration sites. Full lines: coplanar arrangement of water and base. Half-dashed: perpendicular arrangement of water with respect to the plane of the base. Port and Pullman (1973*a*). **X.** *a*. Adenine water bridges. *b*. Out-of-plane hydration of adenine. Port and Pullman (1973*a*).

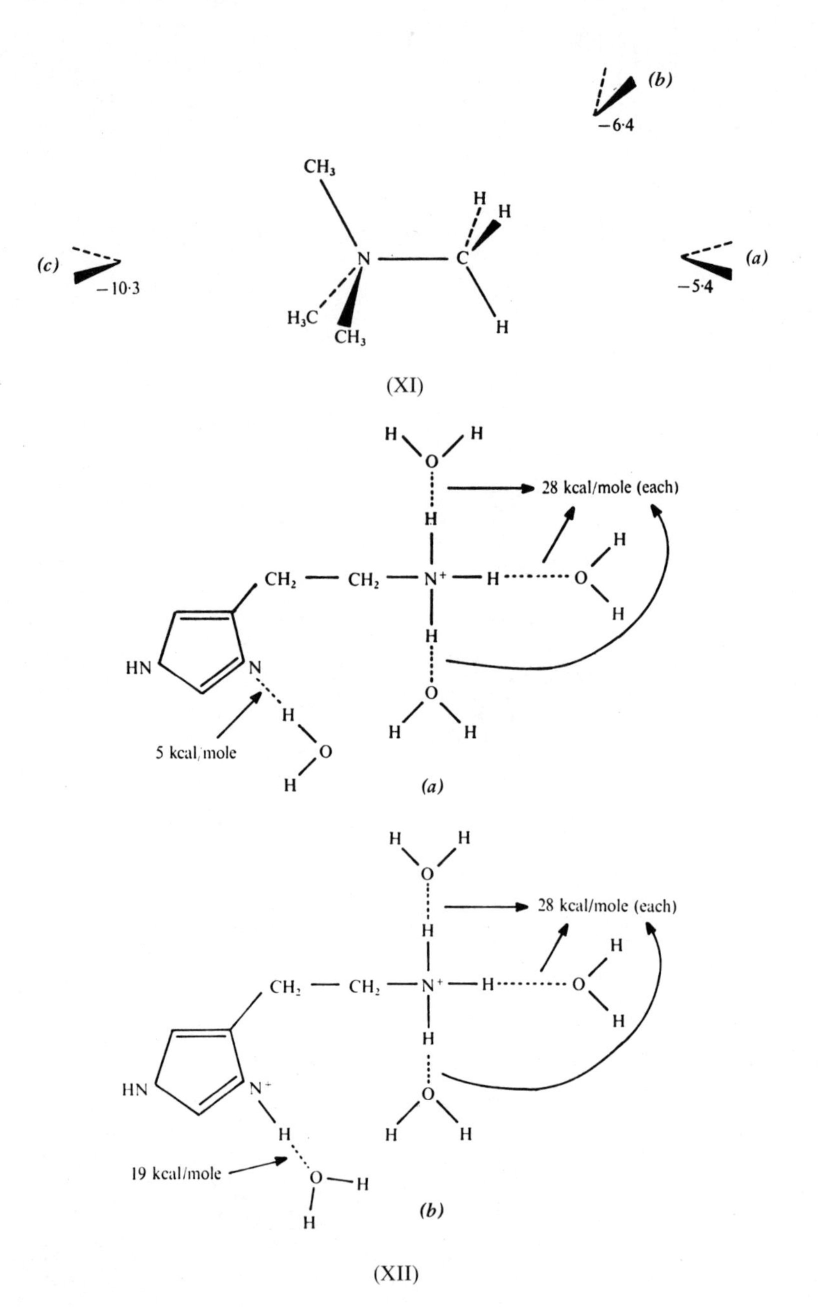

Fig. 14-7 (*continued*)

XI. Stabilization energies in the best positions of a water molecule bound to a tetramethylammonium ion: (*a*) direct approach towards a CH_3 group along the prolongation of an NC bond, (*b*) bisecting approach along the bisectrix of two NC bonds; (*c*) axial approach on an NC axis [opposite to approach (*a*)]. Pullman and Ambruster (1974).

XII. Hydration sites in (*a*) mono- and (*b*) diprotonated histamine. Energies in kcal/mole calculated ab initio (STO_3G). Pullman and Port (1974).

binding energies for various hydration sites (solute groups). These energies can be qualitatively used in deducing which solute hydration sites on larger, more complex solutes will strongly interact with water.

Some general observations can be made from the calculations described in Figure 14-7 that bear on a number of unsubstantiated speculations made in the past about solute–water interactions. Among these are:

1. A number of preferred hydration sites (solvent energy minima) are located about the solute species that can differ markedly in interaction energy (up to 6 kcal/mole).
2. For a given hydration site, there can be a few different solvent orientations, relative to the solute, that can, in themselves, correspond to unique energy minima.
3. The solute–water molecule interaction is largely a hydrogen bonding type of phenomenon in that polar groups tend to interact most strongly with water. Neither CH hydrogen bonding, nor π binding from resonance groups appear to correspond to an appreciable probability of hydration.
4. The actual number of preferred hydration sites seems to be directly proportional to the number of polar groups in the solute molecule and inversely related to the number of aliphatic groups.
5. A water molecule can still form reasonably strong hydrogen bonds with a carbonyl group already participating in a intra- or inter solute molecular hydrogen bond. Thus, bifracated hydrogen bonding involving water is predicted.
6. Aromatic-OH groups should be more hydrated than aliphatic ones. This is in agreement with the conclusions of Watt and Leeder (1968) (based on the construction of differential isotherms of water adsorption on various wool samples), indicating that serine complexes one water molecule, whereas tyrosine complexes two. A similar conclusion was reached by Kuntz (1971) by NMR measurements of "unfrozen" water on rapid freezing of a solution of polypeptide.
7. Ionized carboxyl and ammonium groups interact more strongly with water molecules than their neutral counterparts. There is some experimental evidence to support these calculations for the carboxyl group (Subramaniam and Fischer, 1972; Hnagewyj and Ryerson, 1963; Watt and Leeder, 1968; Breuer and Kennerley, 1971; and Kuntz, 1971). The experimental support for the -$(NH_3)^+$ more strongly interacting with water than -NH_2 is not clear cut and in fact partially nonsupportative. Fraenkel and Kim (1966) concluded that water molecules form three strong hydrogen bonds with the -$(NH_3)^+$ group, whereas Breuer and Kennerley (1971) and Kuntz (1971) did not find a significant difference in hydration between polylysine and polylysine hydrobromide.

8. For solute molecules that are predominantly planar, in-plane solute–water interactions are generally of stronger binding energy than out-of-plane interactions.
9. Aliphatic substitution in an ionic group leads to a decrease in hydration energies with respect to those of the unsubstituted ion strongly reflecting the spreading of charge over the whole molecular periphery. This is illustrated for NH_4^+ to $N\text{-}(CH_3)_4^+$.
10. The magnitudes of solute–water interaction energies are sufficiently large so as to readily suggest that these types of interactions can significantly alter free-space conformational populations.

Weintraub and Hopfinger (1974) have attempted to take into account some of the dynamic aspects of the solvation process by identifying *modes of solvation.* Succintly, these modes of solvation are specific supramolecular solute–solvent complexes that have their own characteristic conformational properties, and, depending on the strength of the solute–solvent interaction energy, are assumed to persist in time to some extent. By averaging over all solvation modes according to the time-persistence weights, a set of time-averaged conformational properties could be achieved. The results of applying this theory to acetylcholine and some of its congeners suggest that the time-persistence weights are critical to the estimation of the conformational populations. This indicates that more work should be done in estimating the time-persistence weights before quantitatively applying this theory of solvation modes.

Beveridge and Schnuelle (1974) have developed the complete statistical thermodynamics for the model shown in Figure 14-1*e* subject to $\varepsilon_1'(\mathbf{s})$ held constant. They have appropriately termed this theory the "supermolecule-continuum model." The major difficulty in the development of the theory resides in configurational averaging over the solvent molecules. Beveridge and Schnuelle (1974) have proposed three configurational averaging techniques: the site, cell, and shell methods. Results of applying the supermolecule-continuum model to ion hydration is given as part of the next section. The supermolecule-continuum model has not yet been applied to bio-organic solute–solvent systems because of computational restrictions arising from the enormous size of the associated conformational-configurational hyperspaces.

IV COUNTERION INTERACTIONS

The structural and corresponding biologic significance of counterion interactions involving groups on a solute molecule are discussed in Chapter 12. Unfortunately, little theoretical modeling of such interactions has taken place. This is in part a result of the difficulty of developing energy functions, be they empirical or quantum mechanical, to describe interactions involving solvated heavy ions. Also, it has only been recently that experimental probes of counterion structural modifications on the molecular level have shown sufficient progress to entice the corresponding theoretical development.

The work of Tiffany and Krimm (Tiffany and Krimm, 1968, 1969) on the interaction of counterions with polypeptides with charged side chains has suggested the possibility of a conformational randomization of the polypeptide only when the charged side chains interact with heavy counterions. To test this hypothesis, Hilter, et al. (1972) modeled the charged polypeptide–heavy counterion interaction by simply placing, at random, point charges of opposite sign in the vicinity of the charged side chain groups and minimizing, as a function of backbone bond rotations, the total interaction energy. The results of these calculations support the Tiffany-Krimm hypothesis, although the treatment of counterion interactions is very crude.

A few quantum mechanical investigations (Pullman, 1974; Perricaudet and Pullman, 1973; Perricaudet, 1973) have been made on the nature of ion binding to cyclic depsipeptides with particular emphasis on enniatin B. Figure 14-8 is an illustration of enniatin B in its complex with K^+. In order to estimate the relative binding energies of Na^+ and K^+ to enniatin in aqueous solution, it is necessary to know the individual binding energies of the ions for the amide and ester groups as well as for water. Table 14-4 lists the equilibrium binding energies and geometries, defined in Figure 14-9, for the three molecular species interacting with Na^+ and K^+.

A comparison of the relative binding properties of the amide and ester groups suggests the following for both cations

1. The intrinsic affinity of the amide carbonyl for cations is appreciably larger than that of the ester carbonyl and larger than the corresponding affinity for water.
2. The equilibrium distances of approach are practically the same for the ester and water, but shorter for the amide.
3. The Na^+ ion shows a stronger binding affinity for carbonyl groups than K^+.

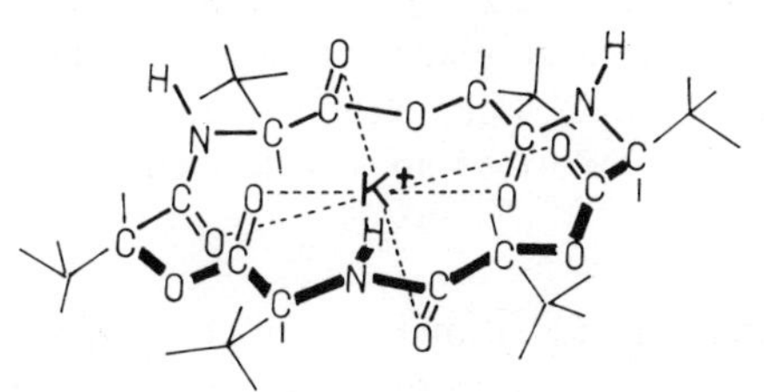

Fig. 14-8 Schematic structure of enniatin B in its complex with K^+ (in fact in enniatin B, the three NH groups are substituted by CH_3). Pullman (1974).

Table 14-4 Binding Characteristics at Equilibrium for Na^+ and K^+ Interaction with Amide, Ester, and Water

	Na^+					K^+				
	$-\Delta E$	d	θ	τ_C	τ_O	$-\Delta E$	d	θ	τ_C	τ_O
Amide	49.7	1.95	35	0	0	35.1	2.35	35	0	0
Ester	38.8	1.99	35	0	0	25.5	2.40	35	0	0
Water[a]	40.7	1.99	0		0	27.9	2.40	0		0
[b]	25.2	2.25	0		0	17.5	2.69	0		0
[c]	0.62	1.13				0.62	1.12			

ΔE, binding energy (kcal/mole); d, minimal distance of approach (Å); θ, angle of the oxygen-cation direction with the CO bond (with the HOH bisectrix for water); τ_C, angle for out-of-plane rotation (CM^+ fixed); τ_O, angle for out-of-plane rotation (OM^+ fixed). All angles in degrees.
[a] Present computation.
[b] Near Hartree-Fock limit.
[c] Ratio b/a.
Pullman, 1974.

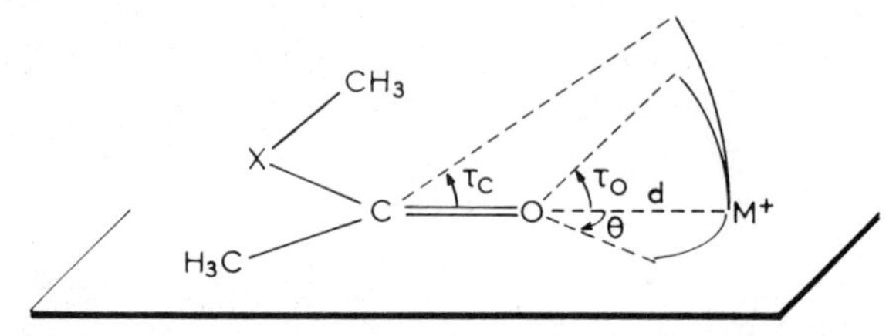

Fig. 14-9 Approach and rotation parameters, X = NH for amide; O for ester. Pullman (1974).

4. Rotation of the cation out of the plane of the peptide and ester bond is very easy for both ions. This was also found to be the case for complexes of N-methyl-acetamide and methyl-acetate with Li^+ and Na^+ (Kostetsky et al., 1973).
5. The rotation of the cation out of the plane of the amide or ester bond has a stronger effect on the binding energy than the in-plane rotation; the Na^+ interaction shows a sharper decrease than the K^+ interaction with increasing angle. However, both in- and out-of-plane rotations are highly flexible up to 30° deviations from planarity.

An explanation for the salt-induced reversible transition between polyproline I and polyproline II, which contain only cis and trans residues, respectively, has not yet been found. All the evidence invoked for the existence of cation binding to peptides points to the carbonyl oxygen as the site of binding (Baddiel et al., 1971; Baron and de Lozé, 1972; Wuepper and Popov, 1970; Balasubramanian and Schaikh, 1973). However, such a mode of complexing should lead to an increase in the double-bond character of the NC bond and increase the height of the cis-trans interconversion barrier. Consequently, the often-used "hand-waving" argument of carbonyl ion ligands lowering the CN torsional barrier is probably worthless. Armbruster and Pullman (1974) have quantitatively demonstrated the increase in the NC torsional barrier as a consequence of $C{=}O \cdots Li^+$ and $C{=}O \cdots H^+$ bindings involving formamide. However, they note that protonation of the nitrogen of the CN bond lowers the torsional barrier substantially. This is consistent with a model assumed by Farmer and Hopfinger (1974) in the study of cis-trans interconversion in polyproline. The interaction of Li^+ with the amide nitrogen also lowers the CN torsional barrier. Armbruster and Pullman also note that, when Li^+ is in a bridged configuration between the carbonyl oxygen and amide nitrogen, a near-perpendicular arrangement of the NH_2 and COH groups is preferred. They suggest that such a bridge binding may be the mediating step in the cis-trans isomerization process.

Kim and Rubin (1973) have calculated the free energy of activation for dehydration of primary hydrated ions in the liquid state. They assumed a quasicrystalline solid for the liquid structure wherein each particle is constrained to move within its own free volume and formulated the resultant partition function. The results are given in Table 14-5, in which n^0_{bw} is the initial number of ion-bound water molecules, and n^*_{bw} the remaining number of ion-bound water molecules after n^0_{fw} water molecules go into free

Table 14-5 Standard Free Energy of Activation (ΔG^* in kcal/mole) for Dehydration of Primary Hydrated Ions in Aqueous Solution

Initial state	Activated state							
n^0_{bw}	n^*_{bw}	n^*_{fw}	Li^+	Na^+	K^+	Cl^-	Br^-	I^-
4	3	1	7.6	7.3	6.2	7.7	7.1	
4	2	2	23.4	19.2	15.9	17.6	16.3	
4	1	3	48.4	38.9	31.4	30.1	27.6	
4	0	4	83.1	66.5	50.1	43.2	38.9	
3	2	1						7.0
3	1	2						15.8
3	0	3						25.1

Kim and Rubin, (1973).

solution. The calculated free energy of activation for dehydration of the primary hydrated ions is in the relative order $Li^+ > Na^+ > K^+$ and $Cl^- > Br^- > I^-$ for the same degree of aqueous hydration. Considerably less free energy is required to remove the first water molecule from the complex than second, third, etc. water molecules for all types of ions.

Manynick and Schaeffer (1975) have calculated the formation energies of phosphate–metal solvation complexes using ab initio self-consistent field techniques. They included Li^+, Na^+, K^+, Be^{2+}, Mg^{2+}, Ca^{2+}, H_2O, and Cl^- mobile ions interacting with phosphate units which are approximate, yet reasonable models for phosphate groups in a nucleic acid environment. Their results are consistent with experimental binding trends.

The supermolecule-continuum model has been applied to the determination of ion-hydration thermodynamics (Schnuelle and Beveridge, 1975, 1976). The idea of treating solvent effects with a discrete representation of the solute ion and vicinal solvent molecules coupled with a continuum repre-

Table 14-6 Calculated and Observed Thermodynamic Functions for Ion Hydration (Units of kcal/mole for ΔG and ΔH, and cal/deg.mole for ΔS)

		ΔG_{el}^{i-w}		ΔH_{el}^{i-w}		ΔS_{el}^{i-w}	
Ion	τ_i	calc.	obs.[d]	calc.	obs.[d]	calc.	obs.[d]
Li^+	0.6	−148.4[a]	−122.1	−158.3[a]	−129.7	−33.3[a]	−25.4
		−142.7[b]		−149.8[b]		−24.3[b]	
Na^+	0.95	−114.0[a]	−98.4	−121.5[a]	−103.6	−25.2[a]	−17.5
		−109.2[b]		−113.9[b]		−15.9[b]	
K^+	1.33	−82.7[a]	−80.6	−86.5[a]	−83.4	−13.0[a]	−9.4
		−77.9[b]		−79.7[b]		−6.2[b]	
		−89.6[c]		−90.3[c]		−19.3[c]	
F^-	1.36	−98.8[a]	−89.5	−97.2[a]	−97.8	−11.5[a]	−27.8
		−88.5[b]		−89.8[b]		−5.6[b]	
		−90.2[c]		−97.1[c]		−18.6[c]	
Cl^-	1.81	−61.0[a]	−76.1	−61.4[a]	−80.3	−1.3[a]	−14.0
		−56.2[b]		−54.9[b]		+2.1[b]	

[a] Site method, 20k configurations (running time ~10 minutes per calculation on IBM 370-168).
[b] Cell method, 50k configurations (running time ~25 minutes per calculation on IBM 370-168).
[c] Shell method, 200k configurations (running time ~100 minutes per calculation on IBM 370-168).
[d] Stearn and Eyning (1937).
From Schnuelle and Beveridge (1976).

sentation for bulk solvent has been suggested several times; the first was the study of water and ionic solutions by Bernal and Fowler (1933). Statistical aspects were first introduced with this model by Kirkwood (1939). Contemporary use of this approach is found in theoretical studies of solvated electron systems, particularly by Copeland et al. (1970) and Fueki et al. (1973). Newton (1973) and Moskowitz et al. (1976) have carried out quantum-mechanical studies of hydrated electrons based on the supermolecule-continuum model, but neglecting configurational averaging. Table 14-6 summarizes the results of the application of the supermolecule-continuum model using different levels of sophisticated configurational averaging for five different ions in dilute aqueous solution. One significant observation from Table 14-6 is that increasing the level of configurational averaging does not necessarily yield thermodynamic properties that increase in agreement with those observed. Still, the theoretical values are generally in good agreement with experimental findings, suggesting that the supermolecular-continuum concept is a reasonable representation for ion hydration.

V LIGAND BINDING ON DILUTE POLYMER MOLECULES IN SOLUTION

The thermodynamics of binding, from solution, of relatively small ligand molecules onto polymer molecules that can undergo conformational transitions has been a relatively popular statistical mechanical topic. In general, the solution is considered to be dilute; thus polymer–polymer interactions (second viral coefficient, etc.) may be ignored. Also, the individual monomer units of each polymer chain are allowed to exist in only two states—usually a function of whether or not the ligand is bound. The ligand molecules are also limited to two states. Consequently, the one-dimensional Ising model (linear and circular) can be applied in the statistical mechanics.

A general formulation of the thermodynamics of ligand binding was given a rigorous treatment by Hill (1964), as an example of the thermodynamics of small systems. Hill (1973) focused his general ligand-binding theory to the binding of complementary monomers and oligomers on polynucleotides. The results of the application of the theory are presented in terms of binding iostherms, optical melting curves, calorimetric melting and mixing studies, the role of the activity of the ligand, quasi-phase transitions and polymer phase diagrams. Crothers (1971) and Damle (1970) have attempted to describe certain ligand-binding interactions using particular statistical mechanical models. Shindo (1971) developed a ligand-binding model to describe protein denaturation as a function of the composition of each type of solvent molecule that is capable of binding to protein residues and inducing an order-disorder state change in the residue. The theory is equivalent to the classic Langmuir adsorption model used to describe the behavior of mobile molecules above and on a surface, but with the added flexibility that the surface adsorption sites can undergo changes in state. Hopfinger (1974*b*) used conformational analysis to evaluate the various interaction energies required in the Shindo model (1971) for a few globular proteins. The denaturation curves as a function of solvent composition were computed.

VI THEORIES OF AGGREGATION IN SOLUTION

Aggregation or cluster formation in an imperfect one-component gas can be treated exactly (Hill, 1955). The McMillan-Mayer solution theory (see Chapter 1, Section II) allows the formal methods of imperfect gas theory to be extended without change to a solution under osmotic conditions. Consequently, aggregration of a solute in a solution, in principle at least, can also be treated exactly. Some specific applications of the general theory include macromolecular aggregration (Hill, 1955) and the stacking of bases and nucleosides (Hill and Chen, 1973). Most often, aggregration processes have been formulated on the basis of an *ideal* mixture of aggregates of various sizes. However, nonassociating interactions between aggregates, at least at the level of "hard" or "space-filling" interactions, must be included in any theory employed to predict activity coefficients, osmotic coefficients, aggregation equilibrium constants, free energies, etc. Hill and Chen (1973) have presented the necessary formalism for including nonassociating interactions in an aggregation theory. Lyngaae-Jorgensen (1976) has proposed a theory to describe aggregate crystallites in solution. This theory provides an expression for the chemical potential of the polymer repeat units in the range between the theory for extremely dilute solutions and the Flory-Huggins theory (see Chapter 1, Section III) for solutions with uniform segment distributions. This theory has been successfully applied to poly (vinyl chloride) in tetrahydrofuran (Lyngaae-Jorgensen, 1976).

VII DRUG-RECEPTOR MODELS: THE INTERACTION PHARMACOPHORE

The action of a drug is, at least in part, a consequence of the direct interaction between the drug and its receptor site. In the cases in which a drug acts as a competitive inhibitor with some natural compound, the relative strengths of ligand-receptor bindings are particularly critical to controlling the activity of the drug. Even though there is every indication that drug-receptor binding plays an important role in drug action, little work has been done to model such interactions. This is probably a consequence of the difficulty in defining, in a general way, those molecular properties which measure the nature of the interaction. Fragmental partition coefficients of constituent molecular groups, as discussed in Chapter 3, have been used in correlation with biologic activity. The idea behind such correlations is that hydrophobic groups and hydrophilic groups, respectively, in the drug and receptor will match up to maximize the binding energy. This same concept has been used by Hopfinger in some proprietary industrial studies using the hydration shell model in which the drug is partitioned into hydrophobic and hydrophilic sections. A binding energy can be computed for each partitioning. Initial results in which the binding energies of a particular partitioning over a congeneric series of compounds have been correlated with activity have yielded encouraging results.

Weinstein (Weinstein, 1975; Bartlett and Weinstein, 1975) has extended the basic concept of pharmacophore, which is defined as a pattern of molecular fragments (atoms or groups of atoms) that contribute to the specific interaction of a drug with a biologic receptor, to that of an "interaction pharmacophore." The interaction pharmacophore is the pattern of electrostatic potential generated by an active molecule in its preferred conformation. Weinstein et al. (1974) proposed that the electrostatic potential of an active compound could be studied as a template for the matching interactions with the receptor that determine the geometries of approach and sites of interaction in the early stages of receptor recognition. Thus regions of strongly negative potentials will indicate the fragments of the drug that would interact preferentially with positively charged sites in the receptor, whereas strongly positive regions protruding from the molecule should interact with negatively charged receptor sites. On this basis, Weinstein believes the structural requirements for the interaction with the receptor can be inferred from the potential map of active drugs. These maps, in turn, can then be used as criteria for the ability of different compounds, even those which lack any atom-to-atom resemblance, to mimic the reactive pattern required to match the receptor sites when the drugs are approaching the receptor. The ability to mimic the reactive pattern should enable the drugs to achieve the same orientation at the receptor as the original agonist and bring the approximate fragments to a position from which they can optimally interact with the corresponding receptor sites.

In Chapter 2, the gauche conformation of acetylcholine (Ach) for the O—C—C—N group was established as the active conformer when engaging the muscarinic receptor by noting the corresponding activity of 3-acetoxyquinuclidine (3-AcQ). Figure 14-10*a*,*b* illustrates the interaction pharmacophores of Ach and 3-AcQ, respectively. The high similarity of the interaction

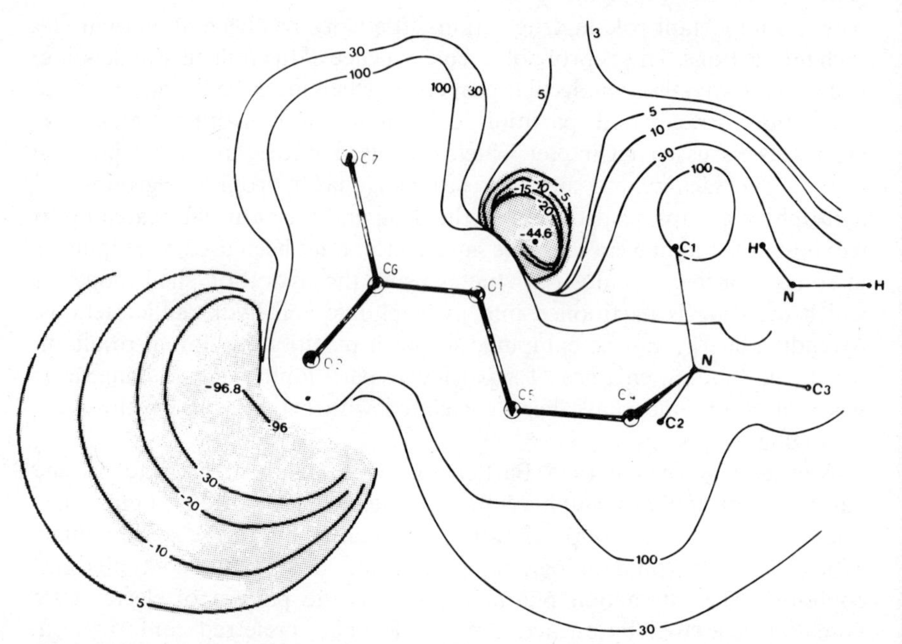

Fig. 14-10(*a*) Interaction pharmacophore of ACh represented by electrostatic potential generated by ACh in interaction with a simulated anionic site (NH_2).

Shaded areas represent regions of negative potential, attractive to positively charged groups. Weinstein et al. (1975).

pharmacophores is obvious and reflects, or accounts for, the similar biologic activities of these two molecules.

Weinstein et al. (1975) have suggested a sequential cholinergic binding sequence based on the interaction pharmacophores. The trimethylammonium cationic group first interacts with the anionic subsite of the receptor and thereby activates the negative region surrounding the ester oxygen. Around this atom, a region of negative potential is formed, which may then interact with the corresponding subsite. Loew et al. (1974) have applied the interaction pharmacophore concept to five morphine-like opiate narcotics. They were able to correlate their predicted extent of interaction with a receptor to available measures of analgesic agonism.

The obvious reservation to the interaction pharmacophore concept is the change in the electrostatic potential field about a drug because of the "bound" water on the drug. One would expect the amount (and location) of the bound water molecules to change (probably decrease) as the drug approaches the receptor site. Thus the interaction pharmacophore might be thought to vary with the distance between the drug and receptor, leading to a path of

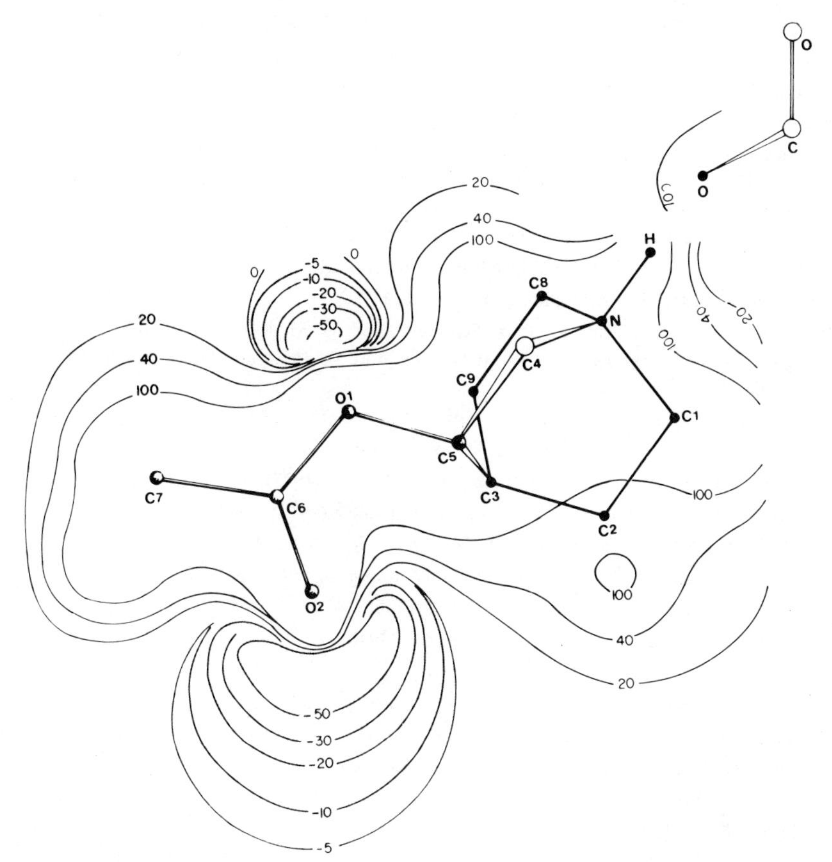

Fig. 14-10(*b*) Interaction pharmacophore of 3-AcQ. The anionic site is simulated by $HCOO^-$. Weinstein et al. (1975).

unique interaction pharamacophores. Pullman (Berthod and Pullman, 1975; Bonaccorsi et al., 1972; Giessner-Prette and Pullman, 1971; Bonaccorsi et al., 1972; Pullman, 1974) have also explored the electrostatic potential in free space about various groups and molecules. This work, however, has not specifically been directed toward explaining drug action.

REFERENCES

Alagona, G., A. Pullman, E. Scrocco, and J. Tomasi (1973). *Int. J. Peptide Protein Res.*, **5**, 251.

Armbruster, A. M. and A. Pullman (1974). *FEBS Lett.*, **49**, 18.

Avignon, M., C. Garrigow-La Grange, and P. Bothorel (1973). *Biopolymers*, **12**, 1651.

Baddiel, C. B., D. Chandhuri, and B. C. Stace (1971). *Biopolymers*, **10**, 1169.

Balasubramanian, D. and R. Schaikh (1973). *Biopolymers*, **12**, 1639.

Baron, M. H. and C. de Loze (1972). *J. Chem. Phys.*, **69**, 1084.

Bartlett, J. and H. Weinstein (1975). *Chem. Phys. Lett.*, **30**, 441.

Bergmann, E. D. and B. Pullman (1973). *Conformation of Biological Molecules and Polymers.* Academic Press, New York. (This volume contains numerous free-space/constant-dielectric calculations).

Bernal, J. D. and F. D. Fowler (1933). *J. Chem. Phys.*, **1**, 515.

Berthod, H. and A. Pullman (1975). *Chem. Phys. Lett.*, **32**, 233.

Beveridge, D. L. (1975). Personnal communication.

Beveridge, D. L., R. J. Radna, G. W. Schnuelle, and M. M. Kelly (1974). In *Molecular and Quantum Pharmacology* (E. D. Bergmann and B. Pullman, Eds.). Reidel-Dordecht, Holland, p. 153.

Beveridge, D. L. and G. W. Schnuelle (1974). *J. Phys. Chem.*, **78**, 2064.

Beveridge, D. L. and G. W. Schnuelle (1975). *J. Phys. Chem.*, **79**, 2652.

Bonaccorsi, R., A. Pullman, E. Scrocco, and J. Tomasi (1972). *Theoret. Chim. Acta*, **24**, 51.

Bonaccorsi, R., A. Pullman, E. Scrocco, and J. Tomasi (1972). *Chem. Phys. Lett.*, **12**, 622.

Brant, D. A. and P. J. Flory (1965). *J. Am. Chem. Soc.*, **87**, 2791.

Breuer, M. M. and M. G. Kennerley (1971). *J. Colloid Interface Sci.*, **37**, 124.

Brown, F. R., III, A. J. Hopfinger, and E. R. Blout (1972). *J. Mol. Biol.*, **63**, 101.

Copeland, D. A., N. R. Restner, and J. Jortner (1970). *J. Chem. Phys.*, **53**, 1189.

Crothers, D. M. (1971). *Biopolymers*, **10**, 2147.

Damle, V. N. (1970). *Biopolymers*, **9**, 353.

Farmer, B. L. and A. J. Hopfinger (1974). *Macromolecules*, **7**, 793.

Forsythe, K. H. and A. J. Hopfinger (1973). *Macromolecules*, **6**, 423.

Fraenkel, G. and J. P. Kim (1966). *J. Am. Chem. Soc.*, **88**, 4203.

Fueki, K., D. Feng, and L. Kevan (1973). *J. Am. Chem. Soc.*, **95**, 1398.

Gibson, K. D. and H. A. Scheraga (1967*a*). *Proc. Natl. Acad. Sci. USA*, **58**, 1317.

Gibson, K. D. and H. A. Scheraga (1967*b*). *Proc. Natl. Acad. Sci. USA*, **58**, 420.

Giessner-Prettre, C. and A. Pullman (1971). *C. R. Acad. Sci.* (*Paris*), **272C**, 750.

Hansch, C., J. E. Quinlan and G. L. Lawrence (1968). *J. Org. Chem.*, **33**, 347.

Hill, T. L. (1955). *J. Chem. Phys.*, **23**, 2270.

Hill, T. L. (1964). *Thermodynamics of Small Systems.* W. A. Benjamin, New York, part II, Chapt. 7.

Hill, T. L. (1973). *Biopolymers*, **12**, 257.

Hill, T. L. and Y. Chen (1973). *Biopolymers*, **12**, 1285.

Hilter, W. A., A. J. Hopfinger, and A. G. Walton (1972). *J. Am. Chem. Soc.*, **94**, 4324.

Hnojewyj, W. S. and L. H. Reyerson (1963). *J. Phys. Chem.*, **67**, 711.

Hopfinger, A. J. (1971). *Macromolecules*, **4**, 731.

Hopfinger, A. J. (1973). *Conformational Properties of Macromolecules.* Academic Press, New York.

Hopfinger, A. J. (1974*a*). In *Peptides, Polypeptides and Proteins* (E. R. Blout, F. A. Bovey, M. Goodman and N. Lotan, Eds.). Wiley-Interscience, New York, p. 71.

Hopfinger, A. J. (1974*b*). *J. Macromol. Sci.-Phys.*, **B9(3)**, 483.

Hopfinger, A. J. and R. D. Battershell (1976). *J. Med. Chem.*, **19**, 569.

Hylton, J., R. E. Christoffersen, and G. Hall (1974). *Chem. Phys. Letters*, (*a*) **24**, 501; (*b*) **26**, 501.

Kim, S. H. and B. T. Rubin (1973). *J. Phys. Chem.*, **77**, 1245.

Kirkwood, J. G. (1934). *J. Chem. Phys.*, **2**, 351.

Kirkwood, J. G. and F. H. Westheimer (1938). *J. Chem. Phys.*, **6**, 506.

Kostetsky, P. V., V. T. Ivanov, A. Yu, A. Ovchinnikov, and G. Schembelort (1973). *FEBS Lett.*, **30**, 205.

Krimm, S. and C. M. Venkatacholam (1971). *Proc. Natl. Acad. Sci. USA*, **68**, 2468.

Kuntz, I. D. (1971). *J. Am. Chem. Soc.*, **93**, 514.

Lipkind, G. M., C. F. Arkhipova, and E. M. Popov (1970). *Strukt. Khim.*, **11**, 121.

Loew, G. H., D. Berkowitz, H. Weinstein, and S. Sregrenik (1974). In *Molecular and Quantum Pharmacology* (E. D. Bergmann and B. Pullman, Eds.). Reidel-Dordecht, Holland, p. 355.

Lyngaal-Jorgensen, J. (1976). *J. Phys. Chem.*, (in press).

Marynick, D. S. and H. F. Schaeffer, III (1975). *Proc. Natl. Acad. Sci. USA*, **72**, 3794.

McCreery, J. H., R. E. Christoffersen, and G. G. Hall (1976*a,b*). *J. Amer. Chem. Soc.* (submitted).

Moskowitz, J. W., M. Boring, and J. Wood (1976). *J. Chem. Phys.* (in press).

Nemethy. G. and H. A. Scheraga (1962). *J. Chem. Phys.*, **36**, 3401.

Newton, M. D. (1973), *J. Chem. Phys.*, **58**, 5833.

Ooi, T., R. A. Scott, G. Vanderkooi, and H. A. Scheraga (1967). *J. Chem. Phys.*, **46**, 4410.

Perricaudet, these 3e' cycle, Paris (1973).

Perricaudet, M. and A. Pullman (1973). *Int. J. Peptide Protein Res.*, **5**, 99.

Port, G. N. J. and A. Pullman (1973*a*). *FEBS Lett.*, **31**.

Port, G. N. J. and A. Pullman (1973*b*). *Theort. Chem. Acta*, **31**, 231.

Port, G. N. J. and A. Pullman (1974). *Int. J. Quantum Chem.* (in press).

Pullman, A. (1974). *Int. J. Quantum Chem.*, **1**, 33.

Pullman, A. (1974). *The Purines: Theory and Experiment, Jerusalem Symposia* (E. D. Bergmann and B. Pullman, Eds.) Academic Press, New York, Vol. 6, p. 1.

Pullman, A., G. Alagona, and J. Tomasi (1974). *Theort. Chim. Acta*, **33**, 87.

Pullman, A. and A. M. Armbruster (1974). *Int. J. Quantum Chem.*, **1**, 58.

Pullman, B. and P. Courrière (1972). *Mol. Pharmacol.*, **8**, 612.

Pullman, B. and G. N. J. Port (1974). *Mol. Pharmacol.*, **10**, 360.

Pullman, A. and B. Pullman (1975). *Quart. Rev. Biophys.*, **7**, 505.

Rein, R., S. Nir, T. J. Swissler, and V. Renugopalakrishnan (1976). *Int. J. Quantum Chem.* (in press).

Rein, R. and V. Renugopalakrishnan (1976). *Biochim. Biophys. Acta* (submitted).

Schnuelle, G. W. and D. L. Beveridge (1975). *J. Phys. Chem.*, **79**, 2566.

Schnuelle, G. W. and D. L. Beveridge (1976). *J. Phys. Chem.* (in press).

Shindo, Y. (1971). *Biopolymers*, **10**, 1081.

Sinanoglu, O. (1974). In *The World of Quantum Chemistry*. (R. Daudel and B. Pullman) Reidel-Dordecht, Holland, p. 265.

Stearn, H. and H. Exrina (1937). *J. Chem. Phys.*, **5**, 113.

Subramanian, S. and H. F. Fischer (1972). *Biopolymers*, **11**, 1305.

Tiffany, M. L. and S. Krimm (1968). *Biopolymers*, **6**, 1767.

Tiffany, M. L. and S. Krimm (1969). *Biopolymers*, **8**, 347.

Tute, M. S. (1971). In *Advances in Drug Research* (N. J. Harper and A. B. Simmons, Eds.). Academic Press, New York, Vol. 6, p. 1.

Venkatachalam, C. M. and S. Krimm (1973). In *Conformation of Biological Molecules and Polymers* (E. D. Bergmann and B. Pullman, Eds.). Academic Press, New York, p. 141.

Watt, I. C. and J. D. Leeder (1968). *J. Text. Ins.*, **59**, 353.

Weinstein H. (1975). *Int. J. Quantum Chem.*, **2**, 59.

Weinstein, H., S. Maayani, S. Srebrenik, S. Cohen, and M. Sokolovsky (1975). *Mol. Pharmacol.*, **11**, 671.

Weinstein H., S. Srebrenik, R. Pauncz, S. Maayani, S. Cohen, and M. Sokolovsky (1974). In *Chemical and Biochemical Reactivity* (E. D. Bergmann and B. Pullman, Eds.). Academic Press, New York, p. 493.

Weintraub, H. J. R. and A. J. Hopfinger (1973). *J. Theort. Biol.*, **41**, 53.

Weintraub, H. J. R. and A. J. Hopfinger (1974). In *Molecular and Quantum Pharmacology*, (E. D. Bergmann and B. Pullman, Eds.). Reidel-Dordecht, Holland, p. 131.

Weintraub, H. J. R. and A. J. Hopfinger (1976). *J. Theort. Biol.* (in press).

Wuepper, J. L. and A. I. Popov (1970). *J. Am. Chem. Soc.*, **92**, 1493.

CHAPTER 15

Ultrastructural Organization

Our structural understanding of a few associating biologic systems has reached a point at which we can describe the molecular arrangement responsible for the observed shape of the composite ultrastructure. The structural understanding of a couple of self-associating, inter-reactive units has allowed us to meaningfully speculate on structure-activity mechanisms of multi-macromolecular systems. This chapter focuses on some of these systems in the hope of reflecting how much we have learned about biomolecular structure, and how very much more still remains unknown.

I MOLECULAR AND ULTRASTRUCTURAL ORGANIZATION OF MYOSIN

A General Structural Features

Myosin is one of the few proteins known at present that has two distinct domains that are covalently linked: one a typical globular conformation with active sites for binding nucleotide and actin, and the other a highly α-helical conformation serving a purely structural role.

All the hydrodynamic properties of myosin indicate a long, thin molecule resembling a rigid rod (Holtzer and Lowey, 1956; 1959). Further evidence for this picture of the solvated molecule is provided by light-scattering data (Holtzer and Lowey, 1959; Holtzer et al., 1962). However, as early as 1957, it was recognized that myosin is not a uniform rod; the curvature in the light-scattering envelope at high angles was best accounted for by assuming a nonuniform particle (Holtzer and Rice, 1957). The conformation and hydrodynamic properties of the tryptic fragments of myosin suggested a rod-like region attached to more globular entities (Cohen, 1961).

The large size and unusual shape of myosin make it a particularly difficult molecule to study by conventional techniques. One way to overcome this restriction has been to selectively degrade myosin into smaller units using proteolytic enzymes. Trypsin (and chymotrypsin) are able to hydrolyze myosin into two types of large fragments called light meromyosin (LMM) and heavy meromyosin (HMM) in reference to their respective molecular weights.

Optical rotatory dispersion (ORD) studies of the meromyosins show that LMM has greater than 90% α helix as compared to about 45% α helix for the HMM (Cohen and Szent-Györgyi, 1957). The wide-angle X-ray diagram for LMM, coupled with simultaneous measurements of the weight and length of several LMM preparations, has led to the conclusion that LMM consists of approximately two polypeptide chains per molecule that adopt a coiled-coil α-helical structure (Holtzer et al., 1962; Lowey and Cohen, 1962). The two α-helical chains of LMM are distinguished from those of other fibrous proteins by an unusually high proportion of polar amino acid residues (Lowey et al. 1969).

Because of its lower helix content, HMM is a much more difficult molecule to analyze than LMM. HMM resembles a typical globular protein in terms of solubility, amino acid composition, and optical rotatory properties, although its viscosity is about ten-fold higher than most enzymes. A double strandedness by direct observation of two subunits in the globular region of myosin and HMM (Slayter and Lowey, 1967; Lowey et al., 1969) has confirmed indirect physico chemical evidence for this organization (Cohen and

Szent-Györgyi, 1960; Lowey and Holtzer, 1959). An excess of trypsin can be used to convert HMM into a less asymmetric molecule, called heavy meromyosin subfragment-1 (HMM S-1) plus large amounts of poorly defined low-molecular-weight components and dialyzable peptides (Mueller and Perry, 1962). Using alternate degradation procedures, the lower-molecular-weight components can be recovered intact and are seen to be part of the coiled-coil α helix that joins the LMM to the (HMM S-1). This subfragment is logically labeled (HMM S-2).

Table 15-1 contains the average length of myosin and several of its subfragments. The overall molecular structure of myosin is summarized in Figure 15-1, which illustrates the position and relative size of the various subfragments. Note that, in the globular region (HMM S-1), the presence of additional subunits is indicated. The following discussion centers around the structural organization of these unusual species.

Table 15-1 Average Lengths for Myosin and Its Subfragments[a, b]

Preparation	Number counted	Peak Length (Å)	Number-average Length (Å)	Weight-average Length (Å)
Rod in myosin	493	1370	1340	1382
Isolated rod	306	1370	1360	1386
Rod in HMM	400	540	528	571
HMM S-2	549	460	474	500
LMM + rod	530	820	1045	1200
LMM	628	730	785	870
Single globules in myosin	216	114	117	121
HMM S-1	210	92	95	89

[a] Lowey et al., 1969.
[b] The averages given in this table were all calculated from the data represented in the histograms. Approximately 25 Å should be subtracted from these dimensions to correct for the accumulation of metal on the molecules during the replication process.

B Noncovalent Subunits in the Myosin Molecule

It had been recognized for many years that myosin exposed to strong denaturing conditions (see Lowey, 1971; Tonomura, 1973; Bourne, 1974 for

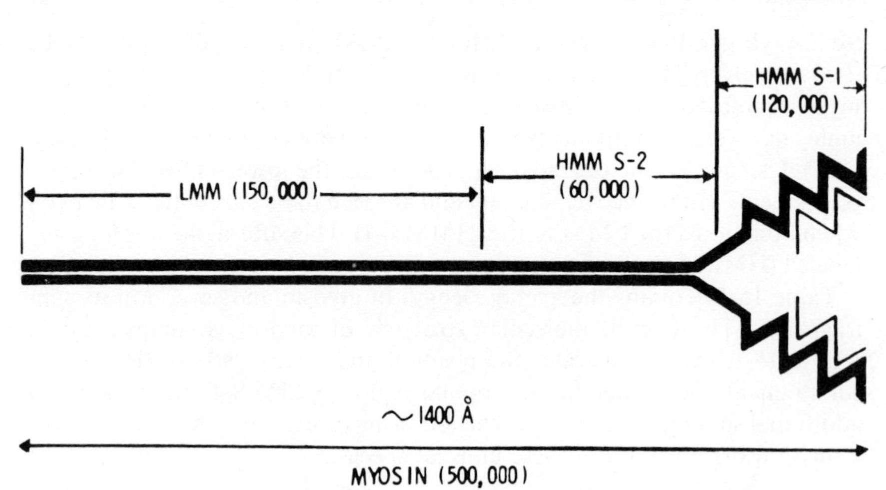

Fig. 15-1 Schematic representation of the myosin molecule. (Reprinted from Lowey et al., 1969, by courtesy of Academic Press, Inc.)

reviews) liberates small amounts of low-molecular-weight protein. It was generally assumed that this minor component was a contaminant. However, the persistence of these small proteins in more elaborate purification procedures made it increasingly difficult to dismiss them as impurities. These noncovalently bound, low-molecular-weight fractions of myosin are referred to as "light chains." Three types of light chains have been identified: Al (M.W. 21,000), A2 (M.W. 17,000), and DTNB (M.W. 19,000). The identification coding is discussed by Lowey (1971).

Ever since the discovery of the light chains of myosin, the problem of relating their structural organization in myosin (and their function in the action of myosin) has existed. The liberation of the A1 and A2 chains is always accompanied by the simultaneous denaturation of myosin. The addition of ATP simultaneously slows the denaturation process and the release of light chains. This suggests that the light chains are perhaps located in the region of the hydrolytic site (Stracher, 1969; Dreizen, 1970). Such a structure would be similar to that found in antibodies, in which the antigen combining site is composed of portions of the light and heavy chains of the γ-globulin molecule (Cohen and Porter, 1964). An ATP analog, which can act as a substrate for myosin, can be bound to the light chains under conditions where hydrolysis is inhibited (Murphy and Morales, 1970). This is additional evidence that the light chains are in the vicinity of the catalytic site. Experiments in which the light chains are separated from the heavy chains in myosin have led to some conflicting results with respect to the integral role of the light chains in activity (Gaetzens et al., 1968; Gershman et al. 1969; Frederiksen and Holtzer, 1968; Dreizen and Gershman, 1970). Nevertheless, it has been shown that, when light chains are separated from heavy chains in some experiments (Stracher, 1969; Dreizen and Gershman, 1970), no activity is detectable in the isolated components. Recombination of components, however, restores a significant portion of the activity. There is also evidence (Sarkar and Cooke, 1970) that the light and heavy chains are synthesized by two different classes of polysomes, with the light chains consequently acting as regulatory proteins in myosin generation.

C Aggregation and Assembly of Myosin Molecules

Myosin in its native environment is organized in a highly specific, cooperative manner, resulting in the functional myosin filament. The molecular structure of myosin has evolved so as to aggregate into a filament assembly. The various solubilities among the subfragments are directly related to the functional role of each region of the myosin molecule. The LMM region is the least soluble; hence it is primarily responsible for anchoring myosin in the case of the thick filament. HMM S-2 is much more soluble than LMM and, correspondingly, has weaker interactions with the backbone of the myosin filament. HMM S-1 is highly water soluble. Its interactions are probably limited to the actin-containing thin filament.

An aqueous environment is essential to the assembly of the thick filament. *PH* and ionic strength are critical in regulating the growth of the thick filament. All the information required to build a thick filament is contained in the myosin molecule, and it is unlikely that another protein(s) participates in the assembly. In other words, myosin seems to aggregate by a self-assembly process.

Josephs and Harrington (1966, 1968*a*,*b*) have estimated the thermodynamic parameters for the formation of myosin filaments based on a rapidly reversible equilibrium between monomer and polymer. Table 15-2 contains a summary of the thermodynamic constants for the association of myosin. One of the major findings from the work of Josephs and Harrington (1968*b*) is the important role played by water in stabilizing the structure of the myosin filament. The myosin molecule has a high percentage of polar residues. As seen in many examples throughout this text, the polar residues most likely lie on the surface of the protein binding to water molecules. As myosin aggregates, protein–protein interactions replace protein–water interactions, and water is released to the extent of about 400 ml per mole of myosin. Such volume changes are quite common in associating systems; for example,

Table 15-2 Thermodynamic Parameters for the Formation of Myosin Filaments[a, b]

Structure	ΔF^{c} kcal	ΔH kcal	ΔS eu	ΔV ml	Effect of electrolytes
Per monomer	−2.2	0	+7.8	+384	Destabilizes
Per polymer	−180	0	650	3.2×10^4	Destabilizes

[a] Reprinted from W. H. Harrington and R. Josephs (1968*b*) by courtesy of Academic Press, Inc.
[b] For the reaction n monomer $\rightleftharpoons$ polymer, where $n = 83$.
[c] At 278°K.

TMV is dehydrated by about 110 ml per subunit (Stevens and Lauffer, 1965) and each G-actin loses 68 ml of water in forming F-actin (Ikkai and Ooi, 1966). The reduced rotational and translational freedom of the molecule in the polymer may be more than compensated for by the increased disorder in the released water, leading to a net increase in entropy, as shown in Table 15-2.

Overall, the filaments of vertebrate skeletal muscle are spindle-shaped objects about 1.6 μ long and 150 Å thick (Huxley, 1963). They show a smooth "bare zone" in the middle of the filament with a rough surface extending away from the center toward either end. It is apparent from the shape of the myosin molecule and the filament that the latter is built by antiparallel aggregation of the myosin monomers, with the rod portion of the molecule forming the base region as shown in Figure 15-2. X-ray diffraction (Huxley and Brown, 1967) and electron microscopy (Huxley, 1957) studies suggest that the myosin molecules in the filament are arranged in the form of a "6/2 helix" with a pitch of 429 Å and a subunit repeat of 143 Å. In other words, the "cross bridges" projecting from the filament are arranged in pairs related by a two-fold axis such that each pair must be rotated 120° and translated 143 Å to generate the helix (see Figure 15-3*a*). The cross bridges probably consist of the HMM S-1 and a portion of HMM S-2. The extent of HMM S-2 contribution to the cross bridges depends on the state of the muscle. Low-angle, equatorial X-ray reflections from striated muscles arise from the hexagonal arrangements of the thin and thick filaments of the myofibril, as shown in Figure 15-3*b*. The relative intensities of these reflections are related to the distribution of mass associated with the filaments. A comparison of the X-ray patterns obtained from relaxed muscles with those from muscles in rigor indicates that a considerable amount of protein passes from the myosin filaments to the actin filaments in going from one state to the other (Huxley, 1968).

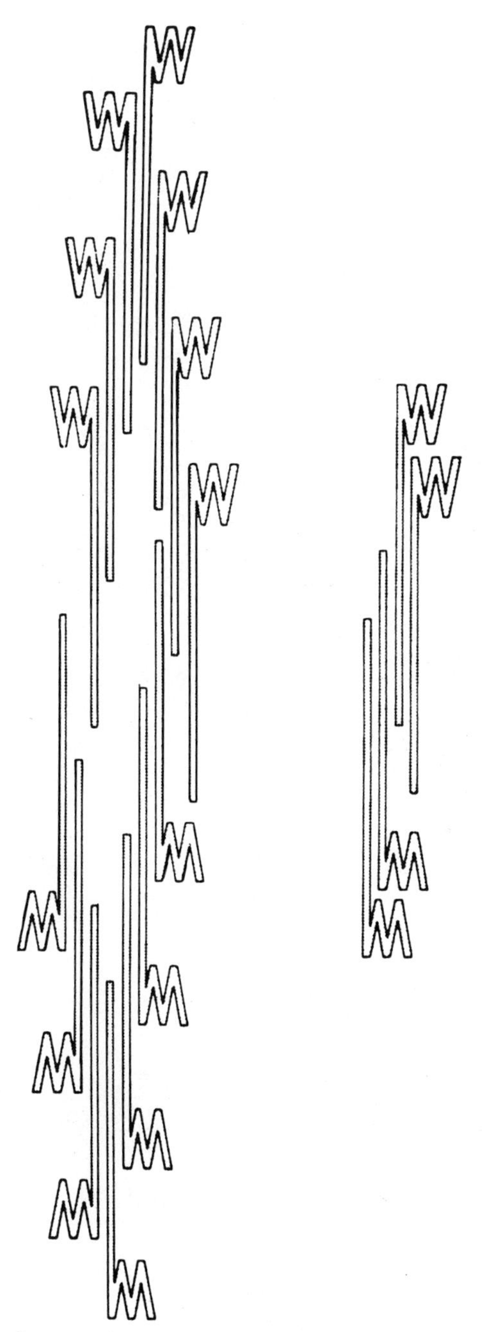

Fig. 15-2 Schematic diagram of myosin molecules illustrating antiparallel growth to form filaments of increasing length. [Reprinted from H. E. Huxley (1965) by courtesy of W. H. Freeman and Co.]

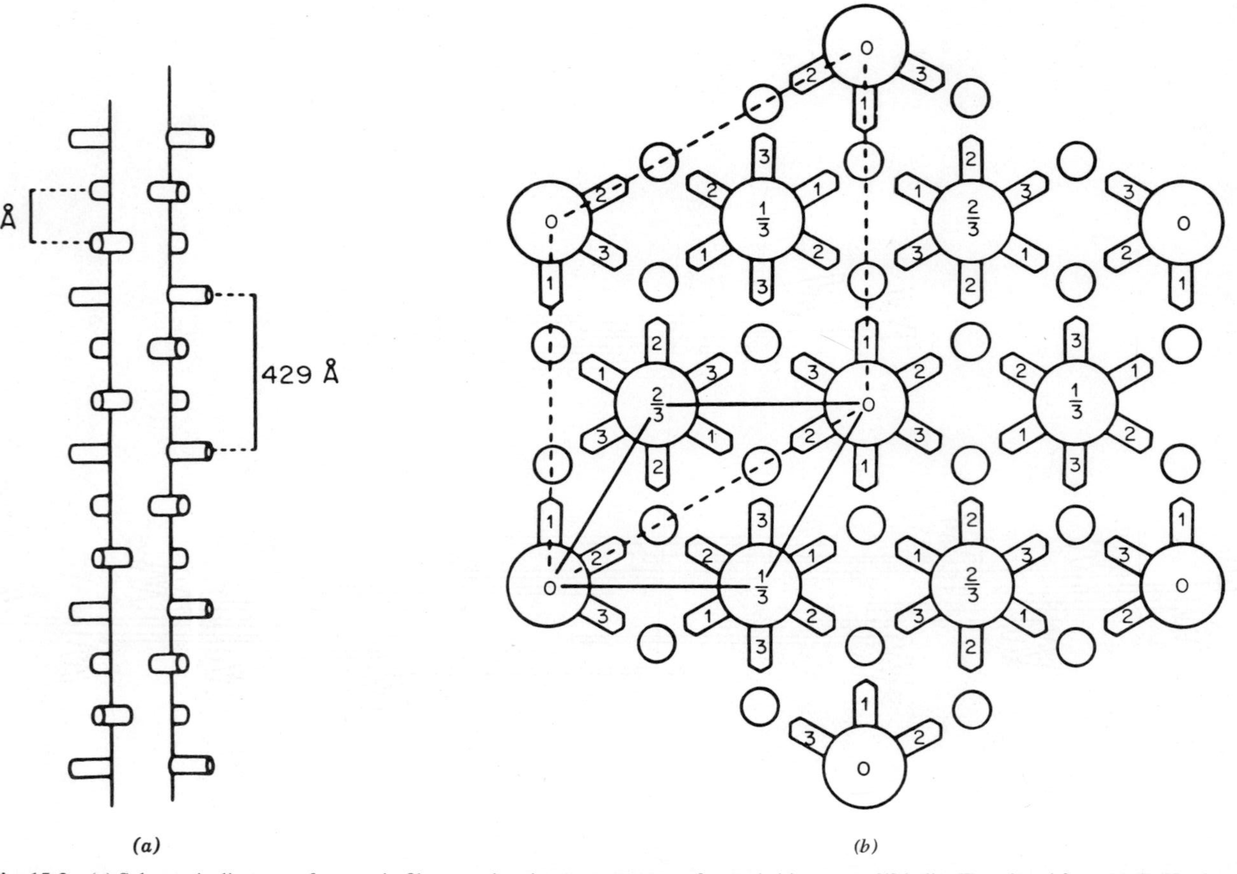

Fig. 15-3 (*a*) Schematic diagram of a myosin filament showing arrangement of cross bridges on a 6/2 helix. [Reprinted from H. E. Huxley and W. Brown (1967) by courtesy of Academic Press, Inc.]. (*b*) Arrangement of filaments in a hexagonal lattice. [Reprinted from H. E. Huxley and W. Brown (1967) by courtesy of Academic Press, Inc.]

II UNIT FIBRIL STRUCTURE OF COLLAGEN AGGREGATES

It has been known for some time (Hodge, 1967) from the electron microscopy of collagen aggregates that molecules in the fibril are related to each other by a stagger of axial translation of multiples of about 670 Å. This value, referred to as D, is the length of a period seen by electron microscopy and X-ray diffraction of collagen fibrils. The collagen molecule has a length of about 4.4D; thus any two overlapping molecules in an ordered region might be related by a stagger of 0D, 1D, 2D, 3D, or 4D. The nonintegral length of the molecule gives rise to "overlap" and "hole" regions of 0.4D and 0.6D, respectively. This structure is shown schematically in Figure 15-4. X-ray diffraction studies by Miller and Wray (1971) and Miller and Parry (1973) have helped to elucidate the way in which collagen molecules in rat tail tendon are packed and give strong support to the concept of specific molecular interactions. Their data suggest that the fibril is a square lattice of five-stranded microfibrils of the type proposed by Smith (1968).

Segrest and Cunningham (1973), using certain assumptions involving the macromolecular structure of collagen, have proposed a model for the surface topography of the monomeric unit. Two possible models for the molecular packing, one hexagonal and the other nonhexagonal, have been inferred. The nonhexagonal packing model is identical to the pentagonal unit fibril first postulated by Smith (1968), whereas the hexagonal model is one of three previously suggested by Segrest and Cunningham (1971).

The hexagonal and the nonhexagonal models share the following attributes: (1) They are consistent with electron microscopic data (Petruska and Hodge, 1964; Kuhn and Zimmerman, 1965). (2) They both are based on the accepted macromolecular structure of the collagen molecules (see Chapters 4 and 5). (3) Both models provide an explanation for the 2/1 ratio of $\alpha 2$ chains in many collagens. (4) They allow a maximum number of quarter-stagger

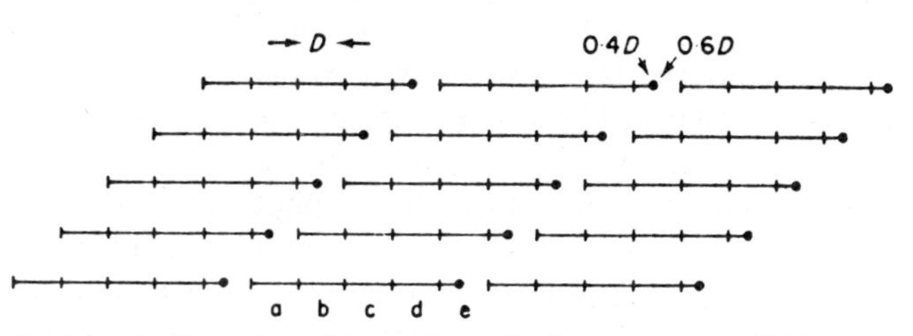

Fig. 15-4 A schematic illustration of the packing of collagen molecules. If the above structure is rolled into a cylinder, the five-stranded microfibril proposed by Smith (1968) is produced. Hulmes et al. (1973).

contacts (Veis et al., 1967), both compatible with the geometrical properties of staggered rods. (5) Finally, both models are sterically compatible with the reported location of hydroxylysine-linked dissaccharide prosthetic groups in skin collagens (Morgan et al., 1970). The nonhexagonal model, shown schematically in Figure 15-5, appears to have greater tensile strength, since it provides the maximum amount of overlap between longitudinally adjacent groups of collagen molecules (Smith, 1968), at least in the absence of covalent intermolecular crosslinking. A second attribute of this model is that a probable natural angle of 108° between $\alpha 1$ chains generates the model proposed by Smith (1968) on other grounds and supported by the X-ray diffraction studies of Miller and Wray (1971) and Miller and Parry (1973). Finally, this model provides a simple evolutionary scheme for development of the respective portion of the native banded pattern of collagen. For these reasons, Segrest and Cunningham (1973) agree with Miller and co-workers about a five-stranded microfibril model similar to the type proposed by Smith (1968) for collagen aggregates.

Hulmes et al. (1973) have applied a relatively unique computer analysis to the amino acid sequence of the $\alpha 1$ chain of collagen in the search of complementary relationships that would explain the stagger of multiples of 670 Å between the rod-like molecules in the fibril. In essence, the sequence of 1011 amino acids was moved passed itself while scoring for complementarity between opposing interchain amino acids on the basis of charge and hydrophobic interactions subject to the constraints imposed by supercoiling. It was found that the interactions are optimized when the chains are staggered by 0D, 1D, 2D, 3D, and 4D, where $D = 234 \pm 1$ residues. The residue repeat derived from this value is 2.86 ± 0.02 Å. The existence of a D separation between interacting residues was shown to be reflected in the actual distribution of large hydrophobic amino acids. Surprisingly, the distribution approximates the pattern $(2D/11)_5$ $(D/11)$ repeated over 4.4D intervals (Hulmes et al., 1973). The distribution of charged residues is less regular and does not show a well-defined periodicity. See Figure 15-6 for the relationship between the periodicity of the hydrophobic interactions and its idealized repeat pattern. The absence of a well-defined periodicity in the distribution of charged residues is perhaps a consequence of the fact that charged groups may be paired with solvent ions as well as with each other and are less restricted in position than hydrophobic groups, since charged interactions take place at the ends of long side chains. Furthermore, the tendency of charges of opposite sign to be located near each other in the sequence not only favours a 0D alignment (which might otherwise be unlikely because of charge repulsion), but also suggests that intramolecular charge interactions are an important feature of collagen structure.

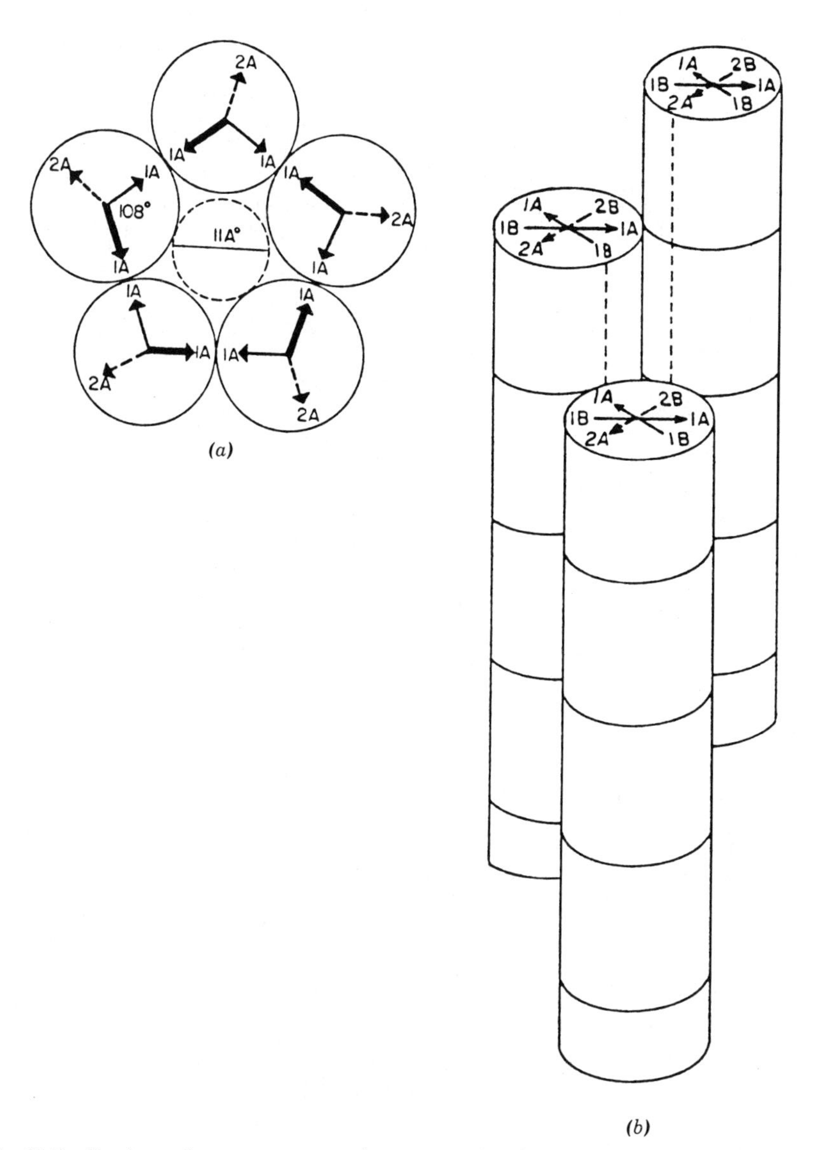

Fig. 15-5 Continuously quarter-staggered, pentagonal unit fibril produced by α 1-L(A) to 1-R(A) interactions. The edge of each α 1-L association edge is designated by a dark arrow and, the A edge of each α 1-R association edge by a light arrow. Segrest et al. (1973).

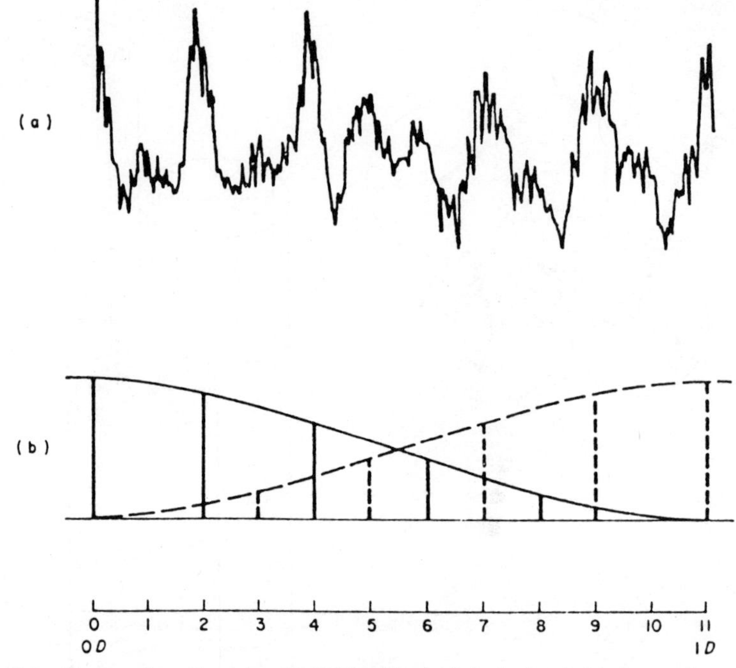

Fig. 15-6 (*a*) An enlarged portion (O-234) residues) of the hydrophobic interaction curve by computer analysis. (*b*) An idealized representation of the curve showing the pattern of which it is composed. Hulmes et al. (1973).

III PROTEIN–PROTEIN INTERACTIONS (QUATERNARY STRUCTURE)

Most proteins organize themselves as aggregates. Most often such aggregates contain identical subunits. Matthews and Bernhard (1973) have tabulated the number of subunits per molecule in a large number of examples.

In a few instances, detailed X-ray diffraction studies of oligomeric proteins have been carried out. Table 15-3 contains a breakdown of the types of pairwise interactions between subunits observed in these oligomeric proteins. The frequency of the types of interactions is surprisingly similar to that which takes place within a typical single subunit. In so far as the van der Waal's contact interactions are generally of a hydrophobic nature, intersubunit organization might be considered to be principally due to hydrophobic interactions. It is reasonable to envision subunit aggregation being driven by a hydrophobic driving force, with the particular sites of interfacial

Table 15-3 Chemical Nature of Subunit Interaction

Protein	Symmetry	Contact	van der Waal's Contacts	H bonds	Ion Pairs
α-Chymotrypsin	2	A	443	9	1
		B	57	6	
Concanavalin A	222	‖ to A		2	
		‖ to B	142	14	6
		‖ to C	174	14	
Hemoglobin					
(oxy)	2	$\alpha_1\beta_1$	110	5	
		$\alpha_1\beta_2$	80	1	
		$\alpha_1\alpha_2$			
		$\beta_1\beta_2$			
(deoxy)	2	$\alpha_1\beta_1$	98	5	
		$\alpha_1\beta_2$	69	1	1
		$\alpha_1\alpha_2$			2
		$\beta_1\beta_2$			1
Insulin	32	OP	111	8	
		OQ	99	2	1

Liljas and Rossmann, 1974.

aggregation a consequence of complementary fit generated through a combination of steric (as reflected by observed symmetries), hydrogen-bonding, and ionic pair interactions. From Table 15-3, it is relatively clear that the symmetries 2,222, and 32 are much preferred to the point groups 3, 4, 5 . . . . The requirements for high symmetry in the subunit organization of viruses, in terms of function, have been discussed by Monod et al. (1965) and Koshland et al. (1966).

Some general patterns, based on secondary chain structures, for subunit interactions have started to emerge. In concanavalin A, glyceraldehyde-3-phosphate dehydrogenase and, to some extent at least, lactate dehydrogenase, a subunit interaction involving pleated sheet structure is seen (Liljas and Rossmann, 1974). When one surface of the sheet constitutes the subunit boundary, a twofold related sheet can be stacked on top of it. Generally, extended main chains in globular proteins are mostly hidden so as not to be susceptible to proteolytic attack, which in subtilisin BPN′ (Robertus et al., 1972), in γ-chymotrypsin (Segal and Powers, 1971), or in the trypsin–trypsin inhibitor complex (Rühlmann et al., 1973), has been demonstrated to involve the temporary formation of a short antiparallel β structure. Dimerization of the protein through the formation of a continuous β structure between subunits as in concanavalin A (Edelman et al., 1972; Hardman and Ainsworth, 1972), insulin (Blundell et al., 1972), and alcohol dehydrogenase (Brändén et al., 1973) is one means of protecting exposed subunit extended chains from proteolytic attack.

The subunits in an associated complex of protein subunits are not always identical or even participating in the same set of functional reactions. There are organized multienzyme complexes in which the products of one reaction step become substrates for the next. Thus it is functionally advantageous to have the enzymes in close and ordered array. An example of such a system in which the structural organization of the complex has been reasonably well identified by electron microscopy is the pyruvate dehydrogenase complex (Reed and Oliver, 1968). Figure 15-7 shows a sequential set of schematic drawings that illustrate the various modes of aggregation in this complex. The total molecular weight of the entire complex is 4,440,000. The core of the complex is a cube in which each of the eight subunit corners contains three individual chains, each of molecular weight 40,000 (see Figure 15-7*a*). The six faces of the cube are each covered by four identical subunits of molecular weight 55,000 that are different from the subunits of the E_2 cube (see Figure 15-7*b*). Each of the 12 edges of the core cube are in contact with two identical subunits of molecular weight 90,000 (see Figure 15-7*c*). These

The E_2 cube of eight trimers. each gray cube is 3 subunits.

(*a*)

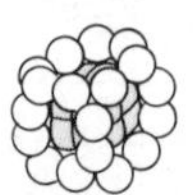

The E_2cube with 24 molecules of E_1(white balls) on the cube edges.

(*c*)

The E_2 cube with 24 molecules of E_3 stippled cubes on the cube faces.

(*b*)

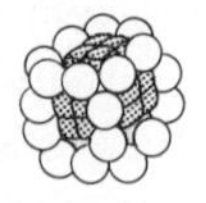

Organization of the total complex: E_1, E_2 ,(inside), and E_3.

(*d*)

Fig. 15-7 Aggregrate assembly of the pyruvate dehydrogenase complex.

subunits are dimers of two different polypeptide chains of similar size. The core cube is not only the foundation, but the organizer of the total complex. Formation of the core cube is spontaneous, and the remaining subunits complex to this cubic core rather than one another. Aspartate transcarbamylase complex formation also involves nonidentical subunits. Aggregation is highly symmetric, possessing 32 point group symmetry. The creation of a central cavity is similar to that of a small spherical virus (Warren et al., 1973).

Similar secondary, or even tertiary binding structures in congeneric sets of proteins do not necessarily lead to similar subunit organizations. This is demonstrated in the dehydrogenases, in which, despite the similarity of the coenzyme binding structures, the subunit interactions involving the binding structures are different in the various enzymes. In alcohol dehydrogenase, the twofold axis in the dimer is located next to the BF strand, resulting in a 12-stranded sheet structure through the dimer. In glyceraldehyde-3-phosphate dehydrogenase and lactate dehydrogenase, the BF strand is in the inside of one subunit. In turn, the coenzyme sites are on the outside of lactate dehydrogenase, whereas they are close to the subunit interfaces in glyceraldehyde-3-phosphate dehydrogenase. The similarity of tertiary structures and differences in subunit aggregation in the dehydrogenases suggests that aggregation organization is a more recent evolutionary event than formation of specific tertiary structure.

Specific functional activity of protein complexes may be the result of deviations from high symmetry as observed in several systems. Significant deviations from high symmetry have been observed in the interactions of nonidentical protein subunits of similar structure. Twofold translation axes are found in hemoglobin (Muirhead et al., 1967). Two dimers in a Bence-Jones protein complex are related through a 60° rotation (Schiffer et al., 1974). Smaller asymmetries are generally observed in aggregates of like subunits. Examples of asymmetry in oligomeric protein complexes are given by insulin (Blundell et al., 1972), chymotrypsin (Tulinsky et al., 1974), concanavalin A (Jack et al., 1971), and hexokinase (Steitz et al., 1973).

Adams et al. (1972) have devised a useful method for describing subunit interactions. Each subunit is portrayed on a two-dimensional surface with as many straight edges as their are subunit contacts. The numbers of individual residues involved in interfacial interactions for each subunit are contained in small boxes which, in turn, are localized in areas delineating particular subunits. Dark solid curves between the boxes of a subunit indicate the topological connectivity of the polypeptide chain. Lines drawn across interfacial edges between residue boxes define the main chain interactions. Such a subunit interaction diagram for dogfish-M4-lactate dehydrogenase is shown in Figure 15-8.

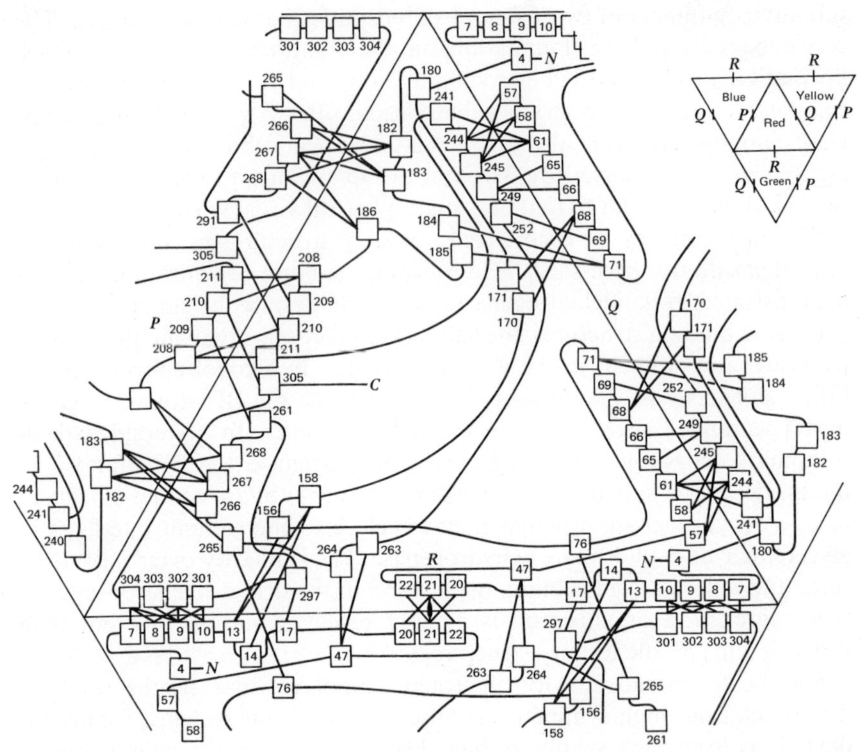

Fig. 15-8 Dogfish M_4 LDH main chain interactions between different subunits ($C\alpha - C\alpha \leqq 6$ Å). Diagrammatic representation of contacts between α-carbon atoms in neighboring subunits. Each subunit is also given a color code: red, yellow, green, and blue. Thus the red subunit contacts the other three by virtue of the Q, R, and P axes, respectively. Adams et al. (1972).

IV PROTEIN–NUCLEIC ACID INTERACTIONS

Interactions between proteins and nuclei acids play an important role in cell chemistry. Establishment of the stereochemistry of these interactions at the atomic level is essential to identifying any set of structure-function relationships. Various nucleotides and nucleosides have been investigated when bound to alcohol dehydrogenase (Brändén, 1973), flavodoxin (Waterpaugh et al., 1972), glyceraldehyde-3-phosphate dehydrogenase (Buehner et al., 1973), lactate dehydrogenase (Adams et al., 1973; Chandrasekhar et al., 1973), malate dehydrogenase (Webb et al., 1974), staphylococcal nuclease (Cotton et al., 1971), phosphoglycerate kinase (Drenth et al., 1971), and ribonuclease-S (Richards et al., 1971; Roberts et al., 1969). The binding of deoxyguanosine to actinomycin can also be considered in this class of interactions (Sobell et al., 1971). Unfortunately, in only the last study has it been possible to determine accurate complex conformations.

Analysis of the dihedral angles of nucleotide structures (Sundaralingam, 1969; Arnott and Hukins, 1969; Arnott and Hukins, 1972; Kim et al., 1973) shows that these are fairly constrained to limited ranges. Furthermore, these ranges are not significantly altered whether a mononucleotide is subject to crystal lattice forces or a polynucleotide is held in a helical form by the π electrons between base pairs. Thus a working assumption is to consider the nucleic acid molecule rigid and the protein flexible in the formation of the protein–nucleic acid complex. This is the case for an actinomycin–deoxyguanosine crystalline complex (Sobell et al., 1971). It is also consistent with NMR results on mononucleotides (Barry et al., 1971).

Imposing the constraint of nucleic acid rigidity in conjunction with electron density maps have led to sets of reasonable dihedral angles for the coenzyme structure of lactate dehydrogenase (Chandrasekhar et al., 1973) and malate dehydrogenase (Webb et al., 1974). Figure 15-9 diagrammatically shows the interactions between the whole of the nicotinamide aderine dinucleotide coenzyme and lactate dehydrogenase in a ternary complex.

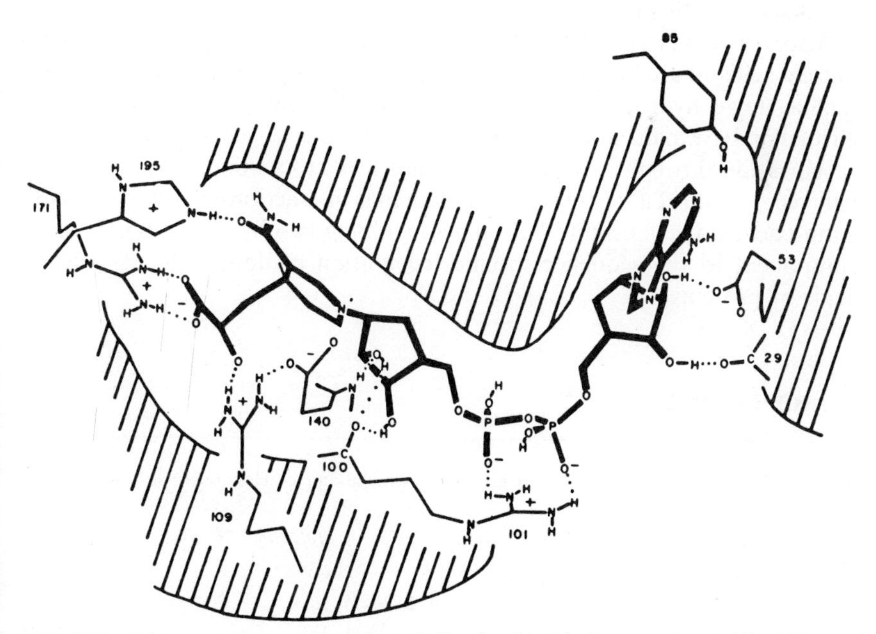

Fig. 15-9 Diagrammatic representation of dinucleotide binding as observed in Liljas and Rossmann (1974).

The binding of nucleotides, and more specifically the phosphate moiety, can have a significant effect on protein structure. In ribonuclease-S (Richards et al., 1971) the imidazole ring of His 119 has four distinct positions controlled by the phosphate binding mode. Lactate dehydrogenase interactions involving the phosphates very likely trigger conformational changes of the "loop" residues (94–114) (Adams et al., 1973). A movement of 12.5 Å in parts of the backbone chain and 23 Å for one guanadinium group is observed. The phosphates in 2′,3′-diphosphoglycerate, when bound to hemoglobin, generate several local conformational changes involving distances of 1–2 Å that lock the molecule into its deoxy quaternary structure (Arnone, 1972). Furberg and Solbak (1974) have investigated single crystals of salts between phosphate diesters and various organic bases simulating arginine and lysine side chains. Such complexes can be considered as models for the contact interactions in nucleoproteins between phosphate groups and basic amino acids. Each complex has a unique hydrogen bonding geometry that may be a necessary intermediate state in generating specific conformational changes in the protein.

Very recently, Sundaralingan and Rao (1975) have edited an extensive work dealing with progress in the study of protein–nucleic acid interactions. Those readers wishing the most up-to-date information on these interactions should consult this reference.

Also, Seeman et al. (1976) have compared the base pairs in double helical nucleic acids to see how they can be recognized by proteins. They conclude that a single hydrogen bond is inadequate for uniquely identifying any particular base pair since this leads to numerous degeneracies. However, specification through two hydrogen bonds leads to a unique means of base pair recognition. These authors postulate that interactions involving specific amino-acid side chains and base pairs involving two hydrogen bonds are a component to the sequence-specific recognition of double helical nucleic acids by proteins.

REFERENCES

Adams, M. M. (1972). *Protein Interactions* (R. Jaenicke and E. Helmreich, Eds.). Springer, New York, p. 139.

Adams, M. J., A. Liljas, and M. G. Rossmann (1973). *J. Mol. Biol.*, **76**, 519.

Arnone, A. (1972). *Nature*, **237**, 146.

Arnott, S. and D. W. L. Hukins (1969). *Nature*, **224**, 886.

Arnott, S. and D. W. L. Hukins (1972). *Biochem. J.*, **130**, 453.

Barry, C. D., A. C. T. North, J. A. Glasel, R. J. P. Williams, and A. V. Xavier (1971). *Nature*, **232**, 236.

Blundell, T. L., G. Dodson, D. Hodgkin, and D. Mercola (1972). *Adv. Protein Chem.*, **26**, 279.

Bourne, G. H. (1974). *The Structure and Function of Muscle.* Academic Press, New York.

Brandén, C.-I., H. Eklund, B. Nordström, T. Bowie, G. Soderlund, E. Zeppezauer, I. Ohlsson, and A. Akeson (1973). *Proc. Natl. Acad. Sci. USA*, **70**, 2439.

Buehner, M., G. C. Ford, D. Moras, K. W. Olsen, and M. G. Rossmann (1973). *Proc. Natl. Acad. Sci. USA*, **70**, 3052.

Chandrasekhar, K., A. McPherson, Jr., M. J. Adams, and M. G. Rossmann (1973). *J. Mol. Biol.*, **76**, 503.

Cohen, C. (1961). *J. Polymer Sci.*, **49**, 144.

Cohen, S. and R. R. Porter (1964). *Adv. Immunol.*, **4**, 287.

Cohen, C. and A. G. Szent-Györgyi (1957). *J. Am. Chem. Soc.*, **79**, 248.

Cohen, C. and A. G. Szent-Györgyi (1960). *Proc. IVth Intern. Cong. Biochem., Vienna*, **8**, 108.

Cotton, F. A., C. J. Bier, V. W. Day, E. E. Hazen, Jr., and S. Larsen (1971). *Cold Spring Harbor Symp. Quant. Biol.*, **36**, 243.

Dreizen, P. (1970). *Trans. N.Y. Acad. Sci.*, , 120.

Dreizen, P. and L. C. Gershman (1970). *Biochemistry*, **9**, 1688.

Drenth, J., J. N. Jansonius, R. Koekock, and B. G. Wolthers (1971). *Adv. Protein Chem.*, **25**, 79.

Edelman, G. M., B. A. Cunningham, G. N. Reeke, Jr., J. W. Becker, M. J. Waxdal, and J. L. Wang (1972). *Proc. Natl. Acad. Sci. USA*, **69**, 2580.

Frederiksen, D. W. and A. Holtzer (1968). *Biochemistry*, **7**, 3935.

Furberg, S. and J. Solbakk (1974). *Acta Chem. Scand.*, **B28**, 481.

Gaetzens, E., K. Barany, G. Bailin, H. Oppenheimer, and M. Barany (1968). *Arch. Biochem. Biophys.*, **123**, 82.

Gershman, L. C., A. Stracher, and P. Dreizen (1969). *J. Biol. Chem.*, **244**, 2726.

Hardman, K. D. and C. F. Ainsworth (1972). *Nature (New Biol.)*, **237**, 54.

Harrington, W. F. and R. Josephs (1968*b*). *Develop. Biol. Suppl.*, **2**, 21.

Hodge, A. (1967). In *Treatise on Collagen. Vol. k, Chemistry of Collagen* (Ramachandran, G. N., Ed.). Academic Press, New York, p. 185.

Holtzer, A. and S. Lowey (1956). *J. Am. Chem. Soc.*, **78**, 5954.

Holtzer, A. and S. Lowey (1959). *J. Am. Chem. Soc.*, **81**, 1370.

Holtzer, A., S. Lowey, and T. Schuster (1962). In *The Molecular Basis of Neoplasia.* University of Texas Press, Austin, p. 259.

Holtzer, A. and S. Rice (1957). *J. Am. Chem. Soc.*, **79**, 4847.

Huxley, H. E. (1957). *J. Biophys. Biochem. Cytol.*, **3**, 631.

Huxley, H. E. (1965). *Sci. Am.*, **213**, 18.

Huxley, H. E. (1963). *J. Mol. Biol.*, **7**, 281.

Huxley, H. E. (1968). *J. Mol. Biol.*, **37**, 507.

Huxley, H. E. and W. Brown (1967). *J. Mol. Biol.*, **30**, 383.

Hulmes, D. J. S., A. Miller, D. A. D. Parry, K. A. Piez, and J. Woodhead-Galloway (1973). *J. Mol. Biol.*, **79**, 137.

Ikkai, T. and T. Ooi (1966). *Biochemistry*, **5**, 1551.

Jack, A., J. Weinzierl, and A. J. Kalb (1971). *J. Mol. Biol.*, **58**, 389.

Josephs, R. and W. F. Harrington (1966). *Biochemistry*, **5**, 3474.

Josephs, R. and W. F. Harrington (1968*a*). *Biochemistry*, **7**, 2834.

Kim, S. H., H. M. Berman, N. C. Seeman, and M. D. Newton (1973). *Acta Crystallogr.*, **B29**, 703.

Koshland, D. E., G. Nemethy, and D. Filmer (1966). *Biochemistry*, **5**, 365.

Kuhn, K. and B. K. Zimmermann (1965). *Arch. Biochem. Biophys.*, **109**, 534.

Liljas, A. and M. G. Rossmann (1974). In *Annual Review of Biochemistry* (E. E. Snell, P. D. Boyer, A. Meister, and C. C. Richardson, Eds.). Annual Reviews Inc., Palo Alto, Vol. 43, p. 475.

Lowey, S. (1971). In *Subunits in Biological Systems* (S. N. Timasheff and G. D. Fasman, Eds.). Dekker, New York, Vol. 5, part A, p. 201.

Lowey, S. and C. Cohen (1962). *J. Mol. Biol.*, **4**, 293.

Lowey, S. and A. Holtzer (1959). *Biochim. Biophys. Acta*, **34**, 470.

Lowey, S., H. S. Slayter, A. G. Weeds, and H. Baker (1969). *J. Mol. Biol.*, **42**, 1.

Matthews, B. W. and S. A. Bernhard (1973). *Annu. Rev. Biophys. Bioeng.*, **2**, 257.

Miller, A. and D. A. D. Parry (1973). *J. Mol. Biol.*, **75**, 441–447.

Miller, A. and J. S. Wray (1971). *Nature* (*Lond.*), **230**, 437–439.

Monod, J., J. Wyman, and J. P. Changeux (1965). *J. Mol. Biol.*, **12**, 88.

Morgan, P. H., H. G. Jacobs, J. P. Segrest, and L. W. Cunningham (1970). *J. Biol. Chem.*, **245**, 5042.

Mueller, H. and S. V. Perry (1962). *Biochem. J.*, **85**, 431.

Muirhead, H., J. M. Cox, L. Mazzarella, and M. F. Perutz (1967). *J. Mol. Biol.*, **28**, 117.

Murphy, A. J. and M. F. Morales (1970). *Biochemistry*, **9**, 1528.

Petruska, J. A. and A. J. Hodge (1964). *Proc. Natl. Acad. Sci. USA*, **51**, 871.

Reed, L. J. and R. M. Oliver (1968). *Brookhaven Symp. Biol.*, p. 21.

Richards, F. M. (1971). *Cold Spring Harbor Symp. Quant. Biol.*, **36**, 35.

Roberts, G. C. K., E. A. Dennis, D. H. Meadows, J. S. Cohen, and O. Jardetzky (1969). *Proc. Natl. Acad. Sci. USA*, **62**, 1151.

Robertus, J. D., J. Kraut, R. A. Alden, and J. J. Birktoft (1972). *Biochemistry*, **11**, 4293.

Rühlmann, A., D. Kukla, P. Schwager, K. Bartels, and R. Huber (1973). *J. Mol. Biol.*, **77**, 417.

Sarkar, S. and P. H. Cooke (1970). *Biochem. Biophys. Res. Commun.*, **41**, 918.

Schiffer, M., R. L. Girling, K. R. Ely, and A. B. Edmundson (1974). *Biochemistry*, **13**, 3816.

Seeman, N. C., J. M. Rosenberg, and A. Rich (1976). *Proc. Natl. Acad. Sci. USA*, **73**, 804.

Segal, D. M. and J. C. Powers (1971). *Biochemistry*, **10**, 3728.

Segrest, J. P. and L. W. Cunningham (1973). *Biopolymers*, **12**, 825.

Segrest, J. P. and L. W. Cunningham (1971). *Nature* (*New Biol.*), **234**, 26–28.

Slayter, H. S. and S. Lowey (1967). *Proc. Natl. Acad. Sci. USA*, **58**, 1611.

Smith, J. W. (1968). *Nature* (*Lond.*), **219**, 157–158.

Sobell, H. M., S. C. Jain, T. D. Sabore, G. Ponticello, and C. E. Nordman (1971). *Cold Spring Harbor Symp. Quant. Biol.*, **36**, 263.

Steitz, T. A., R. J. Fletterick, and K. J. Hwang (1973). *J. Mol. Biol.*, **78**, 551.

Stevens, C. L. and M. A. Lauffer (1965). *Biochemistry*, **4**, 31.

Stracher, A. (1969). *Biochem. Biophys. Res. Commun.*, **35**, 519.

Sundaralingam, M. (1969). *Biopolymers*, **7**, 821.

Sundaralingam, M. and S. T. Rao (1975). *Structure and Conformation of Nucleic Acids and Protein-Nucleic Acid Interactions*. University Park Press, Baltimore.

Tonomura, Y. (1973). *Muscle Proteins, Muscle Contraction and Cation Transport*. University Park Press, Baltimore.

Tulinsky, A., R. L. Vandlen, C. N. Morimoto, N. V. Marri, and L. H. Wright (1973). *Biochemistry*, **12**, 4185.

Veis, A., J. Anesey, and S. Mussell (1967). *Nature*, **215**, 931.

Warren, S. G., B. F. P. Edwards, D. R. Evans, D. C. Wiley, and W. N. Lipscomb, (1973). *Proc. Natl. Acad. Sci. USA*, **70**, 1117.

Watenpaugh, K. D., L. C. Sieker, L. H. Jensen, J. Legall, and M. Dubourdiew, (1972). *Proc. Natl. Acad. Sci. USA*, **69**, 3185.

Webb, L., E. Hill, and L. J. Banaszak (1973). *Biochemistry*, **12**, 5109.

CHAPTER 16

Some Generalizations and Speculations

The discussions in this section are the result of the author's overview of the reported intermolecular processes. The opinions—and they are only opinions—of what might be general characteristics of intermolecular interactions in biologic systems are based on inferences as well as fact in most cases. It is perhaps best to view this chapter as a potpouri of what are reasonable explanations of a set of observations, but certainly not the only (or even the most probable) interpretations.

I SOLUTION AND SOLID-STATE CONFORMATIONS OF SMALL BIOMOLECULES

Most small biologically active molecules adopt preferred conformations in aqueous solution. Often two or more conformational states may be significantly populated, that is, greater than 10%, although one of these states is usually highly preferred over the other(s). In addition, this highly preferred conformation very often corresponds to the observed solid-state conformation.

It is important to put these generalizations concerning conformation into proper perspective. We are referring to the overall conformational characteristics as generated by torsional rotations about bonds that contain large groups on *both* sides of the bond. The observations are not necessarily valid for the conformational behavior of small pendant groups, that is, for example, rotational preferences of methyl groups about their linkage bonds. Generally, a small group connected to the main portion of the molecule through a single bond does not exhibit significant conformational preferences in solution with respect to rotation about the bond over laboratory temperatures. Moreover, those small groups that do possess conformational specificity of some significance in solution, populate conformational states that are only marginally in common with those correspondingly observed in the crystalline state. Still, the tendency for conformational indecision by small groups seems to carry over from solution to the solid state. Those biomolecules which exhibit multiple solid-state conformations, for example, acetylcholine (Chapter 2), differ from one another through different conformations of the small pendant groups that are conformationally flexible in solution. The basic backbone conformation, as specified by torsional rotations about bonds possessing large fragments on both sides of the bond, are usually invariant to external molecular environment for those molecules capable of multiple solid-state conformation.

A general picture emerges that, in solution, conformationally flexible molecules distribute themselves over states that, to a first approximation, are consistent with the isomeric-state model. The force field generated by the solvent medium does *not* appear to seriously modify the distribution of conformational preferences. Roughly, the changes in the population of assessable conformational states varies no more than 10% in going from an aqueous to an alcohol solvent medium. Moreover, the population of the highly preferred conformational state is not modified by more than 5%. That the solid-state conformation is almost always the same as that seen in solution suggests that the intermolecular crystalline force field, like that of the solvent medium, is not strong enough to overcome the conformational dictates of the intramolecular force field. However, the realization of the crystallization process that, generally speaking, excludes solvent molecules from the crystals, indicates that the nucleation and propagation of the crystalline intersolute molecular force field leads to a total lower free energy state than that involving the solvent. Whether this is a consequence of direct interatomic interactions (an enthalpy-driven process) or whether the solute molecules use each other as momentum sinks (an entropy-driven process) is not clear.

II INTERMOLECULAR FORCE FIELDS AND VALENCE BOND GEOMETRY

It is well established that the crystalline environment surrounding a molecule causes perturbative changes in its constituent bond lengths and bond angles. Even larger deviations are seen in the torsional angles about the single bonds of a molecule in different crystalline environments. In fact, different molecular conformations, through torsional rotations, have been observed in several biologic compounds in different crystals. Generally, the deviations in the valence bond geometry have been thought of as inconsequential in structural calculations, with the major emphasis given to the torsional rotations. However, some isolated intramolecular calculations using different observed crystalline valence bond geometries for a single molecule have, in turn, predicted the starting crystalline conformation as the global *intramolecular* energy minimum. This suggests, at least mechanistically, that the intermolecular force field can translate its influence on molecular conformation into the intramolecular force field through the valence bond geometry. In view of this finding, it is reasonable to suggest that a predictive test for estimating how sensitive a molecular conformation might be to different external environments can be achieved by varying valence bond geometry in intramolecular conformational analyses. This may be a particularly useful ploy to use in studying drug-receptor interactions.

Perturbative alterations in valence bond geometry, and consequently, molecular conformation, may have dire consequences in the prediction of protein tertiary structure. Even minor structural changes in the molecular conformation and valence geometry of individual protein residues as a result of interactions with solvent and/or other distant residues (with respect to primary structure) may lead to large changes in the gross tertiary structure of a globular protein. The covalent connectivity of the polypeptide chain may act as a coherent amplifier in magnifying the cumulative small residual changes in molecular structure into a gross molecular reorganization. Consequently, even though we may someday be able to correctly predict the precise sequential distribution of secondary structure in a globular protein, we still may not be able to specify its tertiary structure. It would seem that knowledge of the precise torsional rotational angles resulting from the energy minimization for a specific valence bond geometry is required. Simply knowing if a residue is α helical or not, etc. is not enough. If one thinks about this in terms of the information content of a globular protein, it seems to be a reasonable proposition. Globular proteins perform highly specific and intricate chemical actions. This requires a high degree of organization, that is, information content. The information content must somehow be stored in the molecule. The tertiary structure provides such an information reservoir.

III HYDRATION STRUCTURES ABOUT BIOLOGIC MACROMOLECULES

The hydration shells surrounding macromolecules are often described as "ice-like." This can lead to some confusion, since there are distinct aspects of ice-likeness that have not been clearly distinguished. Ice differs from water in the reduced molecular motion and in the perfect order of its crystal lattice. The motion and the order of water molecules are largely independent. This is shown best by the example of vitreous ice, which preserves the molecular disorder of liquid water in spite of drastic reduction in molecular motion.

There is spectroscopic, calorimetric, and hydrodynamic evidence that the translatory and rotatory motion of water molecules in the hydration shells of biopolymers is reduced with respect to that in liquid water. A reduction of molecular motion is to be expected because of the interaction of water molecules with the polar sites on the surface of the biopolymer. On the other hand, the term "ice-like" also seems to imply to many authors a quasi-crystalline ordering of the hydration shell into a lattice corresponding to ice I structure. The formation of such a lattice has been specifically postulated, even though there appears to be no direct evidence for it.

In fact, the results presented in this book strongly suggest that the most strongly bound water to the biopolymer are "nonfreezable" and, consequently, incapable of ice-like organization. It is likely that the packets of very strongly bound water on the biopolymer are structurally "out of phase" with one another, as well as encroaching external ice structures. The packets of tightly held water are of a sufficiently lower free energy to maintain their structural uniqueness rather than convert into a compromise structure compatible with long-range ice-like structure. Organization-wise, these pockets of tightly bound biopolymer water can be viewed as structural defects that prevent nearest-neighbor water crystallization about biologic macromolecules. Overall, the existence of an inner, noncrystallizable hydration shell may be a general phenomenon of the solvation organization of biologic macromolecules.

IV MOBILE ION INTERACTIONS WITH MACROMOLECULES

Mobile solution ions affect the structure and resultant stability of macromolecules in solution in two identifiable ways: through nonspecific electrostatic interactions in which the strength of ionic perturbant is mainly a function of the sign and magnitude of the ionic charge magnified by its concentration in solution and, second, by site-specific, ligand-like binding interactions to the macromolecule that are largely independent of ionic charge and concentration (provided the actual number of binding sites is less than the number of ions).

The characteristic melt transition temperature, T_m, of macromolecules in solution are generally found to be a linear function of salt concentration where the salt is composed of nonspecific interacting ionic components. An increase or decrease in T_m with increasing nonspecific salt concentration is inversely related to a loss or build-up of charge density on the macromolecules. The nonspecific anion and cation components of a given salt affect T_m in a roughly algebraically additive fashion. The ions tend to fall into a typical Hofmeister series if ranked in order of molar effectiveness as T_m perturbants. This additivity also seems to be roughly preserved in multi-ionic component mixtures. Ionic perturbants that decrease the stability of an ordered conformation of a macromolecule also increase the rate of its conversion to the disordered form under denaturing conditions. Stabilizers have the converse effect.

In general, nonspecific ionic perturbants that destabilize the folded form of a macromolecule in aqueous solution correspondingly increase macromolecular solubility. On the other hand, agents that stabilize the folded conformation by strengthening intramolecular organization decrease macromolecular solubility. Specific attributes of macromolecules that depend on the local maintenance of a special ordered conformation or intermolecular association can be disrupted or stabilized by neutral salts in parallel fashion to their effects on the structural stability of the entire macromolecule.

Site-specific ions, or ionic complexes, that are prequisite to the reactivity of a macromolecule meet the fundamental definition of a drug and probably should be considered as such. At first glance, site-specific ions might appear as relatively unsophisticated drugs, since their interaction with the macromolecule's receptor site requires less information, when translated into the structural chemistry of the interacting pair, than most organic drug–receptor complexes. Often this is probably the case, and it is tempting to consider the first successful drug interactions in the evolutionary path to be those involving site-specific ions. Still, there is increasing evidence from both X-ray and optical spectroscopies that suggests that nature has built ion selectivity into macromolecular binding sites through extremely complex and, presumably, high information-content-requiring, ligand geometries. These sites induce small, but significant, distortions into the complexing orbitals of the visiting ion. Only those ions which can accept the distorted terms of the host site will be able to complex. Moreover, there is reason to believe that the small change in complex energy necessitated by the orbital distortions of the visiting ion may have a significant regulating effect on molecular reactivity.

V AGGREGATE ORGANIZATION AND IONIC INTERACTIONS

A strong and very specific interaction that stabilizes biomolecular aggregates is the *complementary charge matching* between ionic species on different molecules. Since the aggregation process is the nucleus for large-scale molecular organization, it may be safe to infer that complementary ion-charge interactions form the energetic basis for intermolecular assembly in some biologic systems. Although this might be a speculative overstatement, it is not an overstatement to say that the vast majority of known interacting proteins, polynucleotides, and polysaccharides either directly possess ionizable groups like COO^-, N^+H_3, etc., or have developed ligand-binding sites to snatch mobile ions out of solution. Consequently, most interacting biologic macromolecules are macroins under specific solution condition. Moreover, they are highly developed macroins in the sense that the ionic groups sites are uniquely distributed over primary structure. As a result, the secondary and tertiary structures respond during intermolecular interactions to the ion-primary structures by adopting specific spatial structures that presumably have evolved to be best for the particular biologic function.

Bio-macroin intermolecular interactions seem to be more complicated than simply matching up opposite charges on two or more macroins on a one-to-one basis. Rather it appears that biologic macroins contain "clumps" of ionic groups/sites, usually of one type of charge in both sign and magnitude. Oppositely charged "clumps" on different macroins interact to minimize the composite free energy. Thus we use the term *complementary charge matching* as opposed to *opposite charge interaction*. Clearly the "clumps" can be uniformly distributed over molecular structure as is the case for DNA, almost uniformily distributed as in the histones, or quite anisotropic in structural location, as in interacting globular proteins. As one might expect, the proteins offer the largest diversity in the distribution of ions over molecular structure. The fact that bio-macroins do not interact to match up opposite ionic-charge interactions on a one-to-one basis suggests a few possibilities: (1) the most stable complex, as presumably would be achieved by intermolecular ion-charge pairing, is not best for biologic function, (2) some ion sites are needed for functions over and beyond complex stabilization, (3) molecular stereochemistry prevents a higher degree of ion-charge pairing, or (4) some ionic charges that appear wasted from an intermolecular point of view are essential for intramolecular stability.

It is intriguing to speculate why ionic interactions have evolved in biologic systems to become prevalent in intermolecular organizational processes. This is especially the case if one remembers the following basic characteristics of known structural chemistry: the predominant intramolecular, as well as intermolecular, interaction organizing *crystalline* structures of biologic molecules is the hydrogen bond. Even most crystallizable ionic species pack in crystals to form hydrogen-bonding networks over the lattices. A hydrogen bond is relatively weak (3–10 kcal/mole) interaction when compared to an ion-counterion interaction (25–40 kcal/mole). Thus it requires significantly more hydrogen bonds to achieve the interatomic stabilization energy of one ion-counterion bond. In turn, the formation of multiple intermolecular hydrogen bonds requires a significant ordering process. Herein may lie nature's reasons for using a high stability, local interaction—the ion-counterion bond—to build up supramolecular structure. Biologic evolution is an upward struggle against thermodynamic entropy. The use of ion-counterion bonds to achieve a given level of intermolecular stability could be achieved with a corresponding lower level of order than would be required via other interatomic interactions like hydrogen bonds. Thus an evolutionary path using ion-counterion interactions would have lower free energy barriers than paths employing other intermolecular interactions on the basis of entropy. But perhaps even more significant, molecular order usually implies molecular rigidity. The greater the number of intermolecular bonds, the more rigid the complex as a whole, as well as on the basis of individual molecules. Even though we have side-stepped it because of our lack of understanding in this text, biomolecular processes are dynamic and require molecular motion and flexibility. Rigid molecular complexes may simply be incapable of carrying out the chemical actions needed to sustain life. Nature may have found the ideal solution for maintaining stable yet flexible molecular assemblies. The sparse dispersion of very strong interatomic bonds formed by ion/counterion groups located on the ends of flexible chains, for example, the polylysine side chain, yields an assembly strongly held together at a minimal number of contact points to allow significant flexibility. In addition, the ion-counterion bond is quite susceptible to external control. Mobile ions can find their way to the site of the ion-counterion bond and competitively vie with each of the bound ions for interaction with the other type of bound ion. Thus the extent of both complex stability and order can be controlled using outside mechanisms of action.

VI INTERMOLECULAR IDENTIFICATION THROUGH HYDROPHOBIC INTERACTIONS

A troublesome question in dealing with associating biologic macromolecules is how they are able to quickly recognize their organizational role with respect to one another. This is not a random process on the basis of limited kinetic data. The association rates are much too high. The rates are also too high for a set of temporal sequential events as would be implied by series of point-by-point interactions involving formation of hydrogen and/or ionic bonding between the constituent macromolecular species. The information gathered in this book suggests that hydrophobic sightings, which can involve relatively large molecular surface areas (especially for the globular proteins), are the initial means by which many associating biologic molecules identify their self-assembly modes. This should be a well-controlled means for molecules to "see" one another, since the size of the hydrophobic site should be proportional to the ease with which a molecule can pick out potential partners. In turn, the topological intersection "area" of two hydrophobic reactive sites should be proportional to the strength of the association.

More generally, one can envision the hydrophobic association as only the first but nevertheless critical initial step in organized association. Uniquely displaced within the hydrophobic sites are likely to be complentary hydrogen and/or ionic binding groups that impart the specificity to the conjungal molecular reaction. These secondary interactions probably also delineate the precise association energy required for biologic functioning (see the previous section of this chapter).

VII COMPUTER SIMULATION OF MOLECULAR ORGANIZATION

The limited extent of crystallinity, structural flexibility, and heterogeneous molecular composition make the study of associated biomolecular units very difficult. Fixed quantities of structural information using different sources, that is, X-ray IR/Raman, NMR, CD/ORD, etc., can usually be obtained from material samples. However, the amount of information from any single structural-probe technique falls far short of what is needed for any reasonably detailed molecular model.

Some method of quantitatively inter-relating all the bits and pieces of information relevant to the structural complex to maximize the total extent of organizational evidence is needed. One potential means of achieving this goal is through a computer simulation analysis in which the experimental data provides constraints as to the types of models allowed. A variety of simulation-constraint software analysis procedures, usually called "optimization packages," have been developed for a variety of applications in industry. It should be emphasized that the envisioned simulation experiments are more than the typical conformational energy calculations reported in Chapter 14. The calculated energetics of a proposed molecular model would represent an additional constraint to be added to the experimental data base of constraints The method of analyzing the fibril packing of collagen molecules presented in Chapter 15 might be considered as a low-level example of the type of computer simulation system proposed here. Some X-ray structural analysis packages have incorporated constrained optimization procedures and can also be considered as first-generation examples of the structural optimization methods that may evolve.

One can probably guess that the computer simulation approach will very likely provide many possible structures. The key to both limiting the number of molecular models consistent with the constraints as well as quantitatively ranking the "reasonableness" of each proposed model resides in how the various structural constraints are weighted in importance relative to one another. Herein lies the potential shortcoming of the technique. It is up to the investigator to assign the weights to the component bits of structural information in a manner consistent with the reliability and accuracy of the experimental measurements. He may not be able to do this with any confidence.

Index